Uncertainty in Acoustics

Uncertainty in Acoustics

Measurement, Prediction and Assessment

Edited by Robert Peters

CRC Press
Taylor & Francis Group
Boca Raton London New York

CRC Press is an imprint of the
Taylor & Francis Group, an **informa** business

CRC Press
Taylor & Francis Group
6000 Broken Sound Parkway NW, Suite 300
Boca Raton, FL 33487-2742

First issued in paperback 2022

ISBN-13: 978-0-367-49247-2 (pbk)
ISBN-13: 978-1-4987-6915-0 (hbk)

DOI: 10.1201/9780429470622

**Visit the Taylor & Francis web site at
www.taylorandfrancis.com**

**and the CRC Press web site at
www.crcpress.com**

Library of Congress Cataloging-in-Publication Data
Names: Peters, Bob, 1944-2019, editor.
Title: Uncertainty in acoustics : measurement, prediction and assessment / edited by Robert Peters.
Description: First edition. | Boca Raton, FL : CRC Press, Taylor & Francis Group, 2020. | Includes bibliographical references.
Identifiers: LCCN 2019051355 (print) | LCCN 2019051356 (ebook) | ISBN 9781498769150 (hardback ; acid-free paper) | ISBN 9780429470622 (ebook)
Subjects: LCSH: Acoustic models. | Sound—Measurement. | Measurement uncertainty (Statistics)
Classification: LCC TA365 .U48 2020 (print) | LCC TA365 (ebook) | DDC 620.2–dc23
LC record available at https://lccn.loc.gov/2019051355
LC ebook record available at https://lccn.loc.gov/2019051356

Dedicated to Bob Peters

Dr. Bob Peters passed away early on June 22, 2019. Bob was well known for his unbridled enthusiasm for and contributions to acoustics education. I had the pleasure and privilege of working with him for many years through both the Open University and the Institute of Acoustics (IOA).

Bob studied physics at Imperial College, London; graduated in 1965; and stayed on to study for a PhD in physics which included taking the Chelsea College MSc in acoustics and vibration physics, taught by Professor R. W. B. Stephens. Bob remained at Imperial as a research assistant in underwater acoustics until 1969 and obtained his PhD, supervised by Professor Stephens, in 1971.

After three years researching diesel engine noise for CAV Ltd. and two years teaching at Twickenham College of Technology, Bob moved to the North East Surrey College of Technology (NESCOT), where he remained for 23 years. He started and ran diploma and MSc programmes at NESCOT until 2010. Few acousticians will be unaware of Bob's text *Acoustics and Noise Control* (3rd edition, published by Routledge). This was the basis for much of Bob's rewrite of the General Principles of Acoustics Module for the IOA diploma in 2008. While at NESCOT, Bob was a guest lecturer in acoustics for undergraduate and postgraduate courses at many other institutions, including the University of North London, Oxford Brookes University, Imperial College and University College London. For around 25 years, until 2017, he delivered a significant part of the MSc in environmental and architectural acoustics at London South Bank University. He also acted as an examiner for several PhD students around the country.

After retiring from NESCOT, Bob embarked on a new career in consultancy while continuing with a range of teaching activities. For 22 years he worked part time as principal consultant with Applied Acoustic Design (AAD), where his main responsibilities included computer modelling of indoor spaces and acting as an expert witness.

In addition to his involvement with the IOA diploma and certificate courses, Bob worked tirelessly in many other ways for the IOA and contributed enormously to the life of the Institute for over 40 years. Bob chaired the Industrial Noise Group (a forerunner of the current Noise and Vibration Engineering Group [NVEG]) from 1992 to 1997 and was the technical organiser of two of the Institute's autumn conferences on industrial noise, held at Windermere in 1987 and 1989. He also chaired the programme committee for the 1992 Euronoise Conference, held in London.

Bob represented the IOA on several national and international committees. He chaired the Chartered Institution of Building Services Engineers (CIBSE) committees responsible for rewriting the acoustics sections of CIBSE *Guides A* and *B* and was a member of the British Standards Institution (BSI) committee revising BS 6472 on the measurement and assessment of human response to vibration in buildings. At an international level, Bob was the IOA representative on the International Institute of Noise Control Engineering (I-INCE) technical subgroup (TSG3) on noise policies and regulations.

Bob became a corporate member of the IOA (MIOA) in 1977, a Fellow in 1980 and was elected as an Honorary Fellow in 2008. He served as vice-president for group and branches from 1992 to 1998 and received an IOA Distinguished Services Award in 2008. In 2013, he was awarded the R. W. B. Stephens Medal in recognition of his extensive and outstanding work in acoustics education. At the time, Bob commented that this was particularly appropriate because Professor Stephens, as well as being his PhD supervisor, had been his proposer for IOA membership in 1977 and for IOA fellowship in 1980.

Bob was the fifth chair of the IOA Education Committee, serving from 1994 to 1998, and more recently, he took over as the ninth chair in 2015 until standing down owing to ill health in 2017. During these periods, Bob was instrumental in starting the IOA's certificate courses, acting as chief examiner for the certificate of competence in workplace noise risk assessment for several years and creating the distance-learning version of the diploma after its major restructuring in 2008. In 2000, Bob became the first project examiner for the diploma, in which role he undertook to visit every diploma centre each year. He pursued this exhausting schedule every year until 2018.

Despite being nominally an examiner, Bob was not backward in giving fellow examiners a hard time regarding the accuracy, clarity and length of questions, being particularly severe on those setting the examinations for the General Principles of Acoustics Module. Bob became the senior tutor for the tutored distance-learning version of the diploma and was the primary tutor for the St. Albans group and, before the most debilitating stage of his illness, planned to carry on with its successor, the Milton Keynes group.

Bob was a great supporter of the Open University (OU), which necessitated the family acquiring a television and a telephone in the mid-1970s. He was a course tutor for a long period on the second- and third-level courses concerned with pollution control. He helped me as author, particularly on the noise block of the third-level course. Subsequently, Bob found the time to study for an OU arts degree, which he was awarded in 1995.

Bob was the last to acknowledge the severity of his illness or of the side-effects suffered from the intensive treatment he was receiving. While accepting, temporarily, that the loss of an effective immune system as a result of treatment made it impossible for him to travel by public transport, it was clear from telephone conversations with him in early 2019 that he fully expected to be back contributing to the diploma as normal. Those attempting to fill the 'tutor' gap left by Bob are finding that his is a hard act to follow. He will be sorely missed.

Bob must have been one of the most widely recognised figures in the UK acoustics world, regarded with great affection by all who knew him. Here are some quotes from former colleagues and students on hearing of Bob's death:

Stuart Dyne (currently chief examiner for the IOA diploma)

> Bob gave great and long service to the IOA through many activities not least of which was his enthusiastic and authoritative support for education in general and for the Diploma in particular and many of us will have one or more versions of *Acoustics and Noise Control* on our shelves. We will miss that contribution very much, especially at the moderation meetings where he could reliably seek and often find a mark or two to get candidates over the line.

Bridget Shields (professor emeritus at London South Bank University [LSBU])

He was such a lovely guy – and one of best teachers of acoustics in the country. We were so lucky to have him teaching on the MSc [at LBSU] – I think I recruited him as a PT lecturer about 30 years ago and he has been with us ever since. The course would have folded without him. And he did SO much for the IOA nationally and locally too.

David Trevor-Jones (consultant and founder chair of the IOA Certificate of Competence in Environmental Noise Measurements)

A generation of acousticians was taught by Bob, either directly in face-to-face encounters or through his textbook. All who were privileged to know him will have witnessed his profound commitment to acoustics education and experienced his generosity with his time and, also, his own tireless drive to learn as well as to teach.

Jim Griffiths (consultant)

Very sad for me as he started my career at the IOA.

Chris Goff (consultant)

That's really sad to hear, but I'm glad I got the opportunity to be taught by him – he could explain complex subjects in a clear and understandable way which is a very rare thing to be able to do as any kind of teacher (or consultant!), as well as being an extremely helpful and friendly guy. Will be missed!

Alex Krasnic (consultant and member of IOA Education Committee and STEM WG)

I am in a deep state of shock right now as Bob meant a lot to me.

Simon Kahn (consultant and former chair of the IOA Education Committee)

Bob will be greatly missed by many of his students – including me! ... He was a fantastic teacher as well as a fantastic acoustician. As a colleague he would often ask a question that would leave you thinking for weeks. He was both challenging and supportive in all the right ways.

Bob's motivation for producing this book is illustrated by the quotations used to precede his original draft of Chapter 2.

When reporting the result of a measurement of a physical quantity, it is obligatory that some quantitative indication of the quality of the result be given so that those who use it can assess its reliability. Without such an indication, measurement results cannot be compared, either among themselves or with reference values given in a specification or standard. It is therefore necessary that there be a readily implemented, easily understood, and generally accepted procedure for characterizing the quality of a result of a measurement, that is, for evaluating and expressing its uncertainty.

> *[ISO/IEC Guide 98-3:2008, Uncertainty of measurement – Part 3:*
> *Guide to the expression of uncertainty in measurement*
> *(GUM:1995): Introduction, paragraph 0.1]*

and

A measurement result is complete only when accompanied by a quantitative statement of its uncertainty.

> *[Stephanie Bell in the Measurement Good Practice Guide No. 11,*
> A Beginner's Guide to Uncertainty of Measurement, *National Physical*
> *Laboratory, issue 2, 2001)]*

This book is about how to quantify, minimise and evaluate such uncertainties in acoustical measurements. Its publication is a fitting tribute to Bob's passion for acoustics and is dedicated to his memory. It is an honour to have been given the opportunity to contribute as technical editor.

Keith Attenborough

Contents

Chapter 1

Introduction and Concepts

John Hurll

1.1 Introduction

In many aspects of everyday life, we are accustomed to the doubt that arises when estimating quantities, amounts, timings and the like. For example, if somebody asks, "What do you think the time is?" we might say, "It is about 10:15". Use of the word "about" implies that we know that the time is not exactly 10:15 but is somewhere near it. In other words, we recognise, without really thinking about it, that there is some doubt about the time that we have estimated.

We could, of course, be a bit more specific. We could say, "It is 10:15 give or take five minutes". The term "give or take" implies that there is still doubt about the estimate, but now we are assigning limits to the extent of the doubt. We have given some quantitative information about the doubt, or *uncertainty*, of our "guess".

We will also be more confident that our estimate is within, say, five minutes of the correct time than we are that it is within, say, 30 seconds. The larger the uncertainty we assign in a given situation, the more confident we are that it encompasses the "correct" value. Hence, for that situation, the uncertainty is related to the level of confidence.

So far, our estimate of the time has been based on subjective knowledge. It is not entirely a guess, as we may have recently observed a clock or looked at our watch. However, in order to make a more objective measurement, we have to make use of a measuring instrument of some kind; in this case, we can use a timepiece of some kind. Even if we use a measuring instrument, there will still be some doubt, or uncertainty, about the result. For example, we could ask

"Is my watch accurate?"
"How well can I read it?"
"The watch is strapped to my wrist. Am I warming it up?"

So, to quantify the uncertainty of the time-of-day measurement, we will have to consider all the factors that could influence the result. We will have to make estimates of the possible variations associated with these influences. Let us consider some of these.

Is the watch accurate? In order to find out, it will be necessary to compare it with a timepiece or clock whose accuracy is better known. This, in turn, will have to be compared with an even better characterised one, and so on. This leads to the concept of *traceability of measurements*, whereby measurements at all levels can be traced back to agreed references. In most cases, measurements are required to be traceable to the International System of Units (SI system). This is usually achieved by an unbroken chain of comparisons to a national metrology institute, which maintains measurement standards that are directly related to SI units.

In other words, we need a *traceable calibration*. This calibration itself will provide a source of uncertainty, as the calibrating laboratory will assign a calibration uncertainty to the reported values. When used in a subsequent evaluation of uncertainty, this is often referred to as the *imported uncertainty*.

In terms of the accuracy of the watch, however, a traceable calibration is not the end of the story. Measuring instruments change their characteristics as time goes by. They "drift". This, of course, is why regular recalibration is necessary. It is therefore important to evaluate the likely change since the instrument was last calibrated.

If the instrument has a reliable and convincing history, it may be possible to predict what the reading error will be at a given time in the future based on past results and apply a correction to the reading. This prediction will not be perfect, and therefore, an uncertainty on the corrected value will be present. In other cases, the past data may not indicate a reliable trend, and a limit value may have to be assigned for the likely change since the last calibration. This can be estimated from examination of changes that occurred in the past. Evaluations made using these methods yield the uncertainty due to *secular stability*, or changes with time, of the instrument. This is commonly known as "drift".

How well can I read it? There will inevitably be a limit to which we can resolve the reading we observe on the watch. If it is an analogue device, this limit will often be imposed by our ability to interpolate between the scale graduations. If the device has a digital readout, the finite number of digits in the display – and our ability to assimilate them as they change – will define the limit. This also shows that there are human factors involved in the interpretation of measurement results. Another example of this is when a stopwatch is used – the reaction time of the operator may be significantly worse than the inherent accuracy of the stopwatch. This reveals some important issues. First, the measurement may not be independent of the operator, and special consideration may have to be given to operator effects. We may have to train the operator to use the equipment in a particular way. Special experiments may be necessary to evaluate particular effects. Additionally, evaluation of uncertainty may reveal ways in which the method can be improved, thus giving more reliable results. This is a positive benefit of uncertainty evaluation.

The watch is strapped to my wrist. Am I warming it up? Well, it certainly will be at a different temperature to the surroundings – but does this matter? All measuring instruments are, to some degree or other, influenced by the environment to which they are exposed, and it is the designers' task to ensure that such influences are minimised. This is a general point that is applicable to all measurements. Every measurement we make has to be carried out in an environment of some kind; it is unavoidable. So we have to consider whether any particular aspect of the environment could have an effect on the measurement result. The following environmental effects are among the most commonly encountered when considering measurement uncertainty:

Ambient temperature, relative humidity and barometric pressure
Electric or magnetic fields, background charge
Gravity
Electrical supplies to measuring equipment
Presence of interfering objects (e.g. acoustic reflections, magnetic permeability)
Vibration and background noise
Light and optical reflections

Furthermore, some of these influences may have little effect as long as they remain constant, but they could affect measurement results if they are not constant.

It can be seen by now that understanding of a measurement system is important in order to identify and quantify the various uncertainties that can arise in a measurement situation. Conversely, analysis of uncertainty can often yield a deeper understanding of the system and reveal ways in which the measurement process can be improved. The points arising from such an analysis have to be asked, and answered, in order that we can devise an appropriate measurement method that gives us the information we require. Until we know the details of the method, we are not in a position to evaluate the uncertainties that will arise from that method. This leads to a most important question, one that should be asked before we even start with our evaluation of uncertainty: "What exactly is it that I am trying to measure?"

Until this question is answered, we are not in a position to carry out a proper evaluation of the uncertainty. The particular quantity subject to measurement is known as the "measurand". In order to evaluate the uncertainty in a measurement system, we must define the measurand; otherwise, we are not in a position to know how a particular influence quantity affects the value we obtain for it – or even to devise a suitable methodology to conduct the measurement.

1.2 Methodology for Uncertainty Evaluation

1.2.1 Criteria

It has long been recognised that there is a need for an internationally recognised, common approach to evaluation of uncertainty of measurement. To this end, the *Guide to the Expression of Uncertainty in Measurement* (often simply known as the GUM) was first published in 1992. The latest version of this, known as JCGM100:2008, can be obtained from the International Bureau of Weights and Measures (BIPM) web site, at www.bipm.org/en/publications/guides/gum.html. Various useful interpretative documents are also available; in the United Kingdom, the most common one in everyday use is M3003, *Expression of Uncertainty and Confidence in Measurement* – this is available from the United Kingdom Accreditation Service (UKAS) web site, at www.ukas.com.

The processes described in the remainder of this section are consistent, both in methodology and in terminology, with those in the GUM.

1.2.2 Concepts

A quantity Q is a property of a phenomenon, body or substance to which a magnitude can be assigned. The purpose of a measurement is to assign a magnitude to the measurand, the quantity intended to be measured. The assigned magnitude is considered to be the best estimate of the value of the measurand.

The uncertainty evaluation process will include influence quantities that affect the result obtained for the measurand. The GUM refers to these influence, or input, quantities as X, and the output quantity, that is, the measurand, is referred to as Y. As there will usually be more than one influence quantity, they are differentiated from each other by the subscript i. Therefore, there will be input quantities X_i, where i represents integer values from 1 to N, N being the number of such quantities. In other words, there will be input quantities of $X_1, X_2, ..., X_N$.

Each of these input quantities will have a corresponding value. For example, one quantity might be the background noise of the environment – this will have a value, say 40 dBA. A lower-case x represents the values of the quantities. Hence the value of X_1 will be x_1, that of X_2 will be x_2, and so on.

The purpose of the measurement is to determine the value of the measurand Y. As with the input uncertainties, the value of the measurand is represented by the lower-case letter y. The uncertainty associated with y will comprise a combination of the input, or x_i, uncertainties. One of the first steps is to establish the mathematical relationship between the values of the input quantities x_i and that of the measurand y. The measurement process can usually be modelled by a functional relationship between the values of the estimated input quantities and that of the output estimate in the form $y = f(x_1, x_2, \ldots, x_N)$. For example, if sound power W is measured in terms of sound intensity I and area S, then the relationship is $W = f(I, S) = SI$. The mathematical model of the measurement process is used to identify the input quantities that need to be considered in the uncertainty evaluation and their relationship to the total uncertainty for the measurement.

The values x_i of the input quantities X_i will each have an associated uncertainty. This is referred to as $u(x_i)$, that is, "the uncertainty of x_i". These are known as *standard uncertainties*.

Some uncertainties, particularly those associated with the determination of repeatability, have to be evaluated by statistical methods. Others are evaluated by examining other information, such as data in calibration certificates, evaluation of the effects of resolution or long-term drift, consideration of the effects of environment, and so on. In certain cases, special experiments may be needed to evaluate a specific effect.

1.2.3 Types of Uncertainty

The GUM differentiates between statistical evaluations and those using other methods. It categorises them into two types: *Type A* and *Type B*.

Type A Evaluation of Uncertainty

A *Type A* evaluation of uncertainty is carried out using statistical analysis of a series of observations. In the majority of cases, this will involve evaluation of the sample standard deviation resulting from repeated observations and subsequently calculating the standard deviation of the mean value (see Chapter 2 for an explanation of standard deviation). Further information and a worked example can be found in UKAS Document M3003, chapter 4. It should further be noted that if the repeatability data are unreliable (e.g. are based on a very small number of repeat measurements), then it may also be necessary to evaluate the effective degrees of freedom ν_{eff} of the combined standard uncertainty $u_c(y)$. The GUM recommends that the Welch-Satterthwaite equation be used to calculate a value for ν_{eff} based on the degrees of freedom ν_i of the individual standard uncertainties $u_i(y)$. The process for this can be found in appendix B of M3003.

Type B Evaluation of Uncertainty

A *Type B* evaluation of uncertainty is carried out using methods other than statistical analysis of a series of observations. Some examples were shown earlier.

The successful identification and evaluation of these contributions depend on a detailed knowledge of the measurement process and the experience of the person making the measurements. Type B evaluations can be based on information such as that

- associated with authoritative published quantity values,
- associated with the quantity value of a certified reference material,
- obtained from a calibration certificate,
- about drift,
- obtained via special experiments,
- obtained from the accuracy class of a verified measuring instrument,
- obtained from limits deduced through personal experience, and
- obtained from manufacturers' specifications.

1.2.4 Probability Distributions

The input uncertainties associated with the values x_i of the influence quantities X_i arise in a number of forms. Some may be characterised as limit values within which little is known about the most likely place within the limits where the "true" value may lie. A good example of this is the Type B uncertainty due to numeric rounding resulting from the finite resolution of a digital readout. In such cases, it is equally likely that the underlying value is anywhere within the defined limits of plus or minus half the change represented by one increment of the last displayed digit. It can be deduced from this that there is equal probability of the value of x_i being anywhere within the defined range and zero probability of it being outside these limits.

Thus, a contribution of uncertainty from a particular influence quantity can be characterised as a *probability distribution*, that is, a range of possible values with information about the most likely value of the input quantity x_i. In this example, it is not possible to say that any particular position of x_i within the range is more or less likely than any other. This is because there is no information available on which to make such a judgement. The probability distributions associated with the input uncertainties are therefore a reflection of the *available knowledge* about that particular quantity. For this reason, Type B uncertainty evaluations are sometimes referred to as *Bayesian* – that is, based on the concept of "degree of belief". In many cases, there will be insufficient information available to make a reasoned judgement, and therefore a uniform, or rectangular, probability distribution has to be assumed.

If, however, further information is available, it may be possible to assign a different probability distribution to the value of a particular input quantity. For example, uncertainties quoted in a calibration certificate will usually be associated with a Gaussian probability distribution.

1.2.5 Standard Uncertainties and the Central Limit Theorem

When a number of distributions of whatever form are combined, the GUM relies on, apart from in exceptional cases, the resulting probability distribution tending to the Gaussian form in accordance with the Central Limit Theorem. The importance of this is that it makes it possible to assign a confidence level in terms of probability to the combined uncertainty. The exceptional case arises when one contribution to the total uncertainty dominates; in this circumstance, the resulting distribution departs little from that of the dominant contribution.

Consequently, when the input uncertainties are combined, a Gaussian distribution will usually be obtained. The "width" of a Gaussian distribution is described in terms of a *standard deviation*. It will therefore be necessary to express the input uncertainties in terms that, when combined, will cause the resulting Gaussian distribution to be expressed at the one-standard-deviation level. As some of the input uncertainties are expressed as limit values (e.g. the rectangular distribution), some processing is needed to convert them into this form, which is known as a *standard uncertainty* and is referred to as $u(x_i)$. This is accomplished by the use of a divisor that is associated with the particular probability distribution that has been assigned. Some of the more commonly encountered distributions, and their associated divisors, are as follows, where a_i represents the semi-range limits:

Rectangular distribution: $u(x_i) = a_i/\sqrt{3}$
Triangular distribution: $u(x_i) = a_i/\sqrt{6}$
U-shaped (bimodal) distribution: $u(x_i) = a_i/\sqrt{2}$

In the case where a Type B uncertainty has been obtained from a calibration certificate (i.e. imported uncertainty), a coverage factor k will have been used to obtain this expanded uncertainty from the combination of standard uncertainties. It is therefore necessary to divide the expanded uncertainty by the same coverage factor to obtain the standard uncertainty. Therefore, $u(x_i) = U/k$, where U is the quoted expanded uncertainty and k is the associated coverage factor.

The case of repeatability evaluated by Type A means that $u(x_i)$ is simply the standard deviation of the mean, as mentioned earlier.

1.2.6 Sensitivity Coefficients

The quantities X_i that affect the measurand Y may not have a direct, one-to-one relationship with it. Indeed, they may be entirely different units altogether. For example, a dimensional laboratory may use steel end standards (gauge blocks) for calibration of measuring tools. A significant influence quantity is temperature. Because the end standards have a significant temperature coefficient of expansion, an uncertainty arises in their length due to the uncertainty in knowledge of their temperature. In order to translate the temperature uncertainty into an uncertainty in length units, it is necessary to know how sensitive the length of the end standard is to temperature. In other words, a *sensitivity coefficient* is required.

The sensitivity coefficient simply describes how sensitive the result is to a particular influence quantity. The sensitivity coefficient associated with each input estimate x_i is referred to as c_i . It is the partial derivative $\partial f/\partial x_i$ of the model function f with respect to X_i, evaluated at the input estimates x_i. It describes how the output estimate y varies with a corresponding small change in an input estimate x_i. A simple and straightforward approach for such an evaluation is to replace the partial derivative $\partial f/\partial x_i$ by the quotient $\Delta f/\Delta x_i$, where Δf is the change in f resulting from a small change Δx_i in x_i.

1.2.7 Combination of Uncertainties

Once the standard uncertainties x_i have been evaluated and the sensitivity coefficients c_i have been applied, the uncertainties have to be combined in order to give a single value of uncertainty to be associated with the estimate y of the measurand Y. This is known as the *combined standard uncertainty* and is given the symbol $u_c(y)$. It is obtained from the square root of the sum of the squares of the individual standard uncertainties, expressed in terms of the measurand. In accordance with the Central Limit Theorem, $u_c(y)$ takes the form of a normal or Gaussian distribution. As the input uncertainties had been expressed in terms of a standard uncertainty, the resulting Gaussian distribution is expressed as one standard deviation. For a Gaussian distribution, one standard deviation encompasses about 68.3% of the area under the curve. This means that there is about 68% confidence that the measured value y lies within the stated figure for $u_c(y)$.

1.2.8 Expanded Uncertainty

The GUM recognises the need for providing a high level of confidence – referred to as *coverage probability* – associated with an uncertainty and calls it *expanded uncertainty U*. This is obtained by multiplying the combined standard uncertainty by a *coverage factor*. The coverage factor is given the symbol k; thus the expanded uncertainty is given by $U = ku_c(y)$. The GUM recommends that a coverage factor of $k = 2$ be used to calculate the expanded uncertainty. This value of k will give a coverage probability of approximately 95%, assuming a Gaussian distribution.[1]

1.2.9 Evaluation Processes

For practical purposes, it is often convenient to use a spreadsheet such as Microsoft Excel or Apple Numbers for the purpose of uncertainty evaluation. A suggested format for such a spreadsheet is shown in the following table; it provides a systematic and logical presentation of the data.

Symbol	Source of uncertainty	Value	Probability distribution	Divisor	c_i	$u_i(y)$ unit	v_i or v_{eff}
S_s	Uncertainty of applied stimulus	0.2 unit	Gaussian	2	1	0.1	∞
δl_d	Digital rounding of indicator	0.05 unit	Rectangular	√3	1	0.0289	∞
δl_t	Secular stability	0.4 unit	Rectangular	√3	1	0.231	∞
Δt	Temperature effects	5.0°C	Triangular	√6	0.01 unit/°C	0.0204	∞
l_R	Repeatability of indication	0.07 unit	Gaussian	1	1	0.07	4
$u_c(y)$	Combined standard uncertainty	—	Gaussian	—	—	0.2635	>500
U	Expanded uncertainty	—	Gaussian $k = 2$	—	—	0.527	>500

A table such as this is often referred to as an *uncertainty budget*; however, this is not fully correct. An uncertainty budget is the statement of a measurement uncertainty, of the components of that measurement uncertainty, and of their calculation and combination. This implies that the budget should include a narrative that describes where the uncertainties arose from and references where the original data (including any necessary

---

1 In principle, a coverage factor of $k = 2$ provides a coverage probability of 95.45% for a Gaussian distribution. For convenience, this is approximated to 95%, which relates to a coverage factor of $k = 1.96$. However, the difference is not generally significant since, in usual practice, the components and evaluation of uncertainty are based on conservative assumptions.

experimental work) can be found. This kind of narrative should ensure that the data are traceable and, therefore, that others are able to use the information in a consistent and meaningful manner.

1.3 Reporting of Uncertainty

Once the expanded uncertainty has been evaluated for a stated coverage probability (e.g. 95%), the value of the measurand and expanded uncertainty should be reported as $y \pm U$ and accompanied by sufficient information for the user to import into a subsequent uncertainty budget. Examples of such statements include

"The reported expanded uncertainty is based on a standard uncertainty multiplied by a coverage factor $k = 2$, providing a coverage probability of approximately 95%", and
"The reported expanded uncertainty is based on a standard uncertainty multiplied by a coverage factor $k = XX$, which for a t-distribution with $\nu_{\text{eff}} = YY$ effective degrees of freedom corresponds to a coverage probability of approximately 95%".

Statements such as these are flexible and can be amended to suit specific circumstances.

Uncertainties are usually expressed in bilateral terms ($\pm$) either in units of the measurand or as relative values, for example, as a percentage (%), parts per million (ppm), 1×10^x, and so on. In acoustic measurements, some quantities are expressed in logarithmic terms (i.e. in decibels).[2] Combination of relatively small uncertainties expressed in decibels is permissible because $\log_e(1 + x) \approx x$ when x is small and $2.303\log(1 + x) \approx x$. For example, 0.1 dB corresponds to a power ratio of 1.023 and $2.303\log(1 + 0.023) = 0.0227$. Thus, relatively small uncertainties expressed in decibels may be combined in the same way as those expressed as linear relative values. For large values of uncertainty, however, it is recommended that the measurand and its uncertainty be converted into linear units before the evaluation process and converted back to decibel terms afterwards. This is likely to result in uncertainties that are not symmetrical around the reported result; if the asymmetry is significant, then the negative- and positive-going uncertainties should be reported separately.

The number of significant figures in a reported uncertainty should always reflect practical measurement capability. Uncertainty evaluation is based on assumptions and approximations relating to both assigned values and probability distributions; in view of this, it is rarely (if ever)

---

2 Unless indicated otherwise, the abbreviation $\log(\cdot)$ represents logarithm to the base 10.

justified to report more than two significant figures. Conversely, rounding to one significant figure can introduce rounding errors of up to 50% of the uncertainty.

It is therefore recommended that the expanded uncertainty be rounded to two significant figures using the normal rules of rounding. Thus, for example, in the uncertainty table shown in Section 1.2.9, the expanded uncertainty of 0.527 unit would be rounded to 0.53 unit when reporting it with the result. Rounding should always be carried out at the very end of the process, as intermediate rounding can introduce significant cumulative rounding errors.

Chapter 2

Uncertainty in Acoustic Measurements

Robert Peters

Introduction

Inevitably, repeated measurements of any specific sound level, reverberation time or vibration level will yield a range of results, indicating a degree of uncertainty or doubt about the measured value. This chapter aims to introduce the basic ideas and concepts relating to uncertainty. It repeats some of the general ideas in Chapter 1 but offers specific examples of uncertainty in acoustic measurements, including how to minimise and evaluate such uncertainties. Ideas and concepts relating to uncertainty are explored in more detail in the following chapters, each dealing with a different aspect of uncertainty in acoustic measurement, prediction and assessment.

Two of the references listed at the end of this chapter, the International Organization for Standardization's (ISO) *Guide to the Expression of Uncertainty in Measurement* (GUM 1995) and the United Kingdom Accreditation Service's *Expression of Uncertainty and Confidence in Measurement* (UKAS 2012), explain in detail the statistical evaluation of measurement uncertainty leading to an expanded uncertainty, stated to a certain confidence level (usually 95%), as explained in Chapter 1. The third reference, the short book by Stephanie Bell (2001) of the National Physical Laboratory (NPL), in addition to being easier to read, provides a gentler introduction and simplified versions of much of what is in the first two references. Only the fourth reference relates specifically to acoustics and sound levels. It is by Geoff Craven and Nicholas Kerry of University of Salford (2007) and relates specifically to uncertainty in the measurement of environmental noise. All four references are recommended to the reader, but the last two should receive particular attention.

In some cases, the uncertainty is determined not by tests of repeatability but by the resolution of the measuring instrument, that is, by the smallest scale interval available to the observer. A simple example would be measurement of the length of a pencil using a ruler graduated in centimetre divisions, which might produce a result such as 11.5 ± 0.5 cm. Another example might be measurement of a very steady sound level from a loudspeaker using an old analogue sound level meter calibrated in 1-dB steps – an example is given in the Craven and Kerry (2007) good practice guide, where the sound level is constant within 1 dB or so. Such a measurement might produce a result such as 55 ± 0.5 dB. In these situations, although the measurement should be repeated two or three times to confirm results and check for mistakes, further repetitions will not yield any further useful information simply because the result is limited by the resolution of the measurement device.

The situation changes if measurement resolution is increased – by using a ruler with millimetre divisions (or maybe better, such as vernier callipers, micrometers or optical measuring devices) or by using a digital sound level meters displaying sound levels to 0.1 dB. Repeated measurements will then reveal a spread of measurement results from which an estimate of uncertainty may be obtained, as explained later in this chapter.

Uncertainty and Variability

Understanding causes of uncertainty is about understanding what factors associated with the measurement procedure might vary and so cause the result to vary if we were to repeat the measurement whether immediately, the next day, during the next month or over a longer period. The two most obvious sources are variability in the quantity being measured and variability in measurement procedures, but other sources are those associated with the sound propagation between source and receiver and those associated with the receiver and the environment, including the effect of changes in weather conditions.

Variability in the Source (The Quantity Being Measured)

Variability in the quantity being measured includes variability in the sound emission from the sound source of plant and machinery. This can include obvious expected changes arising from changes in machine operating conditions, such as machine speed, load or feed material, all of which information must be included in the measurement report. Even if operating conditions remain constant, the noise emission from a machine might change with time owing to changes in temperature, change in mains electrical supply or changes arising from maintenance conditions or wear and tear.

Other sources of noise are associated with transportation, community and leisure noise. For surface transport, factors such as the vehicle flow rates' average speed, road and rail track conditions affect noise levels and will vary.

Any of these changes might result in changes not only to the overall decibels relative to Z (dBZ, i.e. unweighted) sound level of the machinery but also the spectral content in octave or one-third octave bands. Also, it should be noted that, usually, uncertainties in octave band levels, particularly for the lower-frequency bands, are significantly higher than those in combined A-weighted decibel (dBA) levels. This can be seen in tables of typical uncertainty levels found in standards on sound power determination, such as BS 7345.

Sometimes, particularly for environmental noise, it is the variability with time of the overall dBA level, expressed in terms of parameters such as L_{Amax}, L_{Aeq}, L_{A10} and so on, that is the purpose of the measurement, and each of these different parameters may be subject to a different degree of uncertainty.

For example, suppose that environmental noise measurements were taken over a period of 25 minutes at about 150 m from a dominant noise source at a greenfield site. The measurements were repeated on 10 occasions over a period of two months with similar wind and weather conditions. The environmental noise parameters L_{Aeq}, L_{A10}, L_{A90} and L_{Amax} were measured on each occasion.

Question: Which of these parameters should show the greatest variability, and which the least?

The results of the survey are shown in Table 2.1.

The maximum noise levels (L_{Amax}) show the biggest range and variability and the L_{A90} values the lowest.

Variability in Measurement Procedures

The measurement uncertainties arising from sound measuring equipment, most common of which is the sound level meter, is the subject of Chapter 3. Often, if the sound level meter is Class 1 (known as *Type 1* in earlier instruments), calibrated properly and used in accordance with the manufacturer's instructions, the uncertainty arising from the sound level meter will be small compared with other sources of uncertainty, particularly those associated with the source of sound, as discussed earlier. There will be cases, however, where the uncertainties associated with the sound level meter make a significant contribution to the overall level of uncertainty and must be included.

Table 2.1 Example data for environmental noise indices

Measured parameter	Lowest dBA	Highest dBA	Range dBA	Standard deviation[a]
L_{Aeq}	44.8	49.0	4.2	1.4
L_{A10}	46.2	50.7	4.5	1.4
L_{A90}	41.2	44.8	3.6	1.3
L_{Amax}	51.4	64.4	13.0	4.0

[a] The standard deviation is a measure of the variability of the sample. It was explained in Chapter 1 and will be discussed again later in this chapter.

In addition to uncertainty arising from the instrument itself, there are uncertainties arising from the way it is used. These could include the influence of sound-reflecting surfaces such as the body of the observer close to the microphone; changes in microphone position, height and orientation; the effect of wind over the microphone; selection of measurement position and the measurement sampling strategy, which would involve where, when, for how long and how frequently sound levels are measured. Another possible source of uncertainty is, of course, the operator of the sound level meter, who might vary the way the meter is set up and operated on different measurement occasions.

Variability Associated with Sound Propagation Conditions

The main factors affecting sound propagation between source and receiver are the distance between source and receiver; ground conditions, that is, whether sound reflecting or absorbing; refraction of sound due to wind and temperature gradients; shielding of sound due to sound screens or barriers, whether manmade or naturally occurring as a result of local topography; air absorption and scattering; and the presence of sound reflecting and scattering surfaces, particularly those close to either source or receiver. Variability in any of these factors leads to differences in sound levels measured at the receiver and therefore to uncertainty in the measured result.

Distance

According to the simplest theory, which works reasonably well in practice, the sound pressure level from a point sound source reduces at a rate of 6 dB for every doubling of distance from the source. This means that small changes in source-to-receiver distance are much more important for small distances than for large ones. For example, a change of 1 m in distance will produce a much greater effect on the sound level at the receiver for a microphone located 10 m from the source than for one at 100 m from the source – so careful and precise placement of the microphone is very important when measuring close to the sound source.

The other sound propagation effects will also vary with distance from the source – so, for example, ground attenuation and air absorption are measured in decibels per metre from the source – and so variability with distance can cause changes in the measured sound levels.

Weather and Environment

The main aspects of the weather that affect sound propagation include wind speed and direction and its variation with height above of the ground, air temperature and its variation with height above the ground, humidity and the presence of rain. Any changes in these aspects will lead to a change in sound propagation and therefore to changes in measured sound levels.

The environment includes not only the weather but also the ground and terrain over which sound travels, and its topography and the presence of trees, bushes and other vegetation, as well as the level of background noise. The sound-absorbing and -reflecting behaviour of the ground can change with the weather and through the seasons of the year. Variation in some of these factors can cause changes in the generation of sound at the source, during its propagation to the receiver, and at the receiver position by their effect on the sound level meter. For example, rain can affect not only the generation of road traffic noise at the source but also sound propagation as well as affecting the sound level meter used to measure the sound level.

Changes in these various influences can affect the frequency content of the measured sound level as well as its overall level. Techniques for identifying and separating the influence of some of these factors, such as day-to-day variations, the sound level meter and the operator are discussed in Chapter 5.

A Simple Example of Uncertainty in Measurement: How Long Is This Pencil?

Consider a new, unused pencil; that is, each end is flat and unsharpened, so the ends are well defined. If I hand you a pencil and a ruler marked with centimetre divisions and ask you how long the pencil is, you might, after taking the measurement, say that it is about 12 cm long, or between 11 and 12 cm long, or 12 cm to the nearest centimetre. If I then ask you to say whether it is closer to 11 cm than to 12 cm, you might come to the conclusions that the proper evaluation of the measurement is 11.5 ± 0.5 cm. In this case, the preciseness of the measurement is limited by the measuring instrument. Although I may ask you to repeat the measurement a couple of times just as a check, there would be no need for further repetition in this case – we will always get the same result.

Now suppose that I supply you with a ruler with millimetre divisions and ask you to repeat the measurement. It might be assumed that you would be able to measure to the nearest half millimetre and state a measurement result of 117.5 ± 0.5 mm.

If it was important enough and we were persistent enough, we could employ more precise measuring instruments and procedures with greater degrees of precision, using, for example, vernier callipers, micrometers, a microscope or a laser measuring device.

Now, under these conditions, we might find that repeating the measurements with the same or different observers might yield a spread of results for a variety of reasons. Expansion of the material of the pencil might become a significant factor affecting the measured length, and it might necessary control the temperature of the pencil during measurement.

So far, we are assuming that it is always the same pencil that is being measured, but it might be that we are measuring different pencils from the a nominally identical batch perhaps manufactured to a certain length tolerance, say, for example, to within 10%. Similarly, in acoustics, we might be carrying out repeated sound insulation measurements on the same building element or on a range of similar elements, or of the sound absorption coefficient on a particular sample of sound-absorbing material or on a range of samples.

This simple example has been inspired by a much more thorough and complete explanation about how to estimate the uncertainty in the measurement of the length of a piece of string in the book by Stephanie Bell (2001), which is commended to the reader. This example also illustrates the statistical procedures described later to provide a 95% confidence limit on the uncertainty. Another good illustrative example relating to the uncertainty in the measurement of the height of a flagpole is given in UKAS (2012).

Estimating and Reporting Uncertainties in Measurements

This section repeats much of what is written in Chapter 1 but in a much less mathematical way and, moreover, gives example acoustical contexts. As part of good scientific and engineering practice, we are always encouraged to repeat measurements first to check for mistakes but also because the spread of results from repeated measurements gives us the information we need to obtain a better estimate of the value of the measured quantity as well as a way of estimating its uncertainty.

Mean Value

The best estimate of a measured quantity is usually taken as the average or arithmetic mean of the results of repeated measurements, and the more times the measurement is repeated, the better is the estimate.

Spread and Range

The spread of the results around the mean value forms the basis of ways of quantifying the uncertainty of the measurements. The simplest way to record variability, and so the uncertainty of measurement results, is to state the range, from lowest to highest, of the measured values. However, there is a risk that this may give an unrepresentative indication if it is based on just a few measurements that could include a single 'outlying' result, and there is also more of a chance that the mean will not lie at the centre of the range. More measurements will give a better indication of the spread.

Standard Deviation

The usual way of specifying the spread or variability of the uncertainties of results is to calculate and state the standard deviation of the sample of repeated results. The standard deviation is a way of showing how, on average, the measurement results differ, or deviate, from the mean value. It is the square root of the mean squared deviation of the measurements and is defined mathematically as follows:

$$s = \sqrt{\frac{\sum_{i=1}^{n} (x_i - \bar{x})^2}{n - 1}}$$

where s is the standard deviation, and x_i is the ith measurement of n samples and is the average or mean value of the measurement results.

Example: *Measured SEL Values of Trains*

The sound exposure level (SEL, or L_{AE}) contains information about the sound energy, usually A-weighted, of a discrete noise event such as produced by an individual train, aircraft or motor vehicle. It is the A-weighted sound pressure level which, if occurring over a period of one second, would contain the same amount of A-weighted sound energy as the event.

The continuous equivalent sound level of noise from trains over a period of time (a 16-hour daytime period, for example) can be predicted from the SEL values and the number of train events. Measuring the SEL value of each type of train on a particular railway line means that average levels of train noise can be predicted rather than having to overcome the difficulties of directly measuring the train noise levels over several hours, including filtering out the effects of other sources of noise.

To gather data for a report to accompany a planning application to build dwellings on land adjacent to a railway, the SEL values of 50 trains were measured using a Type 1 sound level meter located 10 m from a railway line in south-west London over a period of five hours.

During the approach of each train, observed either visually or aurally, the sound level meter was switched manually from standby to measurement mode, and the measurement was terminated when the train noise was no longer audible above background noise. The weather was dry and sunny, with little or no breeze. Although not measured, wind speeds were considered to be much less than 5 m/s. The microphone of the sound level mater was satisfactorily calibrated before and after the measurements and was fitted with a windshield during the measurement period. The measuring position was about 15 m from a railway station, where some of the trains stopped and others did not. Passenger trains using the line could be divided into four types: non-stop travelling eastwards, non-stop travelling westwards, stopping (i.e. slowing down) and starting (i.e. accelerating).

For example, the SEL (dBA) measured for 11 non-stopping west-bound trains were as follows:

Train number	1	2	3	4	5	6	7	8	9	10	11
SEL, dBA	95.4	95.9	93.7	95.4	96.7	96.7	89.8	89.8	87.9	96.5	97.0

The arithmetic average of these values is 94.1 dBA, the range from lowest to highest is 9.1 dBA, and the standard deviation is 3.3 dBA.

Similar measurements of SEL were made for each train type. The results are summarized in Table 2.2.

Table 2.2 Example data for the SEL of train pass-bys

Train type	Number in measurement sample	Mean value of SEL, dBA	Lowest to highest	Range	Standard deviation
Nonstop to the west	11	94.1	87.9–97.0	9.1	3.3
Nonstop to the east	12	93.4	88.4–97.9	9.5	3.5
Stopping	12	84.6	82.6–87.1	4.5	1.4
Starting	15	84.4	81.8–87.7	5.8	1.9

The noise levels from non-stop trains are much higher than those from trains that were stopping, but the direction of travel made only a small and probably insignificant difference. The standard deviations can be used to estimate the uncertainty of the calculated average noise level from trains over the required daytime, night-time or evening periods.

Vibration dose values (VDVs) were also measured for each train events and mean and standard deviation obtained, from which, knowing the total number of trains of each type passing during the daytime and night-time periods, it was possible to predict VDVs over the entire day and night periods together with estimates of uncertainty.

The main sources of uncertainty were variability in the noise levels owing to individual trains and operation of the sound level meter. Over the relatively short measurement period and the short distance between source and receiver, it was unlikely that variability in the weather, the environment or sound propagation was a significant contributor to the overall level of uncertainty.

If levels of train noise are required at longer distances from the track or over much longer periods of time, then these factors are likely to become much more significant. In such cases, the options are to carry out more extensive measurements further from the rail track over a much longer period of time or to estimate train noise levels and their uncertainties by constructing an *uncertainty budget*, which uses the uncertainties in existing train noise data but builds in the contributions to uncertainty from variations in these factors based on theory, published data and personal experience. Uncertainty budgets were mentioned in Chapter 1 and feature again later in this and other chapters.

Repeatability and Reproducibility

If repeated measurements are carried out by the same person or by persons who are part of the same team using the same instrument and measuring technique each time, then the resulting data are tests of *repeatability*. If the measurements are repeated by other teams following nominally the same measurement procedure but using their own different measuring instruments and their own, maybe slightly different interpretation of the measurement procedure with the purpose of comparing the variation between the results obtained by different teams, then the resulting data are tests of *reproducibility*.

Therefore, the repeatability of a measurement is the closeness of the measurement results repeated by the same person or team under the same conditions. The reproducibility of a measurement is the closeness of measurement results repeated by different persons or teams under maybe different conditions. The data variation associated with reproducibility will always be greater than that associated with repeatability.

Random and Systematic Uncertainties

Random uncertainties are those which change in a random way each time a measurement is repeated. They can never be eliminated from a measurement result, but their magnitude can be estimated by repeating measurements under the same conditions – the more repetitions, the closer will be the mean to the 'true' value.

Systematic uncertainties are those which do not change when measurements are repeated and so cannot be detected by repeatability tests. Systematic uncertainties are sometimes detected as a result of comparisons with results of measurements carried out in a different way or by reproducibility tests and 'round-robin tests'. Once they have been detected, these types of uncertainty can be allowed for or eliminated. An error in calibration is an example of a systematic uncertainty. For example, this might cause a sound level meter to read 1 dB high until the error is detected, when it is easily corrected.

Type A and Type B Uncertainties

The GUM (1995) and UKAS (2012) classify uncertainties into two types. *Type A* uncertainty estimates are based on statistical analysis (standard deviation) of repeated measurements – to take into account random factors and fluctuations that affect the measured result. *Type B* uncertainty is estimated from all other, non-random factors such as calibration certificates, manufacturers' specifications, calculations, published information, the experience of the person carrying out the measurements and common sense.

Statistical Frequency Distributions: Gaussian and Rectangular Distributions

It is useful to be able to link the estimated uncertainty in a measurement to a statement of the probability (or confidence) that any given measurement will lie within certain limits. To do so, it is necessary to know how the uncertainties in a measurement (the deviations from the mean value) are distributed around the mean value, that is, the so-called frequency distribution of the values.

Two example frequency distributions are the Gaussian or normal distribution and the rectangular distribution. These distributions are useful because they can be applied to the two most common situations that arise when dealing with measurement uncertainties: (1) when either the range of measurements is limited by the smallest measurement interval of the measuring instrument or all we know is the range between the lowest and highest measurement values and (2) where the uncertainty is estimated from the repeatability or reproducibility of the measurement.

The Gaussian distribution applies when the variations in a quantity are subject to random fluctuations. It is common to assume that the fluctuations in a measurement displayed by repeatability tests are random, or approximately random, so the theory of the Gaussian distribution can be applied.

The general idea behind the theory of the normal or Gaussian distribution of uncertainties is that there are many different sources that contribute to the total uncertainty in a measurement. Some of these will be positive (i.e. cause an increase in the measurement result), and some will be negative, and some will be small in magnitude and some large in magnitude. If it is assumed that these different uncertainties are independent of each other and uncorrelated, random in magnitude and direction (positive or negative), then there will be an equal probability that they will be positive or negative, so it is much more likely that they will cancel each other out to produce a small total than that they will combine to produce a large positive or negative total.

In a very simple analogy, suppose that a coin is tossed many times and that each time it is heads there is a score of +1 and a score of –1 for tails. It is assumed that, on any one occasion, there is a 50% chance of heads and a 50% chance of tails. According to the Gaussian statistics, the net score from many tosses is much more likely to be zero, or a positive or negative number close to zero, than to be a large (positive or negative) number.

According to Gaussian theory applied to the distribution of uncertainties, there is a 68% chance that any one measurement value will be within one standard deviation of the mean value and a 95% chance that it will lie within two standard deviations of the mean. This is the basis of the method of stating uncertainty with a 95% confidence level.

The rectangular distribution may be applied when we do not have any repeatability data but only a range between maximum and minimum values, maybe set by the minimum display range of the instrument. In this case, it is often assumed that measurement values are uniformly distributed between the maximum and minimum values. The Gaussian or normal distribution and the rectangular distributions are illustrated graphically in Figure 2.1.

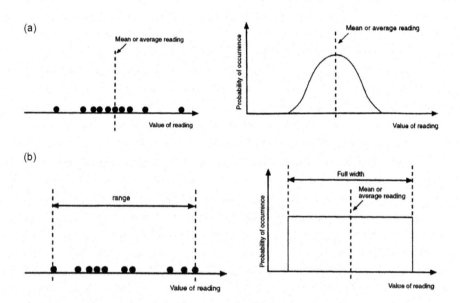

Figure 2.1 (a) 'Blob' plot of a set of values with a normal distribution around a mean value and a corresponding plot of a normal distribution of the probability of occurrence which peaks at the mean value. (b) Blob plot of a set of values with a rectangular distribution and a corresponding plot of a rectangular distribution of the probability of occurrence which is the same at every value.

(Reproduced with permission from Bell S, A Beginner's Guide to Uncertainty of Measurement, Measurement Good Practice Guide No. 11 (Issue 2), National Physical Laboratory, Teddington, UK, 2001.)

There are other theoretical probability distributions which can be applied to measurement uncertainty analysis. Readers should consult the first two references listed at the end of this chapter for more details.

Standard Uncertainty u

The standard uncertainty of a specific source of uncertainty in a measurement is the value of the uncertainty that corresponds to plus or minus one standard deviation from the mean value. When combining different sources of uncertainty, it is important that they are all expressed in terms of a standard uncertainty, that is, to the same level of uncertainty before they are combined, so that we are comparing and combining 'like with like'.

Type A uncertainties: Calculate the standard deviation u from the mean value s of N repeated measurements, and, assuming random (Gaussian) statistical distribution of the values, convert to a standard uncertainty u by dividing by $\sqrt{N} - 1$:

$$u = \frac{s}{\sqrt{N} - 1}$$

Type B uncertainties: If only a range of possible values R is known, then assume a rectangular statistical frequency distribution, and calculate standard uncertainty u using

$$u = \frac{R/2}{\sqrt{3}}$$

Statistical Method for the Estimation of Uncertainty

The method outlined briefly in this section is described in more detail in ISO/IEC Guide 98–3: 2008, *Uncertainty of Measurement*, Part 3: *Guide to the Expression of Uncertainty in Measurement* (GUM 1995), and in *Expression of Uncertainty and Confidence in Measurement*, Edition 3 (UKAS 2012).

Each source of uncertainty must be expressed, quantitatively, in the same way, called a *standard uncertainty*, before the sources can be combined. This is based on a spread corresponding to plus or minus one standard deviation from the mean for a standard (or Gaussian) distribution.

For Type A uncertainties, the standard deviation s of a sample of repeated measurements is calculated. For this type of uncertainty, the standard uncertainty u is obtained by dividing the standard deviation by the square root of the number of samples: $u = s/\sqrt{N}$.

There may be other, Type B uncertainties, for which we have only a range of uncertainty. In this case, we may assume a rectangular distribution of uncertainty, that is, that all values of uncertainty within the range are equally probable, and the standard uncertainty is obtained by dividing half the range by $\sqrt{3}$ (i.e. by 1.732).

When all the sources of uncertainty have been expressed in terms of a standard uncertainty, they may be combined to obtain the value of the combined standard uncertainty using the 'square root of the sum of the squares' method:

$$u_c = \sqrt{\frac{(u_1^2 + u_2^2 + u_3^2 + \cdots + u_N^2)}{N}}$$

The combined standard uncertainty is expressed in terms of a range of plus or minus one standard deviation from the mean value.

The final stage in the process is to express the measurement uncertainty in terms of a confidence limit. This is done by multiplying the combined standard uncertainty by a coverage factor k to obtain the expanded uncertainty U, that is, $U = ku$.

The most commonly used value for k is 2, which gives a combined uncertainty expressed with 95% confidence. The relationship between the coverage factor k and the coverage probability is given in Table 2.3. The process is displayed in Figure 2.2.

Application of Uncertainty Estimates to Determination of Compliance

The following example relates to the determination of compliance with the lower action value of 80 dBA of the 2005 control of noise at work

Table 2.3 Relationship between coverage factor and coverage probability

Coverage factor k	Coverage probability, %
1	68
1.28	80
1.65	90
1.96	95
2.58	99
3.29	99.9

Source: From BS EN ISO 12999–1: 2014.

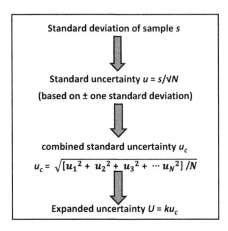

Figure 2.2 Process for determining expanded uncertainty using the statistical method.

regulations. In the example, the personal daily nose exposure level of an employee has been measured. The result of 10 measurements is a mean level (L_M) of 78 dBA and a standard uncertainty u of ±2 dBA. Using a coverage factor of two gives an expanded uncertainty U of ±4 dB with a 95% confidence interval. This means that the noise exposure level lies between 74 and 82 dB with a confidence of 95%.

In this case, however, to determine compliance with the lower action value (L_s) of 80 dBA of the 2005 control of noise at work regulations, it is only the upper limit that is of interest. According to appendix M of M3003 (edition 2, January 2007), this means using a one-tailed distribution with a coverage factor of 1.64 rather than 2 (paragraph M2.10).

Accordingly, $U = 1.64 \times 2$, and the upper limit is 78 + 3.28 = 81.28 dBA, so we can say that the noise exposure level will not exceed 81.28 dBA with a confidence limit of 95%.

Note that BS EN ISO 9612 uses the one-tailed distribution approach with a k factor of 1.65 for estimating the uncertainty of noise exposure levels in the workplace.

If I want to determine the confidence level for complying with the limit of 80 dBA, then using paragraphs M2.14 and M2.15, the table of M2.16 and the example of M2.20, I obtain a confidence level of 84%,[1] which means that the measured noise exposure levels will comply with (i.e. not exceed) the level of 80 dBA with a confidence level of 84%.

---

1 $|(L_s - L_M)/u_c| = [(80 - 78)/2] = 1$ which leads to 84% in the table in paragraph M2.16.

Uncertainty Budgets

An uncertainty budget is a set of calculations setting out the stages in the determination of the expanded uncertainty, as explained earlier. Craven and Kerry (2007) in their guide have suggested that a useful approach to drawing up sources of uncertainty for consideration to include in an uncertainty budget is to review factors relating to, for example, the following:

- *Source:* type, age, condition (maintenance), stability/variability, operating conditions (load, speed, work material, etc.), position relative to microphone, weather conditions
- *Transmission path:* distance; weather, wind speed and direction, temperature gradients; ground conditions; screening/shielding and scattering by nearby surfaces – variability of all of these
- *Receiver:* microphone position, nearby reflecting surfaces, instrumentation and measurement procedure, background noise, environmental influences – variability of all of these

Examples of uncertainty budgets are given in the books by Bell and Craven and Kerry (2007), as well as later chapters in this book.

Combining Uncertainties of Compound Quantities

Sometimes the quantity we wish to measure is a combination of other quantities, each of which has to be measured, and the results combined, together with their uncertainties. A simple example is in estimating the area of a rectangular carpet or floor by measuring length and width and multiplying them together. Suppose that both length and width can be measured with an uncertainty of up to 10% of the true values, but with a random distribution between the maximum of $\pm 10\%$. It is possible that the area of the carpet could be either overestimated or underestimated by 20%, but this would be very unlikely, and the most likely outcome would be $\pm\sqrt{10^2 + 10^2} = \pm 14.14\%$, that is, an error of about 14%.

This can be expressed in terms of fractional uncertainties as

$$\Delta S/S = \sqrt{(\Delta L/L)^2 + (\Delta B/B)^2}$$

where ΔS is the fractional uncertainty is area S, ΔL is the fractional uncertainty in length L, ΔB is the fractional uncertainty in width B, and an uncertainty of $\pm 10\%$ corresponds to a fractional uncertainty of ± 0.1.

This approach can be extended to situations where the quantity to be measured is a combination of many other quantities either multiplied or divided together or involving power-law relationships. For example, the quantity to be measured, M, might depend on two other quantities P and R via a power relationship:

$$M = kP^m R^n$$

where k, m and n are constant values.

The combined fractional uncertainty ΔM in the value of M will depend on the fractional uncertainty ΔP in the value of P and the fractional uncertainty ΔR in R as follows:

$$\Delta M/M = \sqrt{m(\Delta P/P)^2 + n(\Delta R/R)^2}$$

Weighting Factors and Sensitivity Coefficients

This analysis shows that in determining the overall measurement uncertainty of a quantity from various component measurements, the combined uncertainty may be more sensitive to uncertainty in some components than in others. Consequently, the different uncertainties that make up the total or combined uncertainty may have to be weighted with a sensitivity coefficient. Here are a couple of examples relating to acoustics.

Example 1
The sound power W produced in a duct of cross-sectional area S by the flow of air at velocity V is to be determined by measuring S and V:

$$W = kSV^5$$

Therefore, $\Delta W/MW = \sqrt{(\Delta S/S)^2 + 5(\Delta V/RV)^2}$. This shows that the sound power is five times more sensitive to a fractional change in airflow velocities than it is to the same fractional change in area.

Example 2
The sound power W radiated by a small sound source is to be determined using a sound intensity meter to measure the sound intensity I at a distance r from the source under free-field conditions.

The equation (uncertainty model) is $W = 4\pi r^2 I$. The uncertainty relationships are

$$\Delta W/W = \sqrt{2(\Delta r/r)^2 + (\Delta I/I)^2}$$

Therefore, in the determination of sound power from a measurement of sound intensity at a given distance, the uncertainty is twice as sensitive to a certain fractional change in distance than it is to the same fractional change in sound intensity.

This is a very simplified approach to determining how the uncertainties in compound quantities should be combined. A more rigorous mathematical treatment, explained in the GUM, requires knowledge of the mathematical techniques of partial differentiation, as was discussed in Chapter 1 and is illustrated in Example 3.

Example 3   Illustrating the use of sensitivity coefficients

The sound intensity I radiated by a small sound source is measured at 2 m from the source and found to be 0.25 W/m^2 with a standard uncertainty of ±0.05 W/m^2. The standard uncertainty in the distance is ±0.1 m. Calculate the sound power radiated by the source, and estimate the uncertainty.

The value of the sound power is obtained using $W = 4\pi r^2 I = 4\pi 2^2 \times 0.25 = 12.6$W.

There are two approaches to estimating the uncertainty. The first method is, in effect, an approximate method which estimates the change in the final result arising from the uncertainty in each variable using the functional relationship between the variables and the final outcome (i.e. the defining mathematical equation) and then combining the component uncertainties using the square root of the sum of the squares method.

The second, more theoretically correct method is to derive the sensitivity coefficient C_j for each variable j by partial differentiation of the model equation for each of the variables and to use these, together with the estimates of uncertainty in each variable, to determine the combined uncertainty U_c from

$$\text{Combined uncertainty } U_c = \sqrt{(C_1 U_1)^2 + (C_2 U_2)^2 + (C_3 U_3)^2 + \cdots + (C_n U_n)^2}$$

Method 1

The measured value of the sound power is obtained using $W = 4\pi r^2 I = 4\pi 1^2 \times 1 = 12.6$ W. Uncertainty arising from variability in measurement of sound intensity I:

I^+: Using values of $I = 0.3$ W/m^2 and $r = 2$ m gives $W^+ = 15.1$ W.
I^-: Using values of $I = 0.2$ W/m^2 and $r = 2$ m gives $W^- = 10.1$ W.
Average $= (W^+ - W^-)/2 = (15.1 - 10.1)(15.1 - 10.1)/2 = 2.5$.

Uncertainty arising from variability in measurement of distance r:

r^+: Using values of $r = 2.1$ m and 0.25 W/m^2 gives $W^+ = 13.9$ W.
r^-: Using values of $r = 1.9$ m and 0.25 W/m^2 gives $W^- = 11.3$ W.
Average $= (W^+ - W^-)/2 = (13.9 - 11.3)/2 = 1.3$.

$$\text{Combined uncertainty } U_c = \sqrt{(2.5)^2 + (1.3)^2} = \pm 2.8 \text{ W}$$

Method 2

Sensitivity coefficients are obtained by partial differentiation with respect to I (that is, $\partial W/\partial I$) and r (that is, $\partial W/\partial r$) of the equation $W = 4\pi r^2 I$:

$$C_I = \frac{\partial W}{\partial I} = 4\pi r^2 = 4 \times 3.142 \times 2 \times 2 = 50.3$$

Similarly,

$$C_r = \frac{\partial W}{\partial r} 8\pi r I = 8 \times 3.142 \times 2 \times 0.25 = 12.6$$

Component Uncertainties

$$U_I = 0.05 \text{W/m}^2$$

$$U_r = 0.1 \text{ m}$$

$$\text{Combined uncertainty } U_c = \sqrt{(C_I U_I)^2 + (C_r U_r)^2}$$

$$= \sqrt{(50.3 \times 0.05)^2 + (12.6 \times 0.1)^2} = \pm 2.8\text{W}$$

The two methods agree. Some further examples, involving the estimation of uncertainties in simple decibel calculations, are given in the annex to this chapter.

GUM Approaches to Estimating Measurement Uncertainties

The GUM (1995) states that

> "if all of the quantities on which the result of a measurement depends are varied, its uncertainty can be evaluated by statistical means. However, because this is rarely possible in practice due to limited time and resources, the uncertainty of a measurement result is usually evaluated using a mathematical model of the measurement and the law of propagation of uncertainty. Thus, implicit in this Guide is the assumption that a measurement can be modelled mathematically to the degree imposed by the required accuracy of the measurement."

This quotation indicates that the GUM gives two alternative approaches (often complementary) for the determination of measurement uncertainties.

The first method is to perform a series of reproducibility tests (often known as *round-robin tests*) which are sufficiently comprehensive to encompass all the variations in factors likely to cause changes in the measurement result. This may be a very time-consuming process and may be impracticable in some cases. The alternative method is to estimate the uncertainties arising from each of the factors that can affect the measurement result and then combine them to produce an estimated total uncertainty – or, in other words, to construct an uncertainty budget. This second approach requires knowledge of exactly how the variability in each of the factors affects the measurement result, that is, knowledge of the functional relationship between all the factors and the measurement result or, in the language of the GUM, an *uncertainty model*. This could be a mathematical equation involving all the factors and the value of the measured quantity. The equation $W = 4\pi r^2 I$ is a good example of an uncertainty model, as was illustrated in Example 3. In many cases, however, insufficient information is available to construct a full model, and the uncertainty budget has to be constructed using a combination of experience, previous results, published information and noise measurements (e.g. to determine how changes in a particular factor affects the measurement result).

Uncertainty in the Prediction and Assessment of Noise Levels

Uncertainty in noise level prediction may arise from

- the input data (e.g. published sound power levels, sound reduction index values or sound absorption coefficients),
- the prediction model (its assumptions and limitations and range of valid input parameters),
- simplifying the real situation to 'fit' the model (user influence on modelling) or
- the calculation method

Round-robin tests are used to compare different software – to compare results of modelling the same situation using different software applications.

Some Example of Uncertainties in Calculation Methods

The *Calculation of Road Traffic Noise* (CRTN 1988) estimates that the uncertainty in its prediction of the road traffic noise $L_{A10.18h}$ parameter is ±2 dB. ISO 9613-2: 1996, *Acoustics: Attenuation of Sound during Propagation Outdoors*, gives an estimated accuracy (A-weighted, for averaged moderate downwind propagation for distances < 1,000 m and average height of propagation < 30 m) of ±3 dB.

BS EN 12354–5: 2009, *Building Acoustics: Estimation of Acoustic Performance of Buildings from the Performance of Elements*, Part 5: *Sounds Levels due to the Service Equipment*, states (in section 5 on accuracy)

> "As a global indication the expanded uncertainty for the single number ratings (A- or C-weighted levels) with a coverage factor of 2 could be estimated as up to 5 dB for the source input data and up to 5 dB for the transmission predictions; assuming these two aspects to be independent the overall expanded uncertainty would thus be up to 7 dB .... More research and comparison will be needed to be able to specify these uncertainties more accurately and in more detail."

Why Is Uncertainty Important? – Uncertainty and Decision Making

All measurements have a purpose. Often, they lead to decisions such as whether or not

- a theory is confirmed,
- sound insulation of building element passes design targets or building regulations,
- planning permission for a development is agreed,
- acoustic properties of a material or measured or predicted noise levels meet performance specifications,
- possible enforcement action against noise producer should be triggered,
- expensive noise mitigation measures are needed, or
- there are health implications for exposed populations.

All these decisions are much better informed by knowledge of measurement and prediction uncertainty.

Note that differing degrees of uncertainty may apply to different decisions and situations – a fairly quick and imprecise set of measurement, for example, may be ±5 dB – may be adequate in some situations, whereas for others a much greater level of accuracy, say ±2 dB or maybe even ± 0.5 dB may be needed.

The UK standard BS 4142 not only specifies procedures for the measurement of the appropriate environmental noise parameters but also specifies the procedure for applying the results of the measurements to the assessment of the impact of the noise. In its approach to the estimation of uncertainty, the first priority is to decide and state whether or not assessment decisions are likely to be affected by measurement uncertainty. The standard also emphasises the need to minimise uncertainties as far as possible and lists ways for doing this.

The Need to Build Repeatability into Measurements

It is obviously a good idea to repeat measurements, but many acoustic measurements are carried out in a very competitive commercial environment to a very tight time and cost budget. We are never going to be in a situation of being able to say to our client, 'I have carried out the measurements that you require, but I would like to repeat them tomorrow in order to be able to estimate uncertainty'. Therefore, estimation of uncertainty including repeatability needs to be considered when specifying scopes of work and drawing up estimates. This is relatively easy in the case of measuring the SEL values of trains illustrated earlier but may be more difficult in other situations, such as carrying out sound insulation measurements.

It might be that the uncertainty in the measurement procedure can be established by repeatability or reproducibility tests on one measurement occasion and then applied to all other measurements carried out in the same way. It will then be important to repeat this set of tests from time to time, or whenever there is a change in the measurement procedure, to make sure that levels of uncertainty have not changed or that the procedure is updated.

Who Is Responsible for Declaring the Uncertainty in Measurement Result?

If it is accepted that the responsibility for a measurement result lies with the operator who carried out the measurements or with the author of the measurement report, then it surely follows that so does the responsibility for the statement of uncertainty. This surely continues to apply even if it is simply claimed that all measurements and uncertainties have been carried out according to an appropriate national or international standard. In this case, it is the responsibility of the operator or report author to be fully acquainted with the requirements of the standard and to ensure that he or she is aware of and reports any unusual measurement conditions that do not comply. Statements such as 'the measurements were carried out "generally" or "broadly" in accordance with the [stated] standard' should be treated with caution unless the areas of non-compliance are stated and explained.

Requirements of Measurement Standards for Measurement Uncertainty

Standards on acoustic measurements issued since about 2000 usually contain a section on uncertainty which outlines what is required for compliance with the standard and often refer to information given in an annex explaining how these requirements might be achieved. In the future, as acoustic measurement standards issued before 2000 become updated and reissued, it is expected that more information and advice about how to estimate measurement uncertainties will become available.

An example of the kind of statement about uncertainty that might appear in a measurement standard is as follows:

> "The uncertainty of results obtained from measurements according to this International Standard shall be evaluated, preferably in compliance with ISO/IEC Guide 98–3. If reported, the expanded uncertainty together with the corresponding coverage factor for a stated coverage probability of 95% as defined in ISO/IEC Guide 98–3 shall be given. Guidance on the determination of the expanded uncertainty is given in the annex."

In some standards, it is explained that, as yet, there is insufficient information to develop a functional model relating the various factors that affect the uncertainty in the measured quantity and that, therefore, it will be impossible to determine an expanded uncertainty in accordance with the requirements of ISO/IEC Guide 98–3. In these cases, the standard either recommends that the uncertainty statement in a report be based on reproducibility tests or supplies an estimate of the likely maximum uncertainty from such tests.

Uncertainty and Decibels: Consequences of Measuring Sound in Decibels

In Chapter 1, John Hurll advised that

> "relatively small uncertainties expressed in decibels may be combined in the same way as those expressed as linear relative values. For large values of uncertainty, however, it is recommended that the measurand and its uncertainty be converted into linear units before the evaluation process and converted back to decibel terms afterwards. This is likely to result in uncertainties that are not symmetrical around the reported result; if the asymmetry is significant, then the negative- and positive-going uncertainties should be reported separately."

Part of the problem is that the arithmetic mean of a set of decibel readings will not be the same as their energy logarithmic average, as measured, for example, by the L_{eq} value.

> "In their book, *Machinery Noise Measurement*, Yang and Ellison (1985) state that if the sound-pressure levels of a set of measurements have a spread (the maximum variation of L_p values) of less than

5 dB, it is common practice to assume that, to a sufficient degree of accuracy, the mean sound pressure level (i.e. the log average) is equal to the average of the arithmetic average of the sound pressure levels."

These authors also state that

"[f]rom over 250 sets of noise data measured from electric machines, it has been found that the distribution of sound-pressure levels around an electric machine on various spherical and hemispherical measuring surfaces in a semi-reverberant space and in an anechoic chamber is approximately normal (Gaussian) distribution and that the difference between the mean sound-pressure level and the linear arithmetic average of sound-pressure levels can be approximately expressed as a function of the spread, S dB, as:"

$$L_{p,\,\text{logavge}} - L_{p,\,\text{arith.avge}} = 0.0184 \times S^{1.56}$$

For a spread of $S = 5$ dB this gives a difference of 0.2 dB.

Table 2.4 shows how the difference derived from this equation varies for a range of values of the spread of decibel values.

Example 4

This example illustrates how to calculate the mean and standard deviations of a sample of five sound levels (48, 49, 50, 51, 52) that differ in 1-dB steps from each other, for both the linear and logarithmic averaging methods. To determine the log average and standard deviation, the decibel levels are first converted back to the sound energy–related parameter $(p/p_o)^2 = 10^{L/10}$

The steps in the calculations are illustrated in the Table 2.5.

Table 2.4 Difference between logarithmic and linear (arithmetic) averages corresponding to increasing spreads in decibel values

Spread of decibel values, dB	Difference between log and arithmetic average values, dB
5	0.2
10	0.7
15	1.3
20	2.0
25	2.8
30	3.7

Table 2.5 Comparison of linear and logarithmic averaging of a set of sound levels

Linear averaging of decibel levels			Logarithmic averaging of decibel levels		
L	d	d^2	$(p/p_o)^2 = 10^{L/10}$	d	d^2
52	+2	4	1.58×10^5	$+5.32 \times 10^4$	2.82×10^9
51	+1	1	1.25×10^5	$+2.05 \times 10^4$	4.28×10^8
50	0	0	1.00×10^5	-5.38×10^3	2.90×10^7
49	−1	1	7.94×10^4	-2.59×10^4	6.73×10^8
48	−2	4	6.31×10^4	-4.22×10^4	1.79×10^9
Mean = 50	SD = 1.6		Mean = 1.05×10^5	SD = 3.79×10^4	

SD, standard deviation.

Thus, for linear averaging of the decibel levels, we have $50\,\text{dB} \pm 1.6\,\text{dB}$. The value of $(p/p_o)^2$ is $(1.05 \pm 0.38) \times 10^5$. Converting these values of $(p/p_o)^2$ back into decibel values, we have

$$\text{Average level} = 10\log(p/p_o)^2 = 10\log(1.05 \times 10^5) = 50.2 \text{ dB}$$

$$\text{Mean} + \text{standard deviation} = 10\log[(1.05 + 0.38) \times 10^5] = 51.6\,\text{dB}$$
$$= 50.2 + 1.4 \text{ dB}$$

$$\text{Mean} - \text{standard deviation} = 10\log[(1.05 + 0.38) \times 10^5]$$
$$= 48.3 \text{ dB} = 50.2 - 1.9 \text{ dB}$$

Thus, we have $50.2 + 1.4/-1.9$ dB.

The calculations have been repeated for samples of five decibel levels which all have the same arithmetic mean values of 50 dB but which differ in 2- and 3- dB steps from each other: (46, 48, 50, 52, 54) and (44, 47, 50, 53, 56). The three sets of samples have spreads, or ranges, of 4 (as above), 8 and 12 dB, respectively.

The results of these calculations are shown in Table 2.6, illustrating the increasing differences between arithmetic and linear average and increasing asymmetry between plus and minus deviations with increasing spreads of values in the sample.

Limitations and Keeping a Sense of Proportion about Uncertainty

According to Stephanie Bell (2001), 'A measurement result is complete only when accompanied by a quantitative statement of its uncertainty'. Although this is true, a sense of perspective is sometimes required.

Table 2.6 A numerical illustration of the increasing differences between linear and logarithmic averaging with increasing spreads of sample values

Steps	Range, dB	Spread, dB	Calculations based on decibel values		Calculations based on p² values, dB			
			Mean, dB	SD, dB	Mean	SD+	SD−	Mean ± SD
1 dB	48–52	4	50 dB	±1.6	50.2	51.6	48.3	50.2 + 1.4/−1.9
2 dB	46–54	8	50 dB	±3.2	50.9	53.2	45.8	50.2 + 2.3/−5.1
3 dB	44–56	12	50 dB	±4.7	51.9	54.9	34.7	51.9 + 3.0/−7.2

SD, standard deviation.

Suppose, for example, that we are presented with a measured sound level stated as 62.4 ± 3.4 dB. The most important information contained in this statement is that the measured level is 62.4 dBA. Ideally, this is the result of a carefully considered and planned measurement procedure carried out as accurately as possible. Also, the ±3.4 dB is the best estimate of the uncertainty of the measurement. It is important as well that the result (62.4 dB) is as accurate as possible.

The estimate of uncertainty attached to the measured value should be accompanied by an explanation of how the estimate was obtained. Ideally, this might be a fully budgeted estimate of expanded uncertainty to a 95% confidence limit carried out in accordance with the methods of the GUM, or it might be more simply derived from repeatability tests, or it might even be a 'rough and ready' estimate arising from the range of a set of measurements. The first is preferable, but even a rough and ready estimate of uncertainty is better than no estimate at all.

References

Craven NJ, Kerry G, *A Good Practice Guide on the Sources and Magnitude of Uncertainty Arising in the Practical Measurement of Environmental Noise*, Edition 1a, University of Salford, Salford, UK, 2007.

International Organization for Standardization *Uncertainty of Measurement*, Part 3: *Guide to the Expression of Uncertainty in Measurement* (GUM) ISO/IEC Guide 98–3:2008. ISO, Geneva, 2008.

Bell S, *A Beginner's Guide to Uncertainty of Measurement*, Measurement Good Practice Guide No. 11 (issue 2), National Physical Laboratory, Teddington, UK, 2001.

United Kingdom Accreditation Service, *Expression of Uncertainty and Confidence in Measurement*, Edition 3. UKAS, London, 2012.

Yang SJ, Ellison AJ, *Machinery Noise Measurement*, Clarendon Press, Oxford, UK, 1985.

A Little History

In many cases, consideration of uncertainties in acoustic measurements helps to decide compliance with standards, limits, manufacturer's specifications and so on. However, there is a long history of using such considerations about uncertainties in measurements in comparison with predictions to check that the theory 'fits the facts'. Since measurements can never be exact, it is necessary always to carefully and critically examine the results and the measurement method. So, for example, if theory A predicts a result of 40 units and theory B predicts 50 units, then we simply carry out the measurement – thereby asking a simple black-and-white question of nature. But what if nature gives us a grey answer to our question – and we obtain a result of 45 units? What do we do? We need to consider the uncertainties in the measurement and try to reduce them or design a better, more accurate measurement procedure before we can reach a conclusion.

The history of measurement of the velocity of sound in air offers an appropriate example. The first recorded measurements were carried out by Marin Mersenne in 1630, by Pierre Gassendi in 1635 and by Robert Boyle in 1660, among others. They used a time-of-flight method in which the difference between seeing the flash from gunfire from a distant cannon and hearing the arrival of the sound at the reception point was measured.

Aristotle (384–322 BC), on the basis of research into theatre design, reasoned that high pitches must travel faster than lower pitches. The measurements of Mersenne and Gassendi were sufficiently accurate to disprove Aristotle's theory. Mersenne used a variety of firearms and musical instruments as sound sources, and Gassendi showed that the high-pitched crack of a musket and the boom of a cannon arrived at the same time and realised that all pitches travel at the same speed.

There are several sources of uncertainty in time-of-flight measurements and in variations of them, the time of the 'there and back' between a sound and its echo. These arise from the effects of wind speed and temperature fluctuations along the sound transmission path and of the limitations of the timing measurements (Mersenne used a pendulum, and Gassendi used a mechanical timing device) and the reaction times of the operators in using them. The accuracy of measurements improved with time such that in 1738 the Academy of Sciences in Paris had measured and published a speed of sound within 0.5% of the value we accept today.

The first theoretical prediction of the velocity of sound was made by Newton in his *Principia* in 1686. This value was 16% different from the accepted measured value but, more importantly, was in disagreement with results of contemporaneous measurements in England and

France, including by Newton himself. Flamsteed and Halley, at New-ton's request, performed several measurements (i.e. repeatability tests), and they attempted to account for every variable in their measure-ments, to try to account for the discrepancy between measurement and theory, and they resurveyed the distance across the open fields over which the experiment was performed. Their measurements, however, continued to be in disagreement with theory. Newton proposed several modifications to his theory in order to try to bring the predicted result into agreement with measured values, but without success, until, about 100 years later, in 1816, Laplace discovered the cause of the problem, which was Newton's assumption that the air transmitting sound waves behaved isothermally rather than adiabatically. Laplace's value was within acceptable agreement with contemporary measured values. The difference between the two assumptions depended on the ratio of the two principal specific heats of air. This is now known to be 1.4 for air but, at the time, led to another series of heat flow experiments to deter-mine this value. These measurements involved an entirely different set of uncertainties – but that is another story.

Eventually, the search for greater control of the environmental factors in outdoor measurements was abandoned in favour of laboratory measurements using the resonance-tube method. This did indeed lead to more accurate measurement, so much so that the material of the tube and its diameter became factors in the measure-ments – but again, that is another story.

Back to more recent times, where in the 1960s the author learned to write reports of scientific experiments at school in A-level physics classes, but without attempts to evaluate uncertainties, or *experi-mental errors*, as they were then called. This changed completely at university, where in lab reports all measurements were expected to be accompanied by a rigorous estimation of experimental errors, and there were textbooks on the subject.

In the 1970s, one of the first Open University courses, the Science Foundation Course, provided a home experiment kit so that students could carry out simple experiments at home (and also at a one-week summer school), and the determination of uncertainties was expected.

The first edition of the GUM was published in 1993, and UKAS Document M3003 was published in 1997. In acoustics, the Craven and Kerry Salford University guide was published in 2001. From the early 2000s, UK standards on acoustic measurements contained a section on measurement uncertainties. Usually reference is made to the GUM, but sometimes more detailed guidance is given about how to estimate uncertainty. A section on uncertainty was intro-duced into the revised version of BS 4142 in 2014.

Finally, during a recent visit to a secondary school, the author was shown around the science labs and was surprised and delighted to see poster displays (Figure 2.3) featuring definitions of terms such as 'uncertainty', 'accuracy', 'resolution', and so on – these topics are now taught as part of school science syllabuses in the United Kingdom.

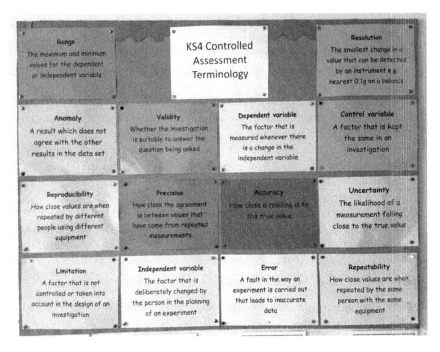

Figure 2.3 Poster observed in a school demonstrating that relevant concepts form part of the current science syllabus.

Annex: Examples of the Estimation of Uncertainties in Decibel Calculations

Introduction

The examples in this annex illustrate the estimation of uncertainties in measured and predicted sound levels where there are uncertainties in more than one variable. The effect of uncertainty in each variable must be estimated before these uncertainties can be combined.

There are two approaches. The first, more laborious method is (1) to estimate the uncertainty in each variable, (2) to use the functional relationship between the variables and the final outcome (i.e. the defining mathematical equation) to estimate the effect that this uncertainty will have on the final outcome and (3) to combine the different uncertainties. In the second method, which is more theoretically correct as well as potentially less laborious, the sensitivity coefficient for each variable is obtained and used, together with the estimates of uncertainty in each variable, to determine the total uncertainty.

The first two examples show that broadly the two approaches are equivalent. Sometimes it is easier to use the second method if the sensitivity coefficients are known or can be determined from the defining mathematical equation, but if not, the more laborious first method can always be used.

These simple examples are intended to show how more complicated situations, involving more variables, can be dealt with. The annex ends with a brief discussion about the difficulties involved in estimating uncertainties with quantities expressed in decibels.

Example A.1   Decibel combination

The noise level at a certain monitoring position arises from the combination of noise emissions from two sound sources operating simultaneously. The sound level at the monitoring position for the first source operating alone is 80 dBA, with an estimated uncertainty of ±2 dBA. When the second source operates alone, the sound level at the monitoring position is 83 dBA, with an estimated uncertainty of ±1 dBA.

What is the expected noise level when both sources are operating together, and what is the estimated uncertainty? Background noise levels are insignificant and may be ignored.

The combined, or total, noise level L_C arising from the two component noise levels L_1 and L_2 may be calculated from the following equation (*uncertainty model* in GUM terminology):

$$L_C = 10 \, \log\left(10^{L_1/10} + 10^{L_2/10}\right)$$
$$= 10 \log\left(10^{8.3} + 10^{8.0}\right) = 84.8 \text{ dBA}$$

Estimation of uncertainty

Method 1
Effect of variation of ±1 dBA in L_1:

If L_1 increases to 84 dB (with no change in L_2), L_C becomes 85.5 dBA.
If L_1 decreases to 82 dB (with no change in L_2), L_c becomes 84.1 dBA.
Therefore, on average, uncertainty arising from variability in L_1 = ±0.7 dBA.

Effect of variation of ±2 dBA in L_2:

If L_2 increases to 82 dB (with no change in L_1), L_C becomes 85.5 dBA.
If L_2 decreases to 78 dB (with no change in L_1), L_C becomes 84.2 dBA.
Therefore, on average, uncertainty arising from variability in L_2 = ±0.7 dBA.

$$\text{Effect on } L_C \text{ of both changes} = \sqrt{(0.7)^2 + (0.7)^2} = \pm\mathbf{1\ dB}$$

Method 2
The sensitivity coefficients may be determined by partial differentiation of
the following equation:

$$L_C = 10 \log\left(10^{L_1/10} + 10^{L_2/10}\right)$$

$$C_{L_1} = \frac{\partial L_C}{\partial L_1} = 10^{L_1/10} \Big/ \left(10^{L_1/10} + 10^{L_2/10}\right) = 0.67$$

$$C_{L_2} = \frac{\partial L_C}{\partial L_2} = 10^{L_2/10} \Big/ \left(10^{L_1/10} + 10^{L_2/10}\right) = 0.33$$

$$U_{L_1} = \pm 1 \text{ dB}$$

$$U_{L_2} = \pm 2 \text{ dB}$$

$$\text{Combined uncertainty } U_{total} = \sqrt{(C_{L_1} U_{L_1})^2 + (C_{L_2} U_{L_2})^2}$$

$$\sqrt{(0.67 \times 1)^2 + (0.33 \times 2)^2} = \pm\mathbf{1\ dB}$$

Thus, the results of using Methods 1 and 2 are in close agreement.

Example A.2   Decibel subtraction

Measurement of the sound level from a machine yields a value of 50 dBA,
with an estimated uncertainty of ±1 dB. When the machine is switched
off, a measurement of the background sound level yields a value of 46
dBA, with an estimated uncertainty of ±2 dB.

Estimate the sound level from the machine after correcting for the influence of background sound, and estimate the uncertainty in the resulting
corrected noise level.

The corrected level L_C is determined from the measured machine level
L_M and the level of background sound L_B using the following equation:

$$L_C = 10 \log\left(10^{L_M/10} - 10^{L_B/10}\right) = 10 \log\left(10^{5.0} - 10^{4.6}\right) = 47.8 \text{ dBA}$$

Estimation of uncertainty

Method 1
Effect of variation of ±1 dBA in L_M:

If L_M increases to 51 dB (with no change in L_B), L_C becomes 49.3 dBA.
If L_M decreases to 49 dB (with no change in L_B), L_C becomes 46.0 dBA.
Therefore, on average, uncertainty arising from variability in $L_M = \pm 1.7$ dBA.

Effect of variation of ±2 dBA in L_B:

If L_B increases to 48 dB, L_C becomes 45.7 dBA.
If L_B decreases to 44 dB, L_C becomes 48.7 dBA.
Therefore, on average, uncertainty arising from variability in $L_B = \pm 1.5$ dBA.

$$\text{Combined effect on } L_C \text{ of both changes} = \sqrt{(1.7)^2 + (1.5)^2} = \pm\,\mathbf{2.3\ dB}$$

Method 2 (using sensitivity coefficients)
The sensitivity coefficients may be determined by partial differentiation of the following equation:

$$L_C = 10 \log\left(10^{L_M/10} - 10^{L_B/10}\right)$$

$$C_{L_M} = \frac{\partial L_C}{\partial L_M} = 10^{L_M/10} \Big/ \left(10^{L_M/10} - 10^{L_B/10}\right) = 1.66$$

$$C_{L_B} = \frac{\partial L_C}{\partial L_M} = 10^{L_B/10} \Big/ \left(10^{L_M/10} - 10^{L_B/10}\right) = 0.66$$

$$U_{L_1} = \pm 1 \text{ dB}$$

$$U_{L_2} = \pm 2 \text{ dB}$$

$$\text{Combined uncertainty } U_{\text{total}} = \sqrt{(C_{L_M} U_{L_M})^2 + (C_{L_B} U_{L_B})^2}$$

$$\sqrt{(0.66 \times 1)^2 + (0.66 \times 2)^2} = \pm\,\underline{\mathbf{2.1\ dB}}$$

Thus, again, the results of using both methods are in close agreement.

Example A.3   Uncertainty in the variation of sound level with distance under free field conditions
It is required to estimate the sound pressure level and associated uncertainty arising from a sound source outdoors under free field conditions at

10 m from the source. The sound power level of the source is 100 dB (re. 1×10^{-12} W), with an estimated uncertainty of ±1 dB, and the estimated uncertainty in the distance is ±1 m.

The sound pressure level is estimated from the following equation:

$$L_p = L_W - 20 \log r - 11$$

where L_p is the sound level at distance r, and L_W is the sound power level of the sound source. This gives an estimated sound level of 69 dB at 10 m from the source.

Estimation of Uncertainty

Method 1
According to the preceding equation, a change of 1 dB in L_W will produce a corresponding change of 1 dB in L_p. To estimate the change in level L_p arising from a change of 1 m in distance, we can use the preceding equation for different distances or, more directly, use a simpler equation derived from it to calculate the change in level ΔL_p arising from a change of distance from r_1 to r_2:

$$\Delta L_p = 20 \log(r_2/r_1)$$

From 10 to 11 m, ΔL_p = 0.83 dB; from 10 to 9 m, ΔL_p = −0.92 dB. On average, this gives ΔL_p = 0.88 dB.

Alternatively, the change from 9 to 11 m gives ΔL_p = 1.74/2 = 0.87 dB, that is, a change of ±0.9 dB. (Note that this is exactly the approach taken in the example on page 13 of the University of Salford good practice guide.)

Therefore, we have an uncertainty of ±1.0 dB arising from uncertainty in the sound power level and of ±0.9 dB from uncertainty in the distance from the source. These can then be combined to estimate the total uncertainty u_t:

$$u_t = \pm\sqrt{(1.0)^2 + (0.9)^2} = \pm\underline{1.3} \text{ dB}$$

Method 2 (Using Sensitivity Coefficients)
Sensitivity coefficient for uncertainty in L_W: $C_W = \partial L_p / \partial L_W = 1$. Sensitivity coefficient for uncertainty in distance r: $C_r = \partial L_p / \partial r = 8.686/r = 0.8686$ (*for* $r = 10$ m) [2]

---

2 Note that $\partial L_p / \partial r = -20(1/r)[1/\ln(10)]$, so for $r = 10$, this is $-2/\ln(10) = -0.8686$.

$$u_t = \pm\sqrt{(1.0 \times 1.0)^2 + (0.8686 \times 1.0)^2}$$
$$= \pm 1.3 \text{ dB (as before)}$$

Example A.4   Uncertainty in measurement or prediction of continuous equivalent noise level $L_{Aeq,T}$

The noise level arising from a machine on a construction site measured at the site boundary bordering the nearest noise-sensitive property is 76 dBA, with an estimated uncertainty arising from variability in machine noise emission of ±1 dB. It is estimated that the machine will be in operation for 6 hours during each 12-hour site working day, with an uncertainty of ±1.5 hours.

Calculate the $L_{Aeq,12h}$ at the property together with the associated uncertainty.

Equation:

$$L_{Aeq,12h} = L_M + 10 \log T_{on} - 10 \log 12$$
$$= 76 + 10 \log(6) - 10 \log(12) = 73 \text{ dB}$$

Method 1
According to the preceding equation, a change of 1 dB in the measured sound pressure level will produce a corresponding change in $L_{Aeq,12h}$ of 1 dB, assuming no change in the machine 'on time'.

To estimate the change in level $\Delta L_{Aeq,12h}$ arising from a change of machine on time, we can use the preceding equation for different on times:

Change from 6 to 7.5 hours on time: $\Delta L_{Aeq,12h} = 10 \log(7.5/6) = +0.97$ dB.
Change from 6 to 4.5 hours on time: $\Delta L_{Aeq,12h} = 10 \log(4.5/6) = -1.20$ dB.
Or change from 4.5 to 7.5 hours on time: $\Delta L_{Aeq,12h} = 10 \log(7.5/4.5) = +2.2$ dB, which gives, after taking an average, $\Delta L_{Aeq,12h} = 1.1$ dB.

Therefore, we have an uncertainty of ±1.0 dB arising from uncertainty in the measured sound pressure level and an uncertainty of ±1.1 dB from uncertainty in machine on time. These can be combined to estimate the total uncertainty u_{total}:

$$U_{total} = \pm\sqrt{(1.0)^2 + (1.1)^2} = \pm\mathbf{1.5\,dB}$$

Method 2 (Using Sensitivity Coefficients)

Sensitivity coefficient for uncertainty in machine sound level, $C_L = \partial L_{\text{Aeq,12h}}/\partial L = 1$. Sensitivity coefficient for uncertainty in on time, $C_T = \partial L_{\text{Aeq,12h}}/\partial T = 4.343/T_{\text{on}} = 4.343/6 = 0.72$. Combining these, $U_{total} = \pm\sqrt{(C_L U_L)^2 + (C_T U_T)^2}$, where $U_L = 1$, $U_T = 1.5$. Thus,

$$U_{\text{total}} = \pm\sqrt{(1.0 \times 1.0)^2 + (0.72 \times 1.5)^2} = \pm\underline{1.5\,\text{dB}}$$

which agrees with the previous estimate.

Suppose now that we wish to include the uncertainty arising from the variability of the output of the sound level meter, which we estimate to be ±1.0 dB, with a sensitivity coefficient of 1.0.

The new total uncertainty will be

$$U_{\text{total}} = \pm\sqrt{(1.0 \times 1.0)^2 + (0.72 \times 1.5)^2 + (1.0 \times 1.0)^2}$$
$$= \pm\underline{1.8\,\text{dB}}$$

This example has related to the estimation of uncertainties in continuous equivalent noise level $L_{\text{Aeq},T}$ of noise in the environment, but a similar approach may be adopted towards noise exposure levels in the workplace. This will be explored in Example 6.

Example A.5   Measurement or prediction of construction site noise

This example continues the construction site noise theme of the preceding example. Uncertainties in three variables, sound power level, distance and machine on time, will be considered. In the interest of brevity, only the second method (involving sensitivity coefficients) will be presented, although the first method could be used if the sensitivity coefficients are not known.

The noise from a construction site at the nearest noise-sensitive property at the site arises from a single machine. The machine is situated 400 m from the property, with an estimated uncertainty in the distance of ±10 m. The sound power level of the machine is 130 dB, with an estimated uncertainty arising from variability in machine noise emission of ±2 dB. It is estimated that the machine will be in operation for 6 hours during the 12-hour site working day, with an uncertainty of ±1.5 hours.

Calculate the $L_{\text{Aeq,12h}}$ at the property, together with the associated uncertainty.

This time the relevant equation is

$$L_{Aeq,12h} = L_W - 20\log(r) - 11 + 10\log T_{on} - 10\log 12$$
$$= 130 - 20\log(400) - 11 + 10\log(6) - 10\log(12)$$
$$= \pm 64 \text{ dB}$$

Sensitivity coefficient for uncertainty in machine sound power level $C_L = \partial L_{Aeq,12h}/\partial L = 1$. Sensitivity coefficient for uncertainty in on time $C_T = \partial L_{Aeq,12h}/\partial T = 4.343/T_{on} = 4.343/6 = 0.72$. Sensitivity coefficient for uncertainty in distance r: $C_r = \partial L_p/\partial r = 8.686/r = 0.02175$ (for $r = 400$ m).

$$U_{total} = \pm\sqrt{(C_L U_L)^2 + (C_T U_T)^2 + (C_r U_r)^2}$$
$$= \pm\sqrt{(1.0 \times 2.0)^2 + (0.72 \times 1.5)^2 + (0.02175 \times 10)^2}$$
$$= \underline{\pm 2.3 \text{ dB}}$$

Example A.6   Uncertainty in measurement of noise exposure levels in the workplace

The noise exposure level of an employee may be estimated by carrying out a noise exposure level assessment, which involves taking a noise level measurement at the employee's ear during the time he or she is performing the noisy task and by estimating the average duration each day the employee spends exposed to this level of noise.

This example has two parts (a) and (b).

(a) For part of each working day, an employee carries out one task during which he or she is exposed to high levels of noise, but for the rest of the day, the employee works in relatively quiet areas where his or her noise exposure is insignificant in comparison with the noise exposure limits set by the UK Noise at Work Regulations. Repeated measurements show that the level at the employee during the noisy activity is 81 dBA, with an estimated uncertainty of ±1 dBA, and the daily average duration of exposure to this activity is 5 hours, with an estimated uncertainty of ±1 hour.

Estimate the daily noise exposure level of the employee over an eight-hour period and the associated uncertainty. It may be assumed that for the remaining hours of each day, the noise exposure of the employee is not significant and may be ignored.

Equation:

$$L_{Aeq,T} = L_{Aeq,T_M} + 10\log T_M/8 = L_{Aeq,T} + 10\log T_M - 10\log(8).$$

With $T_M = 5$, $L_{Aeq,8} = 81 + 10\log(5) - 10\log(8) = 79$ dBA.

Estimation of Uncertainty
Sensitivity coefficients:

$$C_L \frac{\partial L_{\text{Aeq},T}}{\partial L} = 1$$

$$C_T = \frac{\partial L_{\text{Aeq},T}}{\partial T} = 4.343/T_M = 4.343/5 = 0.8686$$

$$U_L = 1.0 \text{ dB}$$

$$U_T = 1 \text{ hour}$$

Total combined uncertainty $U_{\text{total}} = \pm\sqrt{(C_L U_L)^2 + (C_T U_T)^2}$

$$= \pm\sqrt{\left[(1.0 \times 1.0)^2 + (0.8686 \times 1)^2\right]}$$

$$= \pm\underline{\mathbf{1.3 \text{ dBA}}}$$

If, as in Example A.2, we wish to include the uncertainty in measuring the sound levels, say ±1 dB, with a sensitivity coefficient of 1, the new combined uncertainty will be

$$U_{\text{total}} = \pm\sqrt{(1.0 \times 1.0)^2 + (0.8686 \times 1)^2 + (1.0 \times 1.0)^2}$$

$$= \pm\underline{\mathbf{1.7 \text{ dBA}}}$$

(b) The noise exposure level of the employee is increased by having to perform a second work task – in addition to the first one described in part (a) – in a different area of the workplace, where the measured noise level is 90 ± 1 dBA for a period of 1.5 ± 0.5 hours each day. Estimate the daily noise exposure level of the employee over an 8-hour period and the associated uncertainty.

In this example, there are two components to the daily noise exposure pattern, tasks A and B. The contribution of the first task to the daily exposure $L_{8,A}$ was shown in part (a) to be 79 dBA. The contribution of the second task to the daily exposure $L_{8,B}$ is

$$L_{8,B} = 90 + 10\log(1.5) - 10\log(8) = 82.7 \text{ dBA}$$

The total 8-hour noise exposure level is given by

$$L_{8,\text{total}} = 10\log\left(\frac{t_A \times 10^{L_A/10} + t_B \times 10^{L_B/10}}{8}\right)$$

$$= 10\log\left[(5 \times 10^{8.1} + 1.5 \times 10^{9.0})/8\right] = \underline{\mathbf{84.3 \text{ dBA}}}$$

Note that the same result is obtained by combining the two component levels, 79 and 82.7 dBA.

Estimation of Uncertainty

Equation:

$$L_{8,\text{total}} = 10 \log \left(\frac{t_A \times 10^{L_A/10} + t_B \times 10^{L_B/10}}{8} \right)$$

The sensitivity coefficients are obtained by partial differentiation of this equation:

$$C_{L_A} = \frac{\partial L_{8,\text{total}}}{\partial L_A} = (t_A/8) \times 10^{(L_A - L_8)/10} = (5/8) \times 10^{(81-84.3)/10} = 0.29$$

$$C_{t_A} = \frac{\partial L_{8,\text{total}}}{\partial t_A} = (4.343/8) \times 10^{(L_A - L_8)/10} = (4.343/8) \times 10^{(82.7-84.3)/10} = 0.25$$

$$C_{L_B} = \frac{\partial L_{8,\text{total}}}{\partial L_B} = (t_B/8) \times 10^{(L_B - L_8)/10} = (1.5/8) \times 10^{(90-84.3)/10} = 0.70$$

$$C_{t_B} = \frac{\partial L_{8,\text{total}}}{\partial t_B} = (4.343/8) \times 10^{(L_B - L_8)/10} = (4.343/8) \times 10^{(82.7-84.3)/10} = 2.0$$

$$U_{L_A} = 1 \text{ dB}$$

$$U_{t_A} = 1 \text{ hour}$$

$$U_{L_B} = 1 \text{ dB}$$

$$U_{t_B} = 0.5 \text{ hour}$$

Total uncertainty U_{total} is given by

$$U_{\text{total}} = \pm \sqrt{(C_{L_A} U_{L_A})^2 + (C_{t_A} U_{t_A})^2 + (C_{L_B} U_{L_B})^2 + (C_{t_B} U_{t_B})^2}$$

$$= \pm \sqrt{(0.29 \times 1.0)^2 + (0.25 \times 1.0)^2 + (0.70 \times 1.0)^2 + (2.0 \times 0.5)^2}$$

$$= \pm \underline{1.3 \text{ dB}}$$

Assuming that the input uncertainties are standard uncertainties derived from repeatability measurements of sound levels and noise exposure durations, this is the combined standard uncertainty and may be used to calculate an expanded uncertainty by multiplying by a coverage factor.

This example is a simplified version of the worked example presented in annex D of ISO BS EN 9612 as an example of what the standard calls the 'task-based' approach to the determination of noise exposure levels. The statistical background and other aspects of this standard can be explored by reference to papers by Grzebyk and Thiery (2003), Kluge and Ognedal (2005), and Thiery (2003, 2007) listed in the references.

As mentioned earlier in this chapter, BS EN ISO 9612 uses the one-tailed distribution approach, with a k factor of 1.65, for estimation of uncertainty of noise exposure levels in the workplace. A spreadsheet is available with the standard that also allows for the input of the repeatability samples to determine the standard uncertainties of the measured sound levels and noise exposure durations. It is also possible to include the uncertainty arising from the measurement of the sound level meter, and this is also built into the example in annex D of ISO 9612 and the spreadsheet.

Note that the standard suggests allowing a standard uncertainty of 1 dB for a sound level meter, 1.5 dB for a dosimeter, and 1 dB for microphone position (relative to the employee's ear.

Use of Dose Meters to Determine Noise Exposure Levels

Rather than using measurements of sound levels (using sound level meters) and estimated exposure durations, it is also possible to determine noise exposure levels from repeated noise exposure levels readings using dose meters worn by employees, and this approach is the basis of the job-based and whole-day methods illustrated in annexes E and F of ISO 9612. A worked example of the dosemeter approach is illustrated in Chapter 9.

Note 1. There is a typographical error in equation C12 of annex C of the standard: the left-hand side of the equation should be U_i, not U_i^2.

Note 2. The approach of this Example 6 may also be applied to Example 4 on environmental noise, and both cases may be extended easily to more than two components.

Conclusion to Examples A.1–A.6

It is hoped that readers will be able to apply the methods described in these simple examples to estimate uncertainties for other, more complex situations.

Limitations and Difficulties Arising from Applying Uncertainty Theory to Sound Levels in Decibels

There are difficulties in the estimation of uncertainties and confidence intervals using values of sound levels in decibels. This is because the results of repeated sound levels may not follow a normal distribution and because the arithmetic average of the decibel values will differ from their logarithmic or energy average value. This may not matter too much if the spread of measured levels is small, say less than 3 dB, and if a relatively large number of samples are taken, say five or more, but uncertainty estimates based on the methods of the GUM may become unreliable if there is a wide spread of decibel levels and only a small measurement sample. These difficulties are recognised and addressed in ISO 9612, annexes C–F.

Task-Based Methods for Estimating Noise Exposure Levels (Annex D of the Standard)

Annex D of the standard recommends that if the results of a minimum of three measurements of a task (sound level) differ by 3 dB or more, then either the task should be split or at least three additional measurements should be performed.

Job-Based and Whole-Day Methods Using Dose Meters (Annexes E and F of the Standard)

According to standard ISO 9612, a minimum of five measurements is required for each job (though a higher number may eventually be required, according to the standard deviation of the measurements). The following list of references gives further information about the difficulties in estimating uncertainties in decibel measurements, particularly with respect to ISO 9612.

References

British Standards Institution, *Occupational Noise Exposure Engineering Method*, BS EN ISO 9612: 2009. BSI, London, 2009.

British Standards Institution, *Building Acoustics: Estimation of Acoustic Performance of Buildings from the Performance of Elements*, Part 5: *Sounds Levels Due to the Service Equipment*, BS EN 12354–5: 2009. BSI, London, 2009.

Department of Transport and Welsh Office, *Calculation of Road Traffic Noise*. HMSO, London, 1988.

Grzebyk M, Thiery L, 'Confidence intervals for the mean of sound exposure levels', *Am Ind Hyg Assoc J* 64 (2003), 640–645.

Kluge R, Ognedal TA, 'Uncertainty estimation in evaluations of "occupational noise exposure" using combined uncertainty of duration and noise level', in *Proceedings of the INCE Symposium 'Managing Uncertainties in Noise Measurements and Prediction'*. Institute for Noise Control Engineering (INCE), Le Mans, France, 2005.

Thiery L, 'Relationships between noise sampling design and uncertainties in occupational noise exposure measurement', *Noise Control Eng J* 55 (2007), 5–11.

Thiery T, Ognedal T, 'Note about the statistical background of the methods used in ISO/DIS 9612 to estimate the uncertainty of occupational noise exposure measurements', *Acta Acust United Acust* 94 (2008), 331–334.

Chapter 3

Measurement Uncertainty Associated with Sound and Vibration Instrumentation

Ian Campbell

3.1 Legal Considerations

Is it a horse or a mule?

There should be very little uncertainty about this question among those who set about purchasing a working animal. It is easy to tell the difference, particularly for those with an experienced eye who set about making an offer to purchase based on what they have seen before them. This is how things were when four-legged power drove our agrarian economy, and thus it is not difficult to see that in resolving disputes between purchasers and sellers, the rule evolved that the responsibility was with the person making the offer to purchase to determine that he or she was getting what he or she expected. This ancient concept has become part of our law as the principle of *caveat emptor* or, colloquially, "let the buyer beware," and has become a basic principle of law in many English-speaking countries. However, there have been legal refinements over time in that the goods must be correctly represented and be of merchandisable quality, and so on. Moreover, trading practice has evolved, and now, in most cases, the originator and driver of the transaction is the seller, but the principal responsibility to ensure that a need is fulfilled by a transaction remains with the purchaser.

So what has this to do with uncertainty of a sound or vibration measurement? As far as measurement technicians are concerned, the instrument is a black box, and they have little control over what goes on inside it once the signal has passed through the transducer and entered the electronics to emerge as an answer on its display. In many respects, therefore, the uncertainties associated with a measurement are related to the basic design of the instrument, the way it is verified and how it is maintained being as important as the way in which it is operated. Thus, in our modern high-tech world, the *caveat emptor* responsibility needs to be turned on its head. This follows from the fact that the complexity of modern manufactured items makes it almost impossible for the purchaser to be able to verify that the detailed functioning of the instrument is correct in every way and every environment. This makes a lot of sense, as the purchaser will not have access to the highly technical information and test facilities that the seller will most likely have to hand, particularly as the seller would most probably have been the actual designer of the equipment and hence able to undertake these complex verification tasks. So purchasers are not the leaders any more but are responding to the seller and must take the seller's statements as factual, as they have no easy way to verify the instruments for themselves.

These concepts are very important when we come to consider the legal metrology implications of sound and vibration measurements. The designers of the measurement instrumentation are the ones creating the instruments as well as controlling their production and testing – instruments that by their very nature will be used to quantify sound and vibration levels for comparison against legally or commercially required limits. These limits have been decided on by legislators to protect the health, safety and convenience of the public; it follows that exceeding these limits will result in court proceedings with all the consequences that can follow. These entail evidence that the measurements have been correctly made and that the results are of known and acceptable uncertainty.

Most people become involved in sound and vibration measurements because there is a regulation that specifies limits, or desired levels, for the work task or pastime in question. The fundamental human subjective reactions to sound or vibration are generally similar across populations, so the rules in various parts of the world used to control their exposure are also similar. Those who specialise in this field find that using common standards to rate noise and vibration simplifies trade, and hence, economic activity grows. We have therefore a range of standards, regulations and control procedures to define the desirable sound and vibration climate for human activities. From this it follows that we have formalised standards that define both the measurement methods and the accuracy of the measurement instrumentation. This leads us to the primary considerations in the overall uncertainty of the measurements in as far as the instrumentation is concerned. The uncertainties that appear within the measurement instrumentation are generally lower than those associated with the source and transmission path; this does not mean, however, that they do not need to be controlled and accounted for in the overall budget. In many respects, the parameters that affect the overall measurement are under the control of the person supervising the assessment of the levels, but once it has become an electrical analogue of the dynamic time-varying signal, it disappears inside the electronics, only to emerge as a number at the output with very little that the operator can do to affect it. What goes on inside the black box must be understood and controlled if these uncertainties are to be correctly assessed as part of the overall uncertainty budget.

3.2 Performance Specification, Pattern Evaluation and Periodic Verification of Instruments

This is where the instrumentation standards come into the picture. The key requirements are to define how accurate the instrumentation needs to be. But equally important is how this accuracy should be demonstrated, in a legal environment, by independent testing and, finally, what needs to be done to ensure that the instrument remains within the original specification for its working lifetime. This has resulted in a three-tiered level of standards that control these functions.[1] The first tier describes the basic functionality and accuracy of the instrument and the environmental conditions over which it must operate correctly. This is the specification, and it describes two classes of instruments[2]: Class 1 (precision sound level meters) and Class 2 (sound level meters). Both have the same nominal values, but wider tolerances and restricted ranges of environmental conditions are allowed for Class 2 instruments. Meters that do not meet the requirements of the standard are generally known as *sound level indicators*. Following on from the specification comes the need to confirm that any model of instrument designed to the requirements of the specification actually meets its requirements and that this has been assessed and confirmed by an independent authority. This is the *pattern evaluation*, and procedures to do this are set out by the standard's authority; these describe the tests that must be made by an independent national laboratory to prove that the instrument meets the claimed specification. These tests are normally carried out on a sample of about five pre-production samples of the model to prove that the design is sound and stable with time and environmental changes. The final tier is a limited number of tests and calibrations carried out at regular intervals, usually annually or biennially, to confirm that each individual instrument is performing correctly. This is the *periodic verification*. Do not confuse this periodic verification, which is often referred to as *calibration*, with the routine field sensitivity adjustment undertaken at the beginning and end of each measurement sequence. This is dealt with in more detail in Section 7 of this chapter.

3.3 Certificate of Conformance and Legal Metrology

If an instrument has successfully completed both the pattern evaluation and periodic verification requirements, then it can be issued with a certificate of

---

1 For sound level meters, the three parts of the BS EN IEC 61672 standard cover the specification, pattern evaluation and periodic verification in turn. For vibration meters, the three requirements are all listed in part 1 of the BS EN ISO 8041 standard.

2 This relates to the BS EN IEC 61672 sound level meter standard. The BS EN ISO 8041 vibration meter standard only specifies one class of instrument.

both conformance (with the standard) and calibration. The objective of these requirements is to make sure that the uncertainties are understood and well controlled. Results provided by an instrument certified to the standard can then be accepted in legal proceedings without the need for legal professionals to debate the numbers. Within the European Union (EU), there is a unified standards system, and a pattern evaluation granted by a national laboratory in one-member state is valid in all EU member states' courts. In the absence of these requirements, all that a calibration certificate can do is indicate that, at the time and location of the test, the instrument gave the results documented.

The periodic verification procedure is in addition to the routine field calibration and functional tests that a user may perform before and after each measurement sequence. These routine field adjustments, along with the periodic verification procedure, are the responsibility of the user. The manufacturer, or its agents, is responsible for the specification and independent pattern evaluation of the instruments. These inter-related responsibilities are shown graphically in Figure 3.1.

The first step in evaluating the uncertainties associated with a measurement instrument is to understand exactly which standard it is claimed to comply with, has it obtained a certificate of pattern evaluation from a national laboratory, and finally, does it have a certificate of calibration and conformance issued by an accredited laboratory. With this information, it is possible to make some assumptions about the uncertainties of the instrument in question, but without it, considerable effort may have to be made to determine exactly how an instrument performs and over what range of environmental conditions and varying signal dynamics. There are instruments offered for sale that claim compliance with the current standards but do

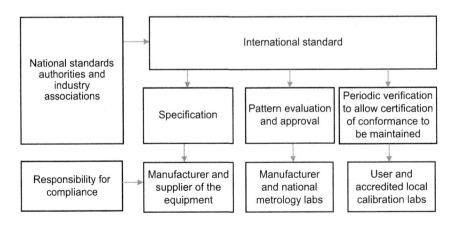

Figure 3.1 Legal metrology compliance responsibility tree.

not have any independent verification of that claim. If the purpose of a measurement is just a matter of obtaining an indication of the sound level, to set up a home hi-fi and so on, then the absence of pattern evaluation will not be a problem. However, if the measurement is in support of a regulatory requirement or is to be the basis of deciding whether a machine meets a contractual requirement, then legal metrology requirements would need to be considered. There are cases where it has been necessary to "confirm" instruments without pattern evaluation; in these cases, the cost of the testing has often exceeded the original purchase price of the instrument.

3.4 National and International Sound and Vibration Meter Standards

The current standards for sound level meters and their associated sound level calibrators are the responsibility of the International Electro-Technical Commission (IEC). Sound level meter standards are published under the title IEC 61672, *Sound Level Meters*. These have been integrated into many different countries' standards catalogues as well as those of the EU, so they appear in legal codes as BS EN IEC 61672, DIN EN IEC 61672, and so on. Sound calibrator standards are similarly published as IEC 60841 and frequency filters as BS EN 61260. Vibration meters are the responsibility of the International Organization for Standardization (ISO), which publishes BS ISO 8041, *Human Response to Vibration: Measuring Instrumentation*. This standard includes procedures for both pattern evaluation and periodic verification, as well as a specification for portable vibration calibrators. There is a full list of the standards in the Annex at the end of this chapter. This list also includes details of the legacy standards that preceded the current issues which also deal with the legal metrology routes for these instruments.

3.5 Tolerance and Uncertainty

Now we can start to consider the impact of tolerances and uncertainties in a measurement, but before doing that, it is worth clarifying when it is a tolerance and when it is an uncertainty. Take the situation where a meter has a tolerance of ±1 dB at 3.15 kHz, and the calibration certificate states that the meter has a deviation of +0.75 dB at 3.15 kHz with an expanded uncertainty of 0.2 dB, so it is within the specification. Naturally, we must take this deviation into account in presenting our results, so because the calibration certificate says that the microphone is 0.75 dB high at 3.15 kHz,[3] we report the result as the measured level with a correction

---

3 Some sound level meters for advanced applications have correction tables to remove relative frequency response errors, and in these cases, these post-measurement corrections would not be

factor of –0.75 dB with an uncertainty of 0.2 dB for 95% coverage. Thus, we can feel confident the reported 3.15 kHz result is within ±0.2 dB of the true value. However, if we did not have the calibration certificate to tell us about the deviation from nominal at 3.15 kHz, but we know that this frequency is important in the measurement, we would have to assume that the true value would be within the tolerance span of ±1 dB. In this case, the tolerance becomes the uncertainty with a rectangular distribution, so the expanded uncertainty increases (half the span divided by the square root of 3).[4] Thus, the reported results become the measured value with an uncertainty of 1.15 dB for the 95% confidence limit. There is a further complication in circumstances where you are not sure if the 3.15-kHz component is present in the measurement results or not; more of this later.

3.6 Evaluation of Uncertainties

The elements we need to consider in an uncertainty budget are

1. Setting of the instrument's sensitivity such that it aligns with the international reference standard and the uncertainty associated with that setting. This relates to the fact that the transducers (microphones or accelerometers) have nominal values, but in production there is a spread of results, and each individual device will have its own sensitivity. Furthermore, this nominal value will change with time, configuration and operating environment. So the use of a field calibrator is required to set the meter to a known traceable national standard, that is, 1 or 10 Pa at 1 kHz (sound pressure level of 94 or 114 dB) for airborne sound and 1 or 10 ms^{-2} at 159.15 Hz for vibration before and after each measurement, backed up by independent calibration of the transducer to confirm its nominal level on an annual or biennial basis.
2. Generation of the electrical analogue and its subsequent processing within the instrument results in consideration of the uncertainties associated with the amplitude, frequency and dynamic performance of both the transducer[5] and the associated electronics.

necessary. These functions do, however, underline the need to make sure that the correction table used for the measurement is the correct one for the microphone in use.

4   U_{k2} = expanded uncertainty; T = tolerance span (±1 dB gives a 2-dB span); thus

$$U_{k2} = 2\left(\frac{0.5T}{\sqrt[2]{3}}\right) = 2\left(\frac{0.5 \times 2}{1.732}\right) = 1.15\,\text{dB}$$

5   The way the transducer interfaces with the source signal can also be important. With vibration transducers, the mounting methods can be significant, and reference should be made to

3. Finally, there are the environmental effects, along with repeatability and reproducibility considerations.

Each of these elements is investigated separately, and they are then combined to give the combined uncertainty of the complete measurement chain. All the items that could affect the uncertainty budget are covered for completeness but looking at the results shows that for systems that meet legal metrology requirements, some are insignificant and therefore can be ignored. However, bear in mind that the effects are cumulative, and the tenths of a decibel here and there all add up in the final result. At the other extreme, where there is limited information from the manufacturer, the uncertainties become much larger.

3.7 Setting the Instruments to the Reference Standard

This first step is the setting of the instrument's sensitivity using a portable field calibration device. This relates to the fact that the transducers (microphones or accelerometers) have nominal values, but in production there is a spread of results, and each individual device will have its own sensitivity. Furthermore, this nominal value will change with time, configuration and operating environment. So the use of a sound calibrator is required to set the meter to a known traceable national standard, that is, 1 or 10 Pa at 1 kHz (94 or 114 dB sound pressure level) for sound and 1 or 10 ms^{-2} at 159.15 Hz for vibration, before and after each measurement, backed up by independent calibration of the transducer to confirm its nominal level on an annual or biennial basis. The standards require that a sound level meter has an associated portable sound calibrator, whilst a field calibrator is recommended for vibration level meters. The periodic verification standards require that the field calibrator is submitted for verification along with the sound level meter. This allows the verification laboratory to check that the

various technical texts to determine the effects of different methods and the associated uncertainties. The impedance matching of the vibration pickup to the instrument also has an effect; piezoelectric types can be charge or integral electronic piezoelectric (IEPE) matched, whilst geophones are low-impedance voltage connections. With charge-type devices, the cable is very important part of the calibration chain and must be the one specified on the calibration certificate or due allowance made for any change in capacitive loading.

With microphones, the situation is much easier; basically, microphones measure a sound pressure level, but then they can be designed to compensate for the effect they have on random (also known as *diffuse field*) or free-field incidence. Thus, there are three basic types of microphones, pressure, random and free field. However, the standards make it clear that the meters should have a free-field response, so we will just consider that case. For specialist applications that require pressure or random incidence results, there are other considerations that need to be investigated.

calibrator is suitable for the meter; for example, a Class 1 meter requires a minimum of a Class 1 sound calibrator, and at the end of the verification process, the two units are paired and quote the correct level to use when setting the instrument in the field. This procedure removes many of the uncertainties associated with variation in levels, load volume and instrument responses. With vibration meters, it is recommended that the field calibration device be submitted with the meter for verification.

Thus, there will be three possible scenarios to evaluate depending on the degree of standardised periodic verification that has been undertaken. First, the ideal situation of a paired calibration; second, individual calibration of the meter and field calibrator; and third, the least favoured situation where there is no individual calibration value on the certificate for the field calibrator and you have to rely on the "nominal" value marked on the calibrator.

3.7.1 Sound Calibrators

The current sound calibrator standards specify and confirm legal metrology requirements for three classes of devices; these are summarised in Table 3.1. The pattern evaluation and periodic verification requirements naturally apply to all three classes of sound calibrator, but fortunately, like sound level meters, they all have the same nominal requirements and only vary in the tolerances allowed and the range of environmental conditions over which they are specified to operate.

Table 3.1 Permitted variation on nominal sound level 160 Hz to 1.25 kHz For full details, see the current issue of BS EN IEC 60942: 2018

Class	Accuracy, dB		Frequency, %	Distortion, %
	Level	Stability		
LS[1]	0.2	0.05	1.0	2.5
1	0.4	0.1	1.0	3.0
2	0.75	0.2	2.0	4.0

1 LS devices, Laboratory Standard, are primarily electro-mechanical devices that are intended for use in climate-controlled calibration laboratories and hence have the tightest tolerances but do not have to meet the wide range of operating environments specified for the field calibrators described in class 1 and 2 of the standards. They are commonly known as "Pistonphones." The Standard requires a LS device to have an individual calibration certificate that states its exact level, and this must be within the range allowed by the standard. Prior to the 2003 version of the standard these devices were known as class 0 calibrators, these often operated at 250 Hz as opposed to the nominal 1 kHz for field calibration devices.

There are now Electronic Pistonphones that meet the LS specifications and operate over the field calibration devices range of environmental conditions.

Elements to Be Considered in the Uncertainty Budget

There are a wide range of influences that can affect the budget, each to a varying degree of magnitude. Bearing in mind that these will be combined in quadrature, small values will not have a significant effect, and it is reasonable to discount them. For completeness, we have considered items here that would perhaps only be considered in a calibration laboratory where precision is paramount, but for a field measurement, they would not be relevant.

UNCERTAINTY OF THE SOUND CALIBRATOR LEVEL

The sound calibrator certificate will give the level, frequency, stability and distortion values for the device, along with the uncertainties associated with them. If there is not a valid calibration certificate and the sound calibrator is marked as compliant with the standard, you can only work on the assumption that the level lies somewhere in the permitted range of ±0.4 dB for a Class 1 device. If there is no level marked on the unit, it cannot comply with the standard, and you just have to go shopping for one that does.

In addition to the actual level, the standard also specifies limits for the short-term stability of the calibrator, and hence, this should also be stated on the documentation along with its uncertainty. Reviewing several recent Class 1 sound calibrator certificates showed typical values of 0.01 to 0.06 dB with uncertainties of 0.02 dB. These could safely be discounted from a field measurement budget.

Associated with the actual level is the distortion produced by the sound calibrator. The standard allows up to 3% of the signal generated by the calibrator to show distortion. In practical terms, this means that for a 94-dB calibrator, around 0.3 dB of the signal can be due to distortion. This is included in the total level certified for the device but is not at the calibration frequency. A typical Class 1 sound calibrator would have a distortion figure of, say 0.5%; then the out-of-frequency component becomes 0.04 dB, which is not nearly so significant. Thus, in theory, there is something to consider here, but in practice, most distortion in sound calibrators is *harmonic*; that is, it is related to the fundamental, so if the fundamental changes, so will the distortion component. Any other distortion would be *random*, such as electrical or acoustic noise present within the calibration cavity, and in this case, it would be independent of the level of the fundamental. Thus, on the basis that distortion is harmonic and stable with a microphone that responds in a similar manner to all frequencies, it need not be considered in the budget – unless, of course, it is close to the limit, you may well see its effect if you calibrate broad band and also in the calibrator fundamental frequency's third octave band.

CORRECT SETTING FOR THE FIELD CALIBRATION CHECK

This will take into consideration the actual sound calibrator level, the volume loading effect of the actual microphone, the pressure to free-field correction, the actual response of the sound level meter at the calibration checkpoint and any offset that is needed to optimise the overall frequency response of the instrument. BS EN 62585: 2012 sets out the points that must be considered when determining this field calibration value. Basic information is given in Figure 3.2, but reference should be made to the standard for full details. When a sound level meter and its associated sound calibrator are periodically verified, the calibration laboratory will consider all these points and quote the correct field setting. The standard makes it clear that this field setting value has no associated uncertainty. If this "correct" value is not quoted, then it is necessary to consider all the points raised individually to determine the correct setting. This all looks quite complicated, but with pattern-evaluated Type 1 meters, the offset from nominal is normally quite small; with non-accredited or Class 2 meters and sound level indicators, however, these offsets can be larger.

REPEATABILITY OF THE APPLICATION OF THE SOUND CALIBRATOR

There will be some uncertainty associated with the repeatability of the field calibration.[6] A few simple tests, using different operators if appropriate, in applying the calibrator should provide the data to determine this correction. An example is shown in Table 3.2, and in this case, both the sound level meter and the sound calibrator have been pattern evaluated and periodically verified, so the fit of the calibrator to the microphone is very good. This example suggests that the uncertainty is so small that it can effectively be ignored, but this may not always be the case, particularly if you decide to derive a single correction figure to cover a range of operators. A particularly difficult situation occurs when calibrating a measurement microphone that has additional weather protection that cannot be removed in the field and additional coupling arrangements are specified by the manufacturer; in such cases, the uncertainty increases considerably.

---

6 The sound calibrator certificate should quote the results from three individual applications of the calibrator on the reference microphone. The standard deviation arising from these three measurements is compounded with the basic uncertainty of the determination of the calibrator's level. Bear in mind that these values are obtained under laboratory conditions using jigs to hold the kit in place. Under field and hand-held conditions, quite different results would be obtained.

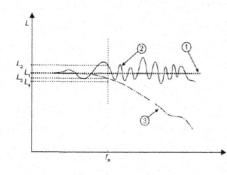

1 Design goal for the free-field response, 2. Free field response, 3. Pressure response

L_1 Reference sound pressure level,
L_2 Indicated level when the instrument is exposed to a free-field sound at the reference level and calibration check frequency.
L_3 Indicated level when the instrument is exposed to a sound pressure at the reference level and calibration check frequency.
L_4 Indicated level when the instrument is exposed to the sound pressure from a sound calibrator producing the reference level at the calibration check frequency. This is the value to be used for setting the instrument in the field.
f_R Calibration check frequency in Hz.

NB The difference (L_3-L_4) can be either positive or negative due to the loading of the sound calibrator by the microphone of the sound level meter.

Figure 3.2 Diagrammatic representation of factors for consideration in optimising adjustment value at the calibration check frequency from BS EN 62585: 2012.

Table 3.2 Repeatability calibration-setting tests with a single operator using a meter in two-decimal mode

Repeat calibrations, single operator			
Run	Result	Average	U,k=1
1	93.73		
2	93.76		
3	93.77	93.76	0.01
4	93.78		
5	93.73		

ENVIRONMENTAL EFFECTS

It may also be necessary to consider environmental variables and to establish whether the values quoted on the calibration certificate are corrected to reference conditions or apply only at the environmental conditions at the time of the periodic verification. If they are corrected to reference conditions, then correction data are available, and the effect under other environmental conditions can be determined. But take care because sometimes correction data only apply to the restricted range of environments found in calibration laboratories. Both the sound level meter and the sound calibrator are affected by the environment in which they operate. Fortunately, for Class 1 devices, the environmental coefficients are relatively small. Sound level meters are usually negative with temperature and barometric pressure, whilst calibrators are positive, so to some extent, they self-compensate. The environmental conditions over which a sound calibrator must operate are given in BS EN 60942, and there is a composite limit of 0.4 dB for Class 1 devices. Again, reference should be made to the standard for the tolerances for the other classes of calibrators. Modern smart calibrators compensate for temperature, altitude and load-volume changes to minimise any corrections necessary; check the device's specification to see data for environmental performance. Thus, if extremes of weather are avoided and there are no significant altitude changes, the impact of the environment on the uncertainty budget is not significant. However, there are still many designs of electro-acoustic and piston-phone sound calibrators that need corrections to account for atmospheric pressure changes; check the manuals and look for the "L" or "C" suffix[7] after the calibrator standard type, as these indicate that the device has a "limited" range of environmental conditions or needs "corrections" when operated over the claimed range.

Example of environmental corrections based on typical environmental performance of measurement microphones and sound calibrators based on data from four pattern-evaluated meters in common use for a change in conditions of 20°C and 5 kPa in pressure.

Self-compensating calibrator	dB/°C	dB/kPa
Microphone	−0.01	−0.01
Calibrator	0.002	0.001
Change in conditions	20	5
Microphone	−0.2	−0.05
Calibrator	0.04	0.005
Combined effect	−0.205	

If we replace the smart calibrator in the preceding with a Laboratory Standard mechanical (LS-M) device or an uncompensated electronic sound calibrator, then in this case the sign of the correction is reversed:

---

7 If a device has been pattern evaluated, then the suffix will be noted on the documentation. There have been examples of non-approved devices being put on the market claiming the standard but without the necessary suffix. Note the 2018 revision of the calibrator standard removes the L and C classifications and only allows barometric correction of classes LS-M and 1-M devices. See the section "Impact of the 2018 Revision" below.

Uncompensated calibrator	dB/°C	dB/kPa
Microphone	−0.01	−0.01
Calibrator	0.005	0.085
Change in conditions	20	5
Microphone	−0.2	−0.05
Calibrator	0.1	0.425
Combined effect	0.275	

Thus, if you choose to make these corrections, then you only have to consider the uncertainty of the correction information, and if not, the correction becomes the uncertainty. The problem is that there is very little information as yet about the uncertainty of the environmental correction figures given, so we come down to reasonable estimates.

Temperature   If the instrument is to be calibrated and operated over a restricted temperature range of, say, ±5°C, the correction is insignificant with a smart calibrator. The sound level meter standard, however, requires the meter to operate over the range −10 to +50°C, and we discuss this more in the section on sound level meters. If the correction information given in the example is available, then corrections can be applied; however, if it is not, then we have to live with the information available. Often temperature range is specified as a maximum deviation over a range, for example, 10–30°C ± 0.5 dB, with no indication as to whether this is a straight line or bell-shaped curve. In such cases, the limit must be taken as an uncertainty with a rectangular distribution.

Barometric Pressure   These corrections can be more problematical, particularly in areas where there are lots of hills and mountains, as these will compound the range of normal pressure changes resulting from weather conditions. In temperate zones, it is unusual for the barometric pressure to vary more than 5 kPa, but there is still the question of altitude changes. A change in altitude of 500 m will cause a change in pressure of approximately −5 kPa, and this, in turn, will cause the sound level generated by the calibrator to change. From the example it can be seen that this will be limited to 0.005 dB in a smart calibrator but to 0.43 dB in an uncompensated device. Thus, when calibrating in the office and then driving to the site to set up and make the field calibration check, you may expect to see a difference due to weather and altitude.

Humidity   There is an insignificant effect on the instrument if condensation is avoided. The danger scenario occurs when the sound calibrator is left in a car overnight and then taken into a warm room. So, even though the numerical value of a correction is close to zero, it can still have an uncertainty. More information about this is provided in following sections.

Microphone Load Volume   Sound calibrators are designed to generate a known sound pressure level in a calibration cavity of known volume. Part of this volume is the equivalent load volume of the measurement microphone, which is added to the calibration cavity volume. It is then necessary to refer to the microphone data to determine its nominal volume; for different models of ½-inch measurement microphones, this would be between 200 and 300 mm^3 and vary by some 5 mm^3 for different examples of a single model.[8] Modern sound calibrators often have automatic compensation algorithms that detect and compensate for the load volume of the microphone, but there are "legacy calibrators" and laboratory standard piston-phone types that do not have these functions. Modern smart calibrators will have very small volume sensitivity, typically 0.0003 dB/mm^3, so with the typical range of microphone volumes mentioned earlier, the correction would also be very small. In the case of legacy calibrators, the corrections can start to be significant, and reference should be made to their manuals to quantify the correction and its associated uncertainty. Among examples of models still in service, the volume-load corrections have been as high as 0.004 dB/mm^3, resulting in a range of corrections of 0.4 dB.

With the electromechanical piston-phone calibrators, it is possible to calculate the change in level precisely, and with some knowledge of the variation in the data for the volume of the calibration cavity and the microphone equivalent volume, some estimates of the uncertainties to be expected can be calculated. With these non-compensated sound calibrators, the relationship is

$$\text{Volume} - \text{load correction } V_k = 20 \log\left(\frac{V}{\Delta V + V}\right)$$

where V is the nominal volume of the calibration cavity, and ΔV is the change in volume from that of the reference microphone. A typical volume for one type of piston-phone is 15,600 mm^3, so now we have

---

8 For calibration of 1-, ½-, ¼- and ⅛-inch measurement microphones, manufacturers provide mechanical couplers to adapt the basic cavity aperture to suit the smaller capsules. The design target is normally to make the smaller microphones mimic the volume of the larger ones to minimise the change in load volume with microphone size – best check the specifications. When adapting to suit larger microphones, additional volume is added, and hence, corrections are required.

a worst-case scenario of 0.06 dB. Bear in mind that piston-phones have a very large calibration cavity, whilst sound calibrators designed for field use are much more compact, and hence, the calibration cavity is much smaller. In these cases, the effect of microphone volume changes will start to become significant.

Power Supplies   The next consideration is the power supply, and in the case of field sound calibrators, this means the batteries. The standard requires, for Class 1 units, the output level to remain within 0.1 dB over the full range of battery voltage. This uncertainty has a rectangular distribution.

Instrument Resolution   The final item for the budget is the resolution of the read-out of the sound level meter. This is usually 0.1 dB. Some meters do have the option to display the second decimal place in place of the hundreds digit, so the improved uncertainty contribution in this case is limited to calibrators operating at 94.0 dB or less. Simple balance between background noise and resolution is the choice in these cases. In a calibration laboratory, this can make a difference, and even in the examples that follow, it can change the result for expanded uncertainty for the paired-meter example from 0.289 to 0.266 dB.

Pressure to Free-Field Correction   As the sound calibrator generates a pressure field, and the standard requires the sound level meter to have a free-field response,[9] it is necessary to add an appropriate correction to the calibrator level. This is because as the wavelength of the sound in a free-field approaches the diameter of the microphone capsule, it will be reflected and refracted around the microphone, causing the sound pressure level to change. For a standard ½-inch microphone, this effect starts to be noticed at 1 kHz, where corrections of around -0.15 dB need to be applied to the actual pressure level generated by the sound calibrator. At 250 Hz, the correction is zero, and it is larger at high frequencies. The correction is normally given in the sound calibrator and/or sound level meter manuals, and if applied, it is only necessary to consider the uncertainty of the correction. If the uncertainty of the correction is not given in the manual, a reasonable estimate is approximately 0.1 dB.

---

9 Although the BS EN standard requires free field response other national standards specify other responses, e.g. American ANSI standards specify random incidence. There are also applications within the UK where random or pressure microphones would be preferable, so check you have the correct microphone and its associated correction.

Background Noise   Any background noise that breaks through into the cali-bration cavity will add to the calibration signal and hence be a source of error or uncertainty. The higher the calibration level, the less likely this is to be a point to consider, but with many calibrators in service with a level of 94 dB, with manufacturers still quoting this level for the field calibration check frequency, it will need to be considered. The basic relationship is as follows:

$$\text{Indicated sound level } L_r = 10 \log\left(10^{L_b/10} + 10^{L_c/10}\right) dB$$

where L_b is the measured background noise less the cavity attenuation, and L_c is the level of the calibrator.

> As the sound calibrator is producing a pure tone and we are meas-uring broadband, the situation can be improved by measuring the A-weighted decibel (dBA) values.
>
> If a frequency filter is available, then checking in the 1-kHz octave or one-third octave band will give even more of an improvement. It obviously would be wise to check that there is no attenuation at the centre frequency of the band when mapping out a procedure to per-form the field calibrations.

Simple measurements in a steady noise level will determine the cavity attenuation that may well be frequency specific. It can then be seen that an 80-dB event will cause a 0.2-dB change in the indicated calibration level. With care, when field calibrations must be made in noisy areas, these problems are easily avoided.

Impact of the 2018 Revision of the BS EN ISO 60942 Standard (Edition 4)
This standard removes the option to use manual corrections for environ-mental parameters. This stems from the improvements in electronics in that it is not difficult to design and support sound calibrators that auto-matically correct for temperature, humidity and barometric pressure. An added benefit here is also that failure to note that the calibrator had a "C" after its type designation means that the user failed to apply the necessary correction and hence increased the uncertainty of the measurement.

The special case of piston-phone calibrators, however, has to be con-sidered because they cannot, like electro-acoustic devices, be easily corrected for changes in barometric pressure. To cover this, the 2018 version of the standard has created a new classification for electromechanical calibrators of

Class LS-M and Class 1-M, with the "M" indicating mechanical method with the need for manual barometric pressure corrections to be applied. No other environmental corrections are allowed.

There has also been a realignment of the way in which uncertainty is treated in this standard. In previous editions, it was included in the tolerance interval; now it is excluded with a maximum permitted uncertainty for the verification laboratory. This makes it look as though there has been a tightening of the tolerance interval, but in practice it has remained very much the same.

Uncertainty Budget for Setting the Field Calibration Level

For all the following examples, we have assumed the use of a self-compensating Class 1 calibrator and a range of environments covering 5–25°C, a 5-kPa change in pressure and no condensation.

PAIRED METER AND CALIBRATOR

The easiest situation occurs when there is a paired associated calibrator and sound level meter having valid accredited calibration certificates. This is in fact the only configuration that conforms to the standard, and hence the legal metrology rules, as the associated calibrator and periodic verification are mandated. So, in this case, there will be a statement, in the calibration certification paperwork, saying what the correct level is for setting the instrument with the sound calibrator. The final calculation of the uncertainty in the setting of the calibration level in this case is therefore as shown in Table 3.3.

SEPARATE METER AND CALIBRATOR CERTIFICATES

If separate accredited calibration certificates for the sound level meter and sound calibrator are available but the results have not been paired, the approach is a little more complex and, of course, open to challenge under legal metrology rules. Most of the details from the individual certificates will read across from those in Table 3.3 into this budget in the same way as the first example to get the revised budget for this new configuration shown in Table 3.4, but we will have to consider the additional uncertainties that arise from the variables discussed earlier. These are

1. When setting the overall sensitivity of the instrument, just one frequency will be used. This is normally 1 kHz, as this is where all the frequency weightings coincide. For historical reasons, measurement microphones have their sensitivity quoted at 250 Hz, and their frequency linearity is

Table 3.3 Typical uncertainty budget for a paired meter and calibrator

Field calibration check uncertainty budget for a paired meter and calibrator

Parameter	Range	Sensitivity	Result	Distribution	Divisor	Contribution	Notes
Imported uncertainty			0.1	Expanded	2	0.050	1
Calibration check level							2
Repeatability			0.01	Normal	1	0.010	3
Temperature	5 to 25°C	−0.01	−0.2	Rectangular	1.732	−0.115	4
Barometric pressure	Δ 5 kPa	0.001	0.005	Rectangular	1.732	0.003	4
Humidity			0				5
Microphone volume							2
Power slm	1	0.05	0.05	Rectangular	1.732	0.029	6
Power calibrator	1	0.05	0.05	Rectangular	1.732	0.029	6
Free-field correction							2
Background noise							7
Resolution of the SLM			0.1	Rectangular	1.732	0.058	8
Combined expanded uncertainty U, $_{k=2}$						0.289	dB

Notes
1 Imported from UKAS calibration certificate, expanded uncertainty, $k=2$
2 Does not apply as correct paired settings given on SLM UKAS calibration certificate
3 From local tests using single operator
4 From manufactures data for microphone and sound calibrator combined
5 From microphone manufacturer's data ensuring no condensation
6 From BS EN IEC 61672-1 maximum deviation over battery life
7 Procedures in place to verify background is at least -20 dB on calibrator level.
8 Level of 114 dB chosen for field calibration to ensure swamping of background noise
NB U,k=2 is twice the square root of the sum of the squares of the individual items in budget

therefore also referenced to this frequency.[10] Class 1 instruments are nominally flat between the 250-Hz reference and 1 kHz, but there can

10 The standard requires a calibration laboratory to have a maximum uncertainty of 0.6 dB in the acoustic measurement of the relative frequency response (100 Hz to 4 kHz), but

sometimes be a small correction to keep the overall frequency response centred around the nominal (see Figure 3.2). This rarely exceeds a few tenths of a decibel. However, with Class 2 meters, there could be a significant correction needed at 1 kHz to keep the overall frequency response balanced in the centre over the range of frequencies for which the instrument is specified. For the example in Table 3.4, we have assumed a Class 1 microphone that is nominally flat, so no offset is needed to balance the response within the acceptance mask, and the only considerations are the uncertainties of the relative frequency response between the reference and calibration check frequencies, for which the typical laboratory figure of 0 dB with an expanded uncertainty of 0.21 dB has been used.

2. Microphone load volume on the calibrator was discussed earlier for standard ½-inch microphones. So, the first step is to determine the sensitivity of the sound calibrator to changes in microphone load volume as well as the nominal volume that the sound calibrator was designed to accommodate. Often this is expressed as the model of microphone for which the sound calibrator was designed. For example, the Nor-1251 Class 1 sound calibrator was designed for use with a 250-mm^3 microphone and has a volume sensitivity of 0.0003 dB/mm^3. If it is used with a 200-mm microphone, there would be a correction of 0.015 dB. If, however, the device was not self-compensating and the load-volume sensitivity increases to 0.004 dB/mm^3, the correction would be 0.2 dB. A bit of research is needed to obtain the information on calibrator and sound level meter performance, and from this, a load-volume effect can be calculated. Let us assume a figure of 0.1 dB for the correction, but we will have to determine an uncertainty associated with this correction. This will be the combined effect of the uncertainty of the measurement of the microphone's volume and the uncertainty associated with the calibrator's volume sensitivity. Not all manufacturers make this information available at present, so a reasonable estimate is 0.1 dB.

3. The sound calibrator produces a sound pressure level in its calibration cavity, and this is then related to the performance of the sound level meter in a free field by means of a correction figure.[11] To

some accredited laboratories can produce results better than 0.25 dB, and this needs to be included in the overall uncertainty of corrections for best-fit responses.

11 Often manufacturers quote a combined figure for the volume and free-field corrections for a specific model of microphone when used in their sound calibrators. For example, the correction figure for the Nor-1251 calibrator with the Brüel and Kjær 4155 half-inch electret measurement microphone is –0.15 dB. So, in this case, the nominal 114 dB of the calibrator becomes 113.85 dB, and we only need to consider an additional correction to take account of any difference needed to centre the response in the acceptance mask.

obtain this information the manufacturer must measure a number of microphones to determine their frequency responses and then check them with the calibrator to see the difference between the pressure and free-field responses. This is normally a negative correction applied to the calibrator's level, and for a standard ½-inch measurement microphone, it is between 0.1 and 0.2 dB, with an expanded uncertainty typically of 0.12 dB. For the example, we have taken a 0.15- dB correction with U_{k2} of 0.12 dB.

Table 3.4 Typical uncertainty budget for independently calibrated meter and calibrator

Field calibration check uncertainty budget with independent calibrator and meter certificates

Parameter	Range	Sensitivity	Result	Distribution	Divisor	Contribution	Notes
Imported uncertainty			0.1	Expanded	2	0.050	1
Calibration check level	1	0.21	0.21	Expanded	2	0.105	9
Repeatability			0.01	Normal	1	0.010	3
Temperature	5 to 25°C	−0.01	−0.2	Rectangular	1.732	−0.115	4
Barometric pressure	Δ 5 kPa	0.001	0.005	Rectangular	1.732	0.003	4
Humidity			0				5
Microphone volume	1	0.1	0.1	Normal	1.000	1.000	10
Power slm	1	0.05	0.05	Rectangular	1.732	0.029	6
Power calibrator	1	0.05	0.05	Rectangular	1.732	0.029	6
Free-field correction	1	0.12	0.12	Normal	1	0.12	11
Background noise							7
Resolution of the SLM	1	0.1	0.1	Rectangular	1.732	0.058	8
Combined expanded uncertainty U, $_{k=2}$						**0.475**	**dB**

Notes
 9 From microphone frequency response, uncertainty at calibration check frequency
10 From microphone and calibrator specifications
11 From sound calibrator specifications, typical FF correction uncertainty

NO ACTUAL CALIBRATOR LEVEL ON CERTIFICATE

In this case, we do not know the actual sound pressure level generated by the calibrator; all we know is the nominal value marked on it or its associated documentation. We would therefore have to accept this value as a base, with the uncertainty being based on the rectangular distribution of the semi-range covered. Thus, for a Class 1 calibrator, the actual level of a nominal 114-dB calibrator could be anywhere between 113.6 and 114.4 dB plus, of course, all the other variables noted in the preceding examples and these have been combined in Table 3.5

If the calibrator's history is not known, it would be unwise to use it without making some simple tests to determine that it is in fact operating

Table 3.5 Typical uncertainty budget where the actual calibrator level is unknown

Field calibration check uncertainty budget with no individual calibrator level on the certificate

Parameter	Range	Sensitivity	Result	Distribution	Divisor	Contribution	Notes
Imported uncertainty	0.5	0.4	0.4	Rectangular	1.732051	0.231	1
Calibration check level	1	0.21	0.21	Expanded	2	0.105	9
Repeatability			0.01	Normal	2	0.010	3
Temperature	5 to 25°C	−0.01	−0.2	Rectangular	1.732	−0.115	4
Barometric pressure	Δ 5 kPa	0.001	0.005	Rectangular	1.732	0.003	4
Humidity			0				5
Microphone volume	1	0.1	0.1	Normal	1	0.100	10
Power slm	1	0.05	0.05	Rectangular	1.732	0.029	6
Power calibrator	1	0.05	0.05	Rectangular	1.732	0.029	6
Free-field correction	1	0.12	0.12	Normal	1	0.120	11
Background noise							7
Resolution of the SLM			0.1	Rectangular	1.732	0.058	8
Combined expanded uncertainty U, $_{k=2}$						**0.655**	**dB**

Notes
12   From BS EN 60942 maximum allowed deviation from nominal level for class 1

correctly, for example, checking it against another calibrator using a known sound level meter as a transfer standard. Again, all the considerations mentioned earlier plus those associated with the use of the sound level meter as a transfer standard would have to be considered in preparing an uncertainty budget for the determined value of the calibrator.

Considering the three scenarios for the basic setting of the instrument's sensitivity, which will be line one in the overall uncertainty budget, we have a range rounded to tenths of a decibel from 0.3 through 0.5 to 0.7 dB.

3.7.2 Portable Vibration Meter Field Calibration Checks

Here the working practice is not as well developed as the case with sound level meters. For sure, vibration transducers are considerably more robust than measurement microphones, but they must work in much more severe environments. They are subject to damage to their connectors and cables and to incorrect setting of the sensitivity of the measurement instrument. Of course, they can suffer damage, particularly geophones that have delicate moving coil elements, but also in industrial situations, temperature and other contaminants can affect the sensitivity.

The current standard for human vibration meters contains an annex that specifies the requirements for the specification, pattern evaluation and periodic verification of a field calibrator for use with these instruments, and these are summarised in Table 3.6. Most of the ones currently on the market are small single-level (10 ms^{-2}), single-frequency (159.155 Hz) devices, but multilevel and frequency devices are becoming available at a cost and size penalty.

When using these devices, it is important to check the manuals for details on how the transducer should be mounted and the orientation in which the device should be operated. There can be complications with tri-axial devices, as not all have tapped mounting points for each axis, and it is necessary to devise mounting methods. Glue is often the solution, and mounting blanks should be used along with the manufacturer's recommended adhesive and associated release agent. Most devices have specified load limits; the standard requires this to be 70 g or more, and with older designs, there may also be de-rating curves that specify how the output changes with increasing mass load that is placed on the device. If this is the case, it is good practice, when sending the field vibration calibrator for periodic verification, to either send the transducer to which it is to be paired at the same time or to let the calibration laboratory know the typical mass loads with which it will be used.

The uncertainty of the initial setup of the vibration meter will comprise components for the expanded uncertainty of the level stated on the calibration certificate; a typical accredited calibration for a device would be between 0.5% and 0.75% for U_{k2} for frequencies between 80 and 160

Table 3.6 Preferred values and limits of error for the mechanical field calibrator (full details are available in BS EN ISO 8041–1: 2017)

Characteristic	Measurement type			
	Hand-arm		Whole body	Low Frequency
Frequency	500 rad/s ±0.5% (79.577 Hz)	1,000 rad/s ±0.5% 159.155 Hz	100 rad/s ±0.5% (15.915 Hz)	2.5 rad/s[a] ±0.5% (0.3978 Hz)
Root mean square (rms) acceleration	10 ms^{-2} ±3%	10 ms^{-2} ±3%	1 ms^{-2} ±3%	0.1 ms^{-2} ±5%

a It is recognized that field calibrators are not currently available at such low frequencies and that vibration pick-up calibration standards do not currently provide calibration methods validated at this frequency. However, to perform reliable measurement of low-frequency whole-body vibration it is desirable to perform calibration checks at a frequency within the frequency range of the measurement. The alternative is either to perform checks at static acceleration (i.e. transducer inversion providing a 2-g change in acceleration) or to test it at frequencies much higher than the measurement range – neither of these options are ideal

Hz. Reference to the specification shows very little influence from the normal range of environmental conditions, so if the correct orientation is maintained and there are no bias effects from the cable or asymmetrical loading, it is only necessary to obtain some data for repeatability to get a basic uncertainty for the calibration. If there is no individual value on the calibration certificate, then it must be assumed that the device is using the full 3% variation allowed by the standard.

With respect to the distortion component, the standard allows up to 5%, so this means that up to 5% of the signal is not at the calibration frequency but is included in the documented output level. Thus, it all depends on how the vibration meter is set with respect to its frequency domain if an error needs to be considered. If the meter is set to measure broadband, then there is no problem, but if it is in a band-limited or narrow-band mode, then either the fundamental or the distortion component may well be attenuated. In the sample budget given in Table 3.7, we have assumed broadband operation; typical portable vibration calibrators that have recently gone through an accredited calibration have returned distortion figures of 0.2% to 1.74%, with uncertainties of 1.05% of the reading.

The load-mass uncertainty would relate to the accuracy of the correction information provided by the calibration laboratory. In the absence of any correction, some laboratories offer calibration at two different loads, and this allows some assumptions to be made about the slope of the correction, and the uncertainty of that slope can be estimated from experience and the uncertainty of the measurements made to determine the load/level slope.

Table 3.7 Typical portable field calibrator uncertainty budgets

| Parameter | Uncertainty budget for vibration field calibration, % | | | | | |
| | No certificate | | | Valid certificate | | |
	U	Divisor	Cont.	U	Divisor	Cont.
Level	3.00	1.73	1.73	0.60	2.00	0.30
Load mass	0.50	1.00	0.50	0.50	1.00	0.50
Distortion	0.00	1.00	0.00	0.00	1.00	0.00
Field conditions	0.25	1.73	0.14	0.25	1.00	0.25
Repeatability	0.25	1.00	0.25	0.25	1.73	0.14
U k=2, %		**3.65%**			**1.30%**	

For this example, we have used a value of 0.5% for the uncertainty of the load correction, but this could easily vary in either direction.

The latest version of the vibration meter standard BS EN ISO 8041-1: 2017 has a very useful annex giving methods for determination of the overall uncertainty of measurement using instruments that comply with that standard.

3.8 Pre-Measurement Checks to Minimise Uncertainties

When any instrumentation is booked out for a measurement task, it is good practice to check that all is well before going to the site. This includes a review of the calibration documentation to make sure that you have the correct kit for the measurement, that calibration dates are valid and that the specification matches the measurement task to be undertaken. Each organisation would have its own procedures for this to ensure that travel time is not wasted – even simple things such as checking that the correct microphone or vibration transducer is as specified on the calibration documentation. Some accessories can introduce uncertainties into the measurement, for example, windscreens and accelerometer mounting methods. In the case of windscreens, extension cables, and so on, they can have an effect and hence are mentioned on the calibration documentation. Their inclusion or omission can influence the uncertainty of the result, and this needs to be understood. For long-term measurements, it is good practice to add to the initial checks prior to going to the site a dummy measurement, that is, if you are going to do a 12 by one-hour survey measurement, set up the kit and then run it in one-minute mode to check that in 12 minutes you have all the information you think you are going to need to cover the ensuing 12-hour report. This is a simple test, but it can save a lot of embarrassment by avoiding requiring you to go back to the site.

Once you get to site, each sound or vibration level meter has some basic pre- and post-measurement checks that the operator must carry out to ensure that the configuration to be deployed is stable and that all the kit is operating correctly. Obvious things such as battery and power supply checks are included, as well as making sure that all the accessory items are the correct ones for the measurement being undertaken, and these should always be documented in the measurement report. The first step is the setting of the instrument's sensitivity using the calibration device. The standards require that a sound level meter has an associated portable sound calibrator, whilst a field calibrator is recommended for vibration level meters, and these have to have the specification in accordance with the standards. There are also cases where the calibration check frequency is not the same as the reference frequency used for calibration of the transducer. Such a case would be an accelerometer from the United States having a 100-Hz reference frequency, whilst the vibration meter has a calibration check frequency of 160 Hz. In such cases, consideration must be given to the difference in response between these two frequencies.

These considerations have given us some data on the uncertainty of setting the instrument ready for a measurement. As an independent check that all is well, the sensitivity setting returned by the sound level meter after the field calibration should correlate with the microphone sensitivity produced during the periodic verification to identify any drift and so on. The sensitivity is normally quoted in millivolts per pascal (a typical value would be 50 mV/Pa[12]) or decibels re 1 V/Pa (for 50 mV/Pa, this would be −26 dB re 1 V/Pa) and is usually visible in the sound level meters calibration setting screen. Differences of approximately 0.25 dB are common due loading of the pre-amplifier on the microphone, but if the differences extend to around 0.5 dB, they should be investigated.

With a single-level, single-frequency sound calibrator, in addition to the mandated check of the dBA value, it is wise to check the other frequency weightings as well. With a 1-kHz calibrator, you would get the same nominal answer for all these measurements. It is also possible to check the maximum values if you avoid errors associated with the click of the switch, so start the calibrator before you start the measurement. Similarly, a check of the peak values should give results close to +3 dB on the calibrator level, particularly if the calibrator frequency is 250 Hz. but with

---

12 One pascal (Pa) of sound pressure level equates to 94 dB (to be precise, 93.98 dB), so when exposed to this sound level, the microphone will be producing an output of 50 mV.

Table 3.8 Change in meter reading for different frequencies on the weighting networks

Weighting network	Nominal attenuation (dB) at each frequency (Hz)[1]								
	63	125	250	500	1kHz	2kHz	4kHz	8kHz	16kHz
A	−26.2	−16.1	−8.6	−3.2	0	1.2	1.0	−1.1	−6.6
C	−0.8	−0.2	−1.3	0	0	−0.2	−0.8	−3.0	−8.5
Z	0	0	0	0	0	0	0	0	0

1 The instrument will have both frequency and amplitude linearity tolerances, so the figures in this table cover only the frequency non-linearity but any meter reading will also include any amplitude linarites. They are however, much smaller than the frequency non-linarites and will be swamped by any change in frequency response due to a damaged microphone.

a 1-kHz calibrator, values may come out below this, but the meter will still be in specification. If the calibrator has an on off switch, simple checks can be made of the $L_{eq,t}$ measurements. Mount the calibrator, and switch it on; then start the $L_{eq,t}$ measurement for, say, 30 seconds. At that point, switch the calibrator off. The meter should be reading the calibration level, say 114 dB. Note this value, and wait another 30 seconds, and the meter then should be reading 111 dB. After another 60 seconds, this should be 108 dB, and so on. For every doubling of time, the meter should be −3 dB of the previous reading until any background noise limits the fall in levels.

If you have a multilevel calibrator, then you can check at different points of the amplitude linearity and on different measurement ranges. With multifrequency calibrators, checks can be made of the weighting networks, including the actual microphone response. When measuring in A-weighted decibels, changing the calibrator frequency from 1 kHz to 250 Hz should cause the reading to change by -8.6 dB. Table 3.8 shows the weighting networks for the common frequencies provided by multifrequency calibrators.

3.9 Uncertainties Associated with the Transducer and Electronic Processing

Having determined the uncertainties associated with setting the instrument at its field calibration check, we have the first entry for overall uncertainty for the measurement and can start to consider the additional uncertainties that come into play under real-world conditions. The actual signal being measured will vary in amplitude, frequency and time domains, and we are now concerned with how well this analogue of the subject signal is processed by the transducer and ensuing electronics, the performance of which will also be influenced by environmental conditions. Thus, we need to consider the uncertainties associated with the tolerances allowed for the deviation of performance from the calibration check level and frequency allowed by the specification. There is therefore a wide range of parameters that need to be considered. Within the microphone itself, there will be changes in its performance with frequency, angle of incidence and environmental conditions, as well as at its maximum rated sound pressure level where its output will begin to distort. Similarly, at the bottom of the range, its self-noise will also affect performance. A similar situation exists with vibration transducers, with added problems with the uncertainty associated with the different methods of mounting it at the test site, including ensuring that the axis is correctly aligned. The electrical analogue of the sound or vibration wave produced by the transducer is passed to the electronics, where it will be quantified and processed to provide the required level. The signal will need to be amplified, frequency weighted, rectified[13] and displayed, the key elements of which are summarised in Figure 3.3. Also, extensive post-processing functions will be required to allow time-varying parameters to be expressed as single numbers, for example, $L_{Aeq,t}$, $L_{A90,t}$, VDV and so on. These functions are controlled by the standards, where tolerances and maximum permitted uncertainty of measurement are specified. The result is a very complex standard with many performance parameters specified and controlled. So, to avoid an overly complex calculation, some simplification of the uncertainty budget is needed to make it practical. The obvious approach here is to concentrate on elements that are most likely to be significant in the measurements.

In addition, we have tolerances and uncertainties, and we must be sure in our minds how we treat them. For example, as discussed earlier, if the instrument has a deviation of 1 dB at 5 kHz and we know that the uncertainty of that deviation is 0.2 dB, how will we treat this in the overall uncertainty budget? If we know that this deviation from the nominal performance is there and that it is a significant item in the measurement, we can correct for it and, hence, only must consider the uncertainty of the correction. If, however, we are not sure if the deviation at 5 kHz is significant in the measurement, we would need to assess how likely it is to be present and include a term for it in the uncertainty budget. This latter situation arises most often when the source contains several frequencies and, hence, the non-linearity of the overall frequency response must be considered rather than at one single frequency.

Before moving on, it is as well to understand that within the standards, the methodology has been to maximise the performance around the critical

---

13 Converted from an alternating-current (ac) waveform to a time-varying direct-current (dc) signal that represents the time history of the signal level. This stage also normally includes the logarithmic conversion of the signal to allow the results to be displayed linearly in decibels.

Figure 3.3 Main elements of a sound or vibration meter. Each stage has its own tolerances and associated uncertainty of determination.

areas of human perception. Thus, we have the tightest tolerances in the
midrange of human perception, with less stringent requirements as we
approach the limits of hearing or vibration perception. Figure 3.4 shows
the range of levels and frequencies that can be heard, where speech is obviously the most important, then moving out through music towards the
limits of audibility. Similarly, with dynamic performance, where the signals
are quasi-steady, there are stricter limits than those which are impulsive or
transitory.[14] In the uncertainty budget, it would seem reasonable to assume
that measurements that are made with steady variations occurring in the
middle of the frequency and amplitude range should have one budget,
whilst those extending over the complete range and containing a wide
range of dynamic signals could have a separate budget. This simplification
would reduce the number of elements, and with careful selection, safe figures for the instrument's contribution to the overall uncertainty of the
measurement could be determined, giving a range over which the uncertainty would exist for most general-purpose measurements.

Returning, then, to the discussion of microphone frequency response,
where there are definite dominant tones, this should be considered as
a special case requiring corrections, and here we should just concentrate on
the situation where the signals are quasi-random. In pattern-evaluated
Class 1 instrumentation, the typical deviations in frequency linearity are
much smaller than the maximum permitted by the standards; in the centre
of the response, these could be as small as ±0.2 dB against permitted tolerance of ±1 dB, moving out to around ±1 dB as you approach the extremes
of the frequency range, where the tolerances are several decibels. In these
cases, where there is good frequency linearity, it is possible to make reasonable estimates of errors. With Class 2 meters, the situation is not quite so
clear, and certainly with instrumentation that has not obtained pattern
approval, the frequency response can sometimes use all, and sometimes
more than, the permitted deviations. In these cases, it is necessary to investigate the frequency response of the instrument in use to decide on the best

---

14 In this case, it is not because impulsive signals are not so well perceived but more because
they are harder for the instruments to capture and process in the time available.

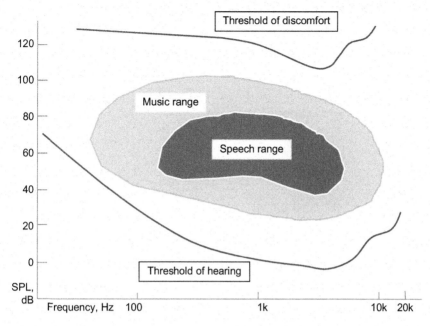

Figure 3.4 Range of human hearing.

approach. The permitted tolerances for complete sound level meters are shown in Figure 3.5. Note that these include the performance of the complete instrument, so the contribution from the microphone and the acoustic effects of the case, front-end accessories and sound level meter electronics must fall within these limits. These results are for incidence of the sound from the reference direction; additional tolerances are allowed if sounds come from other angles.[15]

We can now move on to consider the two scenarios selected: (1) steady midrange signals from close to reference conditions and (2) wide dynamic and frequency range from wider angles of incidence. For each scenario, there are two sets of figures, firstly, based on the maximum permitted uncertainties for the verification of a sound level meter shown in Table 3.9, which would give a worst-case condition, and, secondly, based on the performance obtained by a typical accredited calibration laboratory

---

15 Incidence-angle response is determined at the pattern-evaluation stage and is not part of the periodic verification of the meter. When quoted for a microphone alone, this would just include the preamplifier, but for a complete sound level meter, the case effects are also included.

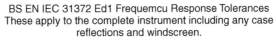

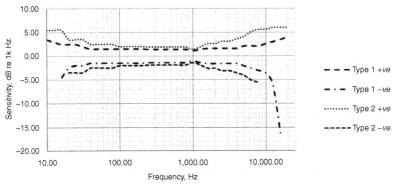

Figure 3.5 Frequency-response tolerances for the complete sound level meter.

Table 3.9 Maximum permitted uncertainties for steady midrange signals

Permitted maximum expanded uncertainty for steady mid-range signals

Parameter	Range	dB U,k=2
Angle of incidence	±30° <4 kHz	0.35
Frequency linearity	50 to 4 kHz	0.6
Amplitude linearity	Δ 10 dB	0.25
Dynamic performance	Quasi steady	0.3
Stability	Long term	0.1
Power supply	–	0.2
Combined expanded uncertainty U,k=2 dB		**0.828**

Notes: Data taken from BS EN IEC 61672

shown in Table 3.10, which should give typical results. In the field, the resulting uncertainties should fall between these two extremes.

So we can see from this analysis that when making the preceding assumptions and dealing with "normal" signals, the maximum uncertainty with which the errors within a sound level meter can be verified would give an uncertainty of 0.83 dB when rounding to two significant figures, and this would represent a worse-case example. However, looking at the uncertainties that would by associated with a pattern-evaluated and periodically

Table 3.10 Typical laboratory uncertainties for steady mid-range signals

Typical UKAS laboratory expanded uncertainty for steady mid-range signals		
Parameter	*Range*	*dB U,k=2*
Angle of incidence	±30° < 4 kHz	0.2
Frequency linearity	50 to 4 kHz	0.23
Amplitude linearity	Δ 10 dB	0.11
Dynamic performance	Quasi steady	0.16
Stability	Long term	0.1
Power supply	–	0.1
Combined expanded uncertainty U,k=2 dB		**0.388**

verified instrument, this figure would drop to 0.4 dB, and this would represent a best-case condition. If the periodic verification data are not available, then one would need to rely on the maximum values specified in the standard, and thus one would end up with a much higher uncertainty.

If the signal extends to cover a wider frequency range and becomes more impulsive, then the uncertainty budget compiled in a similar manner is shown in Tables 3.11 and 3.12. In this case, the worst-case budget gives us 1.43 dB if the standard information is used, but with the additional calibration data, this is improved to 0.94 dB.

Table 3.11 Maximum permitted uncertainties for dynamic wide-range signals

Permitted maximum expanded uncertainty for high dynamic wide-range signals		
Parameter	*Range*	*dB $U_{k=2}$*
Angle of incidence	±90° 10 -8 kHz	0.85
Frequency linearity	10 to 20 kHz	1
Amplitude linearity	<60 dB	0.3
Dynamic performance	Pulse train	0.3
Dynamic performance	Single burst	0.3
Stability	Long term	0.1
Stability	High level	0.100
Power supply	–	0.2
Combined expanded uncertainty U,k=2dB		**1.433**

Notes
Data taken from BS EN IEC 61672

Table 3.12 Typical laboratory uncertainties for dynamic wide-range signals

Typical UKAS laboratory expanded uncertainty for high dynamic wide range signals

Parameter	Range	dB $U_{k=2}$
Angle of incidence	±90° 10 –8 kHz	0.5
Frequency linearity	10 to 10 kHz	0.64
Amplitude linearity	<60 dB	0.17
Dynamic performance	Pulse train	0.224
Dynamic performance	Single burst	0.35
Stability	Long term	0.06
Stability	High level	0.06
Power supply	–	0.1
Combined expanded uncertainty U,k=2 dB		**0.937**

In specialised measurements where very high-dynamic-range impulsive sounds such as gunfire or cartridge-operated tools or where a single frequency dominates the measurement, it would be necessary to look more carefully at the calibration data to determine the performance under these specific conditions. We have also assumed that the frequency non-linearity is not significant, as it is balanced out. If this is not the case, then there will be a need to add a correction to the measurement to take into account the documented performance of the instrument when measuring peaks or the specific frequency in question, and such a correction would then be qualified by a separate statement of uncertainty.

3.10 External Influences on the Measurement

These are all the things that will also have an effect on the outcome of the measurement other than the basic calibration setting and the nature of the signals. Always there are some demarcations here in deciding if an item belongs to either the instrument budget or the others covering the source, transmission path or receiver elements of the overall budget for the complete measurement. The approach taken here is that everything that comes after the sound entering the microphone or vibration transducer is with the instrument, and everything before is somewhere else.

A grey area is the angle of incidence of the sound on the measurement microphone, as this could easily be placed in either budget. In siting the microphone, one of the considerations is to ensure that the measurement location is representative of the reception point and one of these is the angle of incidence of the signal. So, in a noise nuisance investigation where the reception point is defined as the "nearest exposed window," the microphone location could well be 1 m in front, and its orientation should be such that the sound is incident at the reference angle of the instrument; then the angle-of-incidence error is outside the instrument budget. Of course, there can be many situations where this is not nearly so clear, and we discuss this further in the final section of this chapter.

Thus, the specific elements to consider are the reading resolution of the meter, changes in the configuration of the instrumentation between the field calibration check and the start of the measurements and the effects of the environment on the instrument's performance.

3.10.1 Resolution of the Result

This was discussed in the Section 3.1, dealing with the field calibration check, and the same considerations apply here. It is the final-result resolution that applies, so watch out for situations where the instrument may have a resolution of ± 0.1 dB but the data transfer to the final read-out is to ± 1 dB. The entry will be a rectangular distribution of half the span divided by $\sqrt[2]{3}$.

3.10.2 Configuration Changes

These are usually obvious but still tend to get overlooked. Making the field calibration and then installing the meter into a system with extension cable, windscreen and weather protection may well introduce changes that need to be considered. So make the field calibration as near to the complete configuration as possible, and make sure that things such as windscreens and rain protection that have to be removed are properly treated. In many cases, instruments now have correction networks for these items, so make sure they are selected as the manufacturer recommends. Normally, they should be switched off when you make the field calibration check and back on when they are re-installed. In cases where there are no correction networks, reference needs to be made to the manufacturer's specifications for data on the insertion loss of the windscreen and so on, and due allowance must be made in the budget. If for practical reasons it is not possible to check on the site via the extension cable, a good compromise is to check before going to the site so that you know any insertion loss that needs to be accounted for when setting the calibration. If all these items hold true, then there are no extra items for the budget here.

3.10.3 Environment

In these determinations, we have not yet given any consideration to the effects of the environment. In the section dealing with field calibration setting, the environmental effects on the calibrator and microphones were discussed; these are the prime drivers of the performance of the overall measurement chain, as electronics these days are generally fairly stable in temperate environments. With the initial field calibration check, we were dealing with one set of conditions, but these could change and be responsible for different results at the end of measurement check. It is also necessary to consider the environmental effects throughout the measurement, including the impact of wind and rain.

The temperature range considered for field calibration checks of 20°C could easily become 40°C or more, but barometric pressure changes would now be limited to the weather, as the microphone is in one location only. The standards specify the reference conditions and the range of environments over which the instrument's performance must stay within the stated limits, and these apply to the complete instrument, are primarily driven by microphone performance and are summarised in Table 3.13.

Taking the 85–108 kPa and the temperature and humidity data from Table 3.13, we could get a worst-case environmental performance of Taking a more optimistic view, we could consider how the microphone should perform for a type-approved Class 1 system. The data sheets for four different types of WSM2[16] microphone from two different manufacturers

Table 3.13 Range of environmental conditions and specified tolerances for sound level meters

Range of Environments and Performance Limits				
Parameter	Reference	Range	Instrument Class	
			1	2
Pressure	101.325 kPa	85 to 108 kPa	±0.4 dB	±0.7 dB
		65 to 84.9 kPa	±0.9 dB	±1.6 dB
Temperature	23 °C	−10 to +50 °C	±0.5 dB	–
		0 to +40 °C	–	±1.0 dB
Humidity	50 %RH	25 to 90 %RH	±0.5 dB	±1.0 dB

$$U, k_2 = 2 \times \sqrt{\left(\frac{0.4}{\sqrt{3}}\right)^2 + \left(\frac{0.5}{\sqrt{3}}\right)^2 + \left(\frac{0.5}{\sqrt{3}}\right)^2} = 0.94 \text{ dB}$$

16 WSM2 is a definition from BS EN IEC 61094-4:1996 that refers to a "working standard measurement microphone with the 2 designating a ½-inch-diameter microphone." WSM1 would be a 1-inch microphone. These are the microphones used in most pattern-evaluated sound level meters.

were consulted and their average environmental sensitivity determined as a temperature coefficient of –0.007 dB/°C and a barometric pressure coefficient of –0.01 dB/kPa. In both cases, they are negative, which means that the microphone's sensitivity will go down as the temperature and barometric pressure increase[17] above the reference, and vice versa. With respect to humidity, the correction is very small if there is no condensation. A general recommendation is that the change will be less than 0.1 dB over the complete 65% relative humidity span covered by the standards. As it is not possible to correct for these effects, the uncertainty budget needs a term to cover the expected variation in sensitivity caused by changes in the environmental conditions relative to the last field calibration setting. So it is a simple calculation to cover the expected changes due to variations in the environment once the range of the changes has been decided. Thus, as a worst-case example to cover the full range for the microphone,

$$\Delta \text{Temperature } 60°C = 60 \times -0.007 = -0.42 \text{ dB}$$

$$\Delta \text{Barometric pressure } 23 \text{ kPa} = -0.01 \times 23 = 0.23 \text{ dB}$$

$$\Delta \text{Humidity correction} = 0.1 \text{ dB}$$

These values would be treated as a rectangular distribution and hence the root 3 coefficient applies to give

$$U, k_{k2} = 2 \times \sqrt{\left(\frac{0.42}{\sqrt{3}}\right)^2 + \left(\frac{0.23}{\sqrt{3}}\right)^2 + \left(\frac{0.1}{\sqrt{3}}\right)^2} = 0.56 \text{ dB}$$

For different ranges of environmental conditions, then, this can be reworked to take them into account. We still have the effect of the electronics in the sound level meter to consider; pressure and humidity are not normally a problem, but some temperature effects may be seen in the analogue sections. At best, these could be zero, but to keep to the standard, it must not exceed an extra rectangular distribution term of 0.65 dB in the preceding formula. So the range for the expanded environmental uncertainty is 0.56–0.94 dB for this example.

It is also necessary to consider special cases where not all parts of the instrument are in the same environment; for example, when the microphone is mounted outdoors and the instrument is inside a building, there could be big differences in temperature between the microphone and the instrument.

---

17 This is a general statement, but it should be noted that some parameters, other than the basic sensitivity, also change with the environment. The most noticeable is a change in the frequency response with barometric pressure. These effects are small but should be considered in specialised measurements.

In permanently installed systems, all-weather microphone systems often have electrical heating systems to keep the microphone dry and above ambient temperature, hence reducing the range of temperatures. In addition, the electronics are mounted in weather-protected enclosures where there is a heating element. At the opposite end of the scale, precautions have to be taken to ensure that there is not too much thermal gain from direct sunlight that could overheat the electronics.

Wind and rain naturally produce noise or make the trees move, and this is outside the instrumentation budget, but we have to consider the effects on the instruments themselves. The effects of rain hitting the body of a preamplifier or instrument should not induce microphonic noise, but if rain gets into the kit, it will induce faults and result in a malfunction, with the microphone being the weakest link, so it must be kept out. Wind, however, can contain turbulence itself that will induce instability in the microphone grid that will be measured as noise. It is the job of windscreens to interrupt the direct airflow without impeding the alternating pressures that we are trying to measure. The insertion losses of windscreens are normally now reasonably well documented and even accompanied by uncertainty information, so they can be corrected for in the frequency domain. However, the amount they reduce the actual wind noise is not so easy to account for. Usually the larger the windscreen, the more it reduces wind turbulence but the greater is the insertion loss. The effect of wind turbulence is to dynamically modulate the self-noise level, causing uncertainty in the measurement span. In the end, this comes down to a question of professional judgement as to the impact of wind-induced measurement artefacts.

3.11 The Final Budget

Having looked at the ranges of results for each of the key elements of the field calibration check, the instrument itself and the operating environment, we have all the elements needed to construct the final budget. With a variety of possible conditions to consider, we have a range of answers to each of these elements, and it seems reasonable to present the most optimistic and pessimistic views of how they perform. These now need to be combined to give the final answer, and in the two budgets in Tables 3.14 and 3.15, they have been combined to give the range of results expected.

Thus, we can see a range of outcomes that go from an unlikely expanded uncertainty of 0.74 dB through to a cautious 1.83 dB. This assumes a pattern-evaluated and periodically verified instrument, so be aware that it could be worse if these points are in doubt. Taking a view on the significance of each of the points discussed, along with some experience in the kinds of influences that apply in different kinds of measurement, the right data can be selected to build a budget for the instrument's contribution to any individual measurement being undertaken.

Table 3.14 Optimistic view of the uncertainty budget for a measurement

Total instrument uncertainty budget for ideal situation with steady signals	
Parameter	Result, U,k=2 dB
Field calibration check	0.289
Meter steady signals	0.39
Other effects	0.56
Combined expanded uncertainty U, $_{k=2}$	**0.741**

Total uncertainty budget for ideal situation with wide range signals	
Field calibration check	0.17
Meter steady signals	0.93
Other effects	0.56
Combined expanded uncertainty U, $_{k=2}$	**1.099**

Notes
1   From figure 3.3
2   From figure 3.10
3   From figure 3.12
4   Text #3.10.3

When it comes to the repeatability and reproducibility of these measurements, most of the considerations lie outside the actual instruments. If you feed the same signals into the instrument and the environment remains the same, you will get the same answers with very small uncertainty, so you are just down to the permitted variations in power supply and the long-term stability of the meter. The change in performance with power supply variations is 0.1 dB for a Class 1 meter, and the long-term stability is also 0.1 dB, and these values combine to give an expanded uncertainty of 0.14 dB. The environmental changes during a lengthy measurement are more than likely to swamp this result. As far as reproducibility is concerned, we would need to combine the uncertainty budgets for each of the instruments. This would give a more significant figure but again would have to be combined with all the other variables in the measurement.

The examples and topics chosen here are designed to show the range of effects that need to be considered, even if it is just to be able to dismiss them as insignificant for the particular measurement in question. As each measurement project has its own individual challenges, there may well be

Table 3.15 Pessimistic view of the uncertainty budget for a measurement

Total uncertainty budget for worst case with wide range signals	
Parameter	Result, U,k=2 dB
Field calibration check	0.66
Meter steady signals	0.937
Other effects	0.94
Combined expanded uncertainty $U_{,k=2}$	1.482

Total uncertainty budget for worst case with wide range signals	
Field calibration check	0.66
Meter steady signals	1.43
Other effects	0.94
Combined expanded uncertainty $U_{,k=2}$	1.834

Notes
1   From figure 3.5
2   From figure 3.11
3   From figure 3.12
4   Text #3.10.3

items for the uncertainty budget that have not appeared before; we always must be on the lookout for the "unknown unknowns."

These observations have primarily been related to the measurement of sound levels. When it comes to assessment of the uncertainty of human vibration levels, many of the considerations are the same, but be careful of the details. The latest version of the vibration meter standard, BS EN ISO 8041-1: 2017, has a very useful annex giving methods for determination of the overall uncertainty of measurements using instruments that comply with that standard.

3.12 Some Sources of Error in Sound and Vibration Measurements

3.12.1 Microphone Calibration

The microphone response shown in Figure 3.6 is typical of a good-quality WSM2 measurement microphone. Its frequency linearity is very good, and even when extended by the uncertainty of the calibration laboratory, it stays well within the limits. In such cases, it would be reasonable to

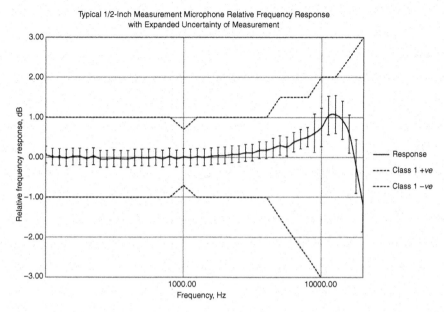

Figure 3.6 Class I acoustic response with typical expanded uncertainties obtained by an accredited calibration laboratory.

assume a good result if the frequency distribution were random. If, however, there was a tone at 12,500 Hz, the meter would read 1 dB high, and a correction would be needed to accurately describe this tone.

In the second example shown in figure 3.7, the microphone response shows a drop in mid-band frequencies, and in this case, it would not be reasonable make the assumptions mentioned earlier. Although this microphone is within the tolerances without a statement of measurement uncertainties, it is not possible to say whether it complies with the standard or not. It would also not be possible to make the assumption that the frequency variations would average out, as there is no certainty that the negative bias around 6.3 kHz would be balanced out by the positive bias around 12.5 kHz.

3.12.2 Calibration Records

Naturally, regular calibration verification is important, and this drives confidence in the numbers derived for the uncertainty budget, particularly if your activities are open to third-party scrutiny following the use of your measurements in a court case or as part of an audit by an accreditation agency or quality-management scheme. The long-term data analysis also

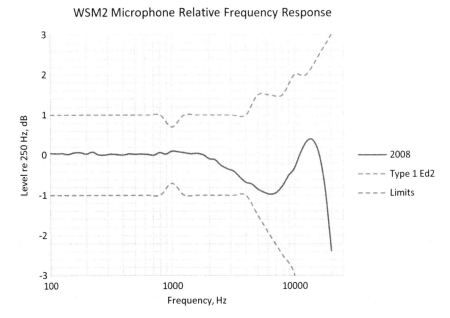

Figure 3.7 Microphone response showing a drop in response at mid-band frequencies.

drives decisions about the frequency of calibration verifications; if an item shows stable results over time, then there are very good reasons to extend the interval between calibrations, although for some applications the maximum period between verifications is specified, and in these cases, this option is not available. Similarly, if the results are showing drift or instability, the calibration interval may need to be shortened.

In the two examples in this section, you can see firstly the overlaid frequency responses of a measurement microphone over a 10-year period (Figure 3.8). The initial smooth response has been maintained, and with time, the interval between calibrations has extended. The second example is for a sound calibrator where the nominal value has been logged for each year, along with the error bars that show the uncertainty given by the laboratory for that calibration (Figure 3.9). For the one lower figure returned for 2004, the laboratory had a higher level of uncertainty, so this effect could very well be due to the laboratory's procedures rather than drift in the device. If you keep this kind of record, it is important that the laboratory reports to you if it resets the calibrator level; in such cases, it is best to keep the record of the adjustments made as well. Table 3.16 is an example of a log of a measurement microphone's open-circuit sensitivity at the reference frequency.

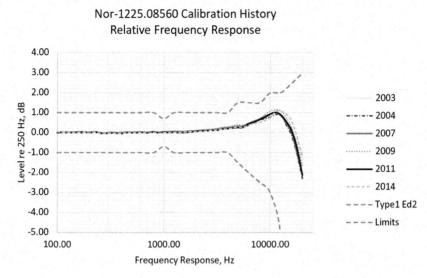

Figure 3.8 Calibration history records drive confidence in the uncertainty budget.

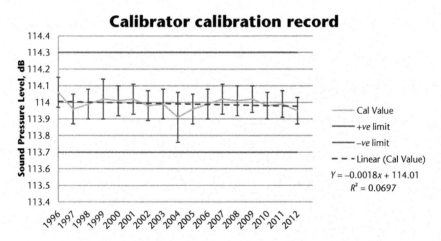

Figure 3.9 History of calibrator verification with the uncertainty of the measurement shown by the error bars.

Table 3.16 Example of the sensitivity calibration log for a ½-inch measurement microphone

Typical Microphone Calibration history

Date	2006	2008	2009	2011	2012	2014	2016
S_{ref}	−25.65	−25.68	−25.59	−25.52	−25.45	−25.49	−25.43
Δ annual	–	−0.03	0.09	0.07	0.07	−0.04	0.06
Δ cumulative	–	−0.03	0.06	0.13	0.2	0.16	0.22
Uncertainty	0.1	0.1	0.1	0.1	0.1	0.1	0.1

3.12.3 Low-Frequency Response of a Microphone

The electrostatic actuator method of relative frequency response calibration is not normally used at frequencies below 100 Hz, as it is unable to detect small holes or acoustic leaks in the microphone and would therefore give an overestimate of the microphone's performance at low frequencies. To provide these data, the microphone must be placed in a closed pressure chamber where both the front and rear vents are exposed to the sound pressure. This enables the microphone to be calibrated down to frequencies of 2 Hz or less. This is important if there is any doubt about the integrity of the diaphragm

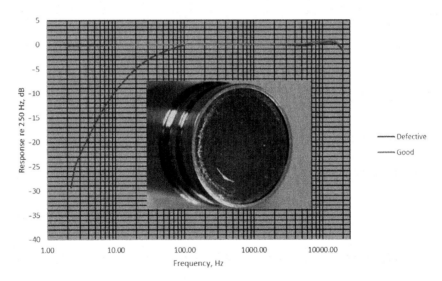

Figure 3.10 Effect on microphone low frequency response from a pinhole following a scratch on the diaphragm.

or the atmospheric equalisation system within the microphone. In Figure 3.10, the microphone would calibrate correctly at both 250 Hz and 1 kHz but is outside the A-weighting tolerances below 63 Hz. The most common cause of acoustic leaks in microphones is pinholes in the diaphragm. It also follows that detritus can enter the microphone via the holes, resulting in an increase in the self-noise of the microphone capsule.

3.12.4 Polar Response

If the sound is not arriving perpendicular to the plane of the diaphragm, there can be considerable deviations at higher frequencies that would need to be dealt with. These are detailed in Table 3.17 along with the maximum permitted uncertainties of measurement allowed by the standard; individual instrument manufacturers may achieve better uncertainties, and reference should be made to their literature for more information on this point. It is important to note that these limits apply to the complete instrument and hence include the effect of the instrument housing and any other front end accessories; performance may well be different if the microphone is operated via an extension cable. For completeness a typical polar response of a ½ inch measurement microphone is shown in Figure 3.11

Making the assessment of the incidence angle takes a bit of experience; however, it can be a major contributor because not only is there the

Table 3.17 Tolerances on the polar response of a sound level meter along with maximum permitted measurement uncertainties

	Maximum value of the difference between the displayed sound levels at any two sound incidence angles within $\pm\Theta$ degrees from the reference direction								
Frequency, kHz	$\Theta=30°$			$\Theta=90°$			$\Theta=150°$		
	Class and uncertainty, dB			Class and uncertainty, dB			Class and uncertainty, dB		
	1	2	U, k=2	1	2	U, k=2	1	2	U, k=2
0.25 to 1	1	2	0.25	1.5	3	0.13	2	5	0.13
>1 to 2	1	2	0.25	2	4	0.23	4	7	0.23
>2 to 4	1.5	4	0.35	4	7	0.23	6	12	0.23
>4 to 8	2.5	6	0.45	7	12	0.43	10	16	0.43
>8 to 12.5	4	–	0.55	10	–	0.58	14	–	0.58

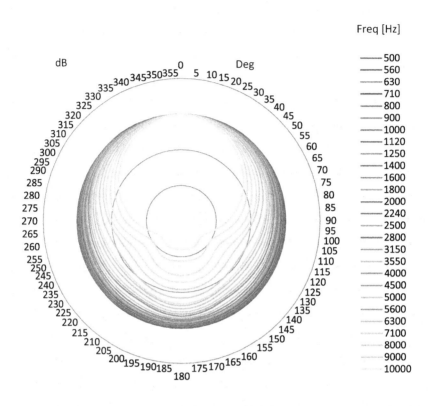

Figure 3.11 Polar plot for a typical ½-inch measurement microphone.

uncertainty of the polar response of the microphone but also there can be uncertainty in the determination of the incidence angle itself. If the sound is not arriving perpendicular to the plane of the diaphragm, then there can be considerable deviations at higher frequencies that would need to be considered. It is worth noting that the standard requires that sound level meters have a free-field response microphone, but in practice, WSM2 microphones are available with their frequency response quantified as either pressure, random or free-field responses. These responses are significantly different at higher frequencies, and this needs to be borne in mind when considering alternative microphones[18] for use on the sound level meter.

18 During pattern evaluation, the make and type of microphone will be registered as part of the approval. Most accreditation authorities will however accept and alternative

The most common situation occurs when a free-field microphone is measuring a source that is coplanar with the horizon, such as an environmental noise study. It should be mounted horizontal to the ground at the specified height and pointing at the source. If it is then to be used with weather protection, it is turned vertical to allow the rain cap to divert the water away from the acoustic ports, and its incidence angle changes from 0 to 90 degrees, and as a result, there will be attenuation of the higher frequencies that needs to be accounted for.

3.12.5 Pressure, Random and Free-Field Microphones

The BS EN standards require the use of free-field microphones, whilst American National Standards Institute (ANSI) standards specify the use of random incidence microphones and this may be a good place for a quick note on the differences between them. Basically, a microphone responds to the sound pressure that is incident on its diaphragm; these microphones are known as *pressure microphones*, and the design objective is to get the pressure response to be as flat as possible across the operating range. These microphones are used in closed-cavity measurements such as hearing aid couplers. If they are used in free space, their response will be as described earlier; that is, the response will vary depending on the angle with which the sound arrives at the diaphragm, and this effect gets more pronounced as the frequency increases. At 0-degree incidence, when the sound source is directly in front of the plane of the diaphragm, the output from the microphone will start to increase as the wavelength of the sound approaches the diameter of the microphone, so using a pressure microphone in a free field will overestimate the higher frequencies. To avoid this, the microphone designer can, by changing the damping and other features in the construction, cause the pressure response to roll off earlier in such a manner that the free-field response is flat to the highest possible frequency. It is now a free-field microphone and now would, of course, under-report higher frequencies if it were used in a pressure measurement situation. There is a halfway house between these two, and that is where the sound arrives at random angles of incidence, as would be the case in a reverberant room. In this case, the response is made up of a combination of the responses given in the earlier situation for the various angles. This results in a response between a pressure and free-field microphone as the design compromises are biased towards rolling of the pressure response to reflect the sum of random incidence responses.

manufacturer's microphone when submitted for periodic verification if it is documented as an equivalent.

As these free-field and random incidence corrections for individual designs of microphones are well defined, it is possible to construct electronic filters that can alter the response of an instrument fitted with a free-field microphone to perform like a random-incidence or pressure microphone. These correction algorithms are designed for the microphone type fitted to the sound level meter; microphones of similar construction, that is, WSM2, will have similar but not necessary the same correction data. WSM1 and WSM3 microphones would be completely different and hence should not be use with these WSM2 correction networks to convert to random-incidence response.

3.12.6 Self-Noise

The lower limit of the measurement range is often controlled by the self-noise generated within the instrument or microphone. The standard specifies that the lower limit of the measurement range is defined and that there is an under-range indicator to warn the operator that he or she is outside the verified measurement range of the instrument, and hence the answer could be misleading. In the design of the meter, the self-noise is normally designed to be at least 10 dB below the lower limit of the amplitude linearity range. Assume this to be 25 dBA; hence. at this point the meter will have a signal at 25 dBA plus a self-noise signal at 15 dBA. This will result in the following display of the sound level:

$$10 \times \log\left[\left(10^{25/10} + 10^{15/10}\right)\right] = 25.4 \text{ dB}$$

So there is a 0.4- dB amplitude linearity error, which would be within the specification at this point but will obviously decrease as the signal level increases. However, the standard accepts that self-noise is age dependent and tends to increase as the instrument ages. As such, there is the possibility that the low-end linearity errors could therefore increase and compromise the under-range indicator's function. For many years, self-noise was measured using a dummy microphone, and in such a case, this just reports the self-noise of the electronics and overlooks the fact that there is also a significant contribution from the self-noise of the microphone itself. So now the standards have added an additional test for self-noise of the complete instrument, and a typical report of instrument noise is given in Table 3.18. Here you can see the electrical noise that is measured on all the frequency weightings and the acoustic self-noise that is just required to be measured in A-weighted decibels. Naturally, to undertake this test, the laboratory must have a test environment where the background noise level is greater than 10 dB(A) below the expected self-noise of the microphone.

Table 3.18 Typical self-noise data from a calibration report

Self-generated noise - IEC 61672-3 Ed.1 #10				
Weighting	Level, dB	Limit, dB	U_{k2}, dB	Comment
A	17.3	18.0	0.49	Microphone installed
A	12.8	14.0	0.49	Equivalent capacity
C	12.8	14.0	0.49	Equivalent capacity
Z	17.4	19.0	0.49	Equivalent capacity

Annex Acoustic Instrumentation Standards

By their very nature, the standards' procedure takes a long time; the concepts of uncertainty and legal metrology are relatively recent refinements in the presentation of measurement results, and hence, not all the standards are up to date yet. Standards for sound level meters are perhaps the leaders in this respect, followed by those for sound calibrators, with those for vibration meters catching up now. The original version of the current sound level meter standard was published in 2002, with part 1, *Specification*, part 2, *Pattern Evaluation*, and part 3, *Periodic Verification*, following in 2005; these are detailed in Table 3.19. There was a revision in 2013 shown in Table 3.20, that has added a few extra tests and has changed the way in which uncertainty is handled in determining the acceptance interval. Instruments manufactured and accredited to the original version of the standard are known as *edition 1 instruments*, and those to the later version

Table 3.19 Previous sound level meter standards, edition 1

Number	Date	Title	Comment
IEC 61672–1	2003	*Electro-acoustics: Sound Level Meters Specification.* Also as BS EN standard.	Limits are quoted inclusive of the uncertainty obtained by the laboratory performing the tests.
IEC 61672–2	2003	*Electro-acoustics: Sound Level Meters Pattern Evaluation.* Also as BS EN standard.	
IEC 61672–3	2005	*Electro-acoustics: Sound Level Meters Periodic Verification.* Also as BS EN standard.	

Table 3.20 Current sound level meter standards, edition 2

Number	Date	Title	Comment
IEC 61672–1	2013	*Electro-acoustics: Sound Level Meters Specification*	Meters to this version are known as edition 2 instruments. In addition to adding extra requirements, the uncertainties of the testing laboratory are now considered in determination of the acceptance limits.
IEC 61672–2	2013	*Electro-acoustics: Sound Level Meters Pattern Evaluation*	
IEC 61672–3	2013	*Electro-acoustics: Sound Level Meters Periodic Verification*	

are *edition 2 devices*. Both versions have the same centre values, but the way in which uncertainty is handled in each instrument is different; they are more strongly controlled in edition 2 meters. It can be taken that a meter that conforms to edition 2 of the standard will also conform to edition 1 but not necessarily the contrary.

With respect to the current sound level meter standards, they make some assumptions about having an associated sound calibrator and that this device is also pattern evaluated and periodically verified. This concept allows the final stage of the sound level meter periodic verification to include a test that matches the meter and calibrator and in so doing optimises the basic sensitivity setting in the field before any measurements are made. If this is not done or a "non-standard" calibrator is used, then there are additional sources of error to be considered along with their associated uncertainties.

There are a lot of instruments in service known as *legacy* instruments designed to earlier national standards. Some of these standards go back to the 1960s, and they often originated as national standards and were latter developed into an IEC standard and these are summarised in Table 3.21. In most cases, the centre values of such things as the A-weighting network, fast and slow time constants and so on. have kept the same basic definitions, but the means of verifying performance have changed over the years. These early standards did not always cover the legal metrology requirements, and hence, there could be differences between manufacturers' implementations of an individual feature. As we approached the end of the 20th century, the uncertainties were starting to become better understood, and framework legal metrology procedures were developed. The OIML[19] developed two documents to specify the pattern evaluation procedures necessary

---

19 Organisation Internationale de Métrologie Légale (International Organisation for Legal Metrology) based in Paris.

Table 3.21 Legacy sound level meter standards

Number	Date	Title	Comment
IEC 60651	1994	*Sound Level Meters Specification*	Originally numbered IEC 651. Also published as BS 5969:1981.
IEC 60804	2001	*Integrating Sound Level Meters*	Added time-averaging functions to IEC 60651 specification.
OIML R58	1998	*Pattern Evaluation of Sound Level Meters*	
OIML R88	1998	*Pattern Evaluation of Integrating Sound Level Meters*	
BS 7580	1997	*Periodic Verification of Sound Level Meters*	Used in the United Kingdom only

OMIL, Organisation Internationale de Métrologie Légale (International Organisation for Legal Metrology).

Table 3.22 Human vibration meter standards[a]

Number	Date	Title	Comment
BS ISO 8041–1	2017	*Human Response to Vibration: Measuring Instruments*	Clarifies and expands on the requirements for pattern evaluation and periodic verification. Contains an annex on the calculation of overall uncertainty of measurement. Subsequent parts covering personal vibration dose meters have been proposed.
ISO 8041	2005	*Human Response to Vibration: Measuring Instruments*	Superseded by 2017 version and withdrawn.
ISO 8041	1990	*Human Response to Vibration: Measuring Instruments*	Superseded by 2005 version and withdrawn.

a There is also the BS 4675–2 series, which deals with vibration instrumentation for rotating machines, but these are outside the scope of this investigation as they do not cover subjective reactions to vibration levels.

Table 3.23 Sound calibrator standards

Number	Date	Title	Comment
IEC 60942	2003	*Electro-acoustics: Sound Calibrators. Also as BS EN standard.*	This standard has annexes dealing with pattern evaluation and periodic verification. Revision pending (see next entry).
IEC 60942	2017[a]	*Electro-acoustics: Sound Calibrators*	There are changes to the type designations and the use of manual correction data.
BS 7189	1989	*Specification for Sound Calibrators*	Superseded by BS EN IEC 60942 and withdrawn.

a The BS EN IEC version is dated 2018.

to obtain legal metrology status for the instruments but it was left to each individual national standards authority to determine the periodic verification requirements. In the United Kingdom and many English-speaking countries, the BS 7580 standard was used to set out what was needed for periodic verification. Many individual regulations refer to the standards numbers to which the measurement instruments should conform; sometimes these laws are not updated, and reference to the old standard numbers still appear in the regulations. The international standards committees have always tried to make sure that there is a thread of continuity running through the various issues of the standards. As a rule, if an instrument

Table 3.24 Filter Standards

Number	Date	Title	Comment
BS EN IEC 61260-1	2014	*Electro-acoustics: Octave Filters Specification. Also a BS standard.*	Replaces 1996 version but not activated until parts 2 and 3 are published.
EN 61260-2 DC	2013	*Electro-acoustics: Octave Filters Pattern Evaluation.*	Draft for public comment.
EN 61260-2 DC	2013	*Electro-acoustics: Octave Filters Periodic Verification*	Draft for public comment.
EN 61260	1996	*Electro-acoustics: Full and Fractional Octave Band Filters*	Superseded and withdrawn.
IEC 225	1966	*Full, ½ and ⅓ Octave Filters for Sound and Vibration Applications*	Superseded and withdrawn.

complies with the current IEC 61672 standard, it will also comply with the earlier legacy standards. Take care, however, because the old standards quote four accuracy "types", whilst the new standard only covers two "classes" of accuracy. So this general rule only applies to IEC 61672 edition 2 meters in Classes 1 and 2, which are like but not the same as IEC 60651 Type 1 and 2 meters, respectively.

In this summary we have just used the international numbers, for example, ISO or IEC, as they are the originators. In most cases, these are then reproduced with dual numbers as both European standards EN and national standards BS, DIN, and so on. For example, IEC 61672–3: 2013 is also known as IEC EN BS 61672–3: 2013 in the United Kingdom and as IEC EN DIN 61672–3: 2013 in Germany.

Chapter 4

Uncertainty in the Prediction of Sound Levels

Roger Tompsett

Uncertainty in Noise Predictions

This chapter identifies and discusses some of the uncertainties in the prediction of airborne sound, particularly in relation to environmental noise. It looks at a variety of standards, calculation procedures, verification work and case studies. It also considers ways of handling and reducing uncertainty so that noise predictions can be used with a known and appropriate level of confidence. Underwater and structure-borne sounds are not covered. These are specialist areas: underwater sound is mainly of interest in the defence sector. Structure-borne sound prediction techniques usually entail analysis of the spread of energy as vibration through the structure of interest. This can include both ground-borne transmission of sound and vibration and the transmission of sound through a building. This of interest to building acousticians and is covered in Chapter X. The terms 'sound' and 'noise' are used largely interchangeably in the literature and throughout this chapter.

What Is Noise Prediction?

Noise prediction is the process of calculating the noise levels expected at one or more locations arising from one or more sources of noise. It is a process of mathematical modelling as distinct from a measurement procedure. The term 'prediction' implies the uncertainty of looking into the future. Whilst a major use of noise prediction is for evaluating the noise impact of proposed projects, many situations are well defined and quantified, so the term 'calculation' may be preferred, particularly when quantifying the noise level received from a known source.

The principal purpose of noise prediction is to provide the information needed when assessing the noise impact of an existing or proposed project. Calculations may also be done as part of a process of 'noise control' or 'mitigation', where a completed project is giving rise to more noise than expected or permitted. Such calculations can help acousticians to understand where the problem is arising and what steps may be most effective in designing a solution.

Noise prediction procedures have been devised for every type of environmental noise, including roads, railways, aircraft, wind farms, industrial sites, minerals extraction sites, construction work and leisure activities. Some procedures are general purpose and can be applied to many types of noise source. Others are specific to a particular type of noise, which may make them easier to use for relevant projects. Whilst some of the earlier procedures could be undertaken by manual calculation using charts, nomograms and the like, in practice they are now all implemented through computer software, much of which is available commercially.

Uncertainty, Risk and Evaluation criteria

Elements of Uncertainty

The public perception of noise predictions is that they are 'theoretical' values and a poor substitute for measurements, which are seen as more reliable and 'true'. Project promoters are therefore wary of alluding to uncertainty. As we shall see in this chapter, the prediction or calculation process itself is rarely the principal source of uncertainty. Where environmental assessments do refer to sources of uncertainty, and even to their magnitude, it is rare for their effects to be quantified in the reported results. Moreover, it has not been easy to learn from experience, as promoters have often considered it to be inadvisable to check the results once the project is complete – this is viewed as a wasteful exercise in looking for problems.

There are four major elements that contribute to uncertainty in noise assessment [1]. These are

1. Uncertainty in the model inputs;
2. Uncertainty in propagation characteristics;
3. Uncertainty in the model structure and formulation; and
4. Uncertainty in the evaluation of impact.

Uncertainty in the evaluation of noise impact (i.e. its effect on people and activities) is a major issue. A large number of noise indexes are in use for rating the impact of noise: they are all far from perfect, especially for rating the impact of noise on an individual person. However, a number of indexes or rating procedures have been standardised for specific purposes and are generally regarded as being acceptably reliable when applied to a larger population, as this allows differences between individuals to be averaged. Even where an evaluation procedure and noise calculation methods are agreed, there are difficulties with reliably assessing the number of people affected. This chapter considers the effects of uncertainty in impact criteria only so far as this influences decisions on noise prediction procedures.

The value of a noise assessment should not be judged by its level of uncertainty alone but rather on its reliability as a tool in decision making, and this judgement should be made according to the specific application and situation under consideration. Every noise prediction study should be made with its ultimate application in mind, and the outputs should have accompanying contextual information that enables decision makers to understand how the information can be used [2].

Given the degree of flexibility and interpretation permitted by evaluation criteria, users may have widely differing expectations of what is involved in conducting a predictive study. Usually, the modelling process should be governed by the requirements of the evaluation criteria that will be used for the project under consideration. The challenge for practitioners is to ensure that the end user appreciates the decision risks associated with the various alternative approaches and to develop an assessment strategy that strikes an appropriate balance between the scale of resources required for the study, against the costs (social, financial and otherwise) and likelihood of an incorrect decision resulting from an inadequate study. For example, at the early sift stage, rapid evaluation techniques may be advised. These could allow unsuitable schemes to be rapidly eliminated from consideration without the need to spend a large amount of time and money to undertake a fully-detailed analysis. Such sifting processes can be as broad-brush as counting the number of properties within various distances of a proposed road alignment. These must be used with care, however, as such techniques could unnecessarily introduce new noise into quiet areas and at the same time failing to serve the local transport needs by skirting residential areas.

A common concern is that a prediction will underestimate the full scale of the noise impact of the project, with associated social and financial consequences. However, perhaps the most frequently-underestimated risk is that of the unnecessary development costs (direct costs as well as those associated with lost development opportunities) arising as the result of an over-estimation of noise impact. This can occur because worst-case approaches are frequently used to address the uncertainty and limitations of practical environmental noise studies.

Assessment Criteria

Leaving aside the question of whether it is desirable to identify the worst case – with all the problems of cost and sustainability this incurs – it is not even certain that we can know what constitutes the worst case. For example, is it better to affect a large number of people by a small amount, or small number of people by a large amount? Or is it better to have a scheme that affects a few people often or a lot of people infrequently?

Should we ensure that a fixed limit is never exceeded, even under extreme or rare weather or operating conditions? Such questions show that there is a considerable element of uncertainty and risk in any assessment, independent of the numerical uncertainty of the prediction method.

Attempts have been made to remove these value judgements by monetising the impact – for example, by assessing the effect of a scheme on property prices or by asking people what sum of money would compensate them for the increased noise. But many schemes can bring benefits in terms of property prices owing to improved access to facilities, a factor entirely independent of environmental impact. Moreover, asking people about amounts of compensation may result in their considering compensation to move house rather than to live with the noise, leading to very high costs. A further problem is how to weigh the advantages and disadvantages of a scheme. It is well recognised that increased noise levels can have a negative effect on people, whilst reduced noise levels may provide little overt benefit. It has been observed that an immediate increase in noise exposure caused by a scheme evokes a much stronger response than the same increase that occurs over several years.

As we shall see in some of the case studies described later, there is a tendency for promoters to overestimate the 'benefits' of their scheme in terms of passenger numbers, usage, flow rates, and so on, as these are important elements of the cost-benefit analysis required for the economic justification of the scheme. This can lead to overprediction of noise levels with consequent negative effects. Public perception that all transportation schemes are excessively noisy leads to negative perceptions and resistance that can inhibit the progress of economically and socially necessary developments.

An appreciation of these issues is crucial to the correct assessment of projects and the willingness of promoters to place greater value on noise prediction studies. Historically, scheme promoters have often regarded noise studies as a 'distress purchase' needed for regulatory purposes but adding little or no value to a project. Fortunately, the value of noise prediction exercises is now being recognised in formal assessment guidance. For example, the UK Highways Agency's *Design Manual for Roads and Bridges*, 2011 edition [3] (DMRB), now requires at the scoping stage the identification of properties likely to undergo a 1 dBA change in the short term or 3 dBA in the long term. It may be possible to fulfil this requirement solely by considering changes in traffic flows, but this approach ignores the possibilities of mitigation, for example, by putting roads in cuttings or by using barrier mounds. Moreover, this approach has the potential to sterilise some areas of land whilst introducing noise into quiet areas. However, an example of updated practice, the Irish National Road Authority's *Good Practice Guidance for the Treatment of Noise during the Planning of National Road Schemes* [4], published in 2014, now advises that noise footprints should be created even at the scoping stage, where detailed road design is probably not available, an approach that is becoming more practicable as computer prediction techniques improve.

Achieving Design Targets

It is common for planning permissions and construction contracts to have a clause along the lines of 'The sound level at point X shall not exceed Y dB $L_{Aeq,T}$'. Much of this chapter, and much of this book, is about the difficulties in achieving such apparently simple objectives and the degree of risk that can be accepted during design.

The most cautious approach would be to ensure a large margin of safety to cover all uncertainty, but this can have excessive effects on cost and sustainability. One study [5] in Ireland found that enormously long and high barriers were being provided for certain road schemes in order to meet an absolute noise limit for isolated properties. Whilst this was arguably essential to meet standards, the long-term cost implication was excessive, and with the other environmental impacts of such long barriers, this approach was arguably unsustainable. This could lead to pressure to relax the target level, which would have an adverse effect on other locations where the original target would have been easy to achieve. Thus, excessive caution in dealing with uncertainty can and does have negative effects.

In another case [6], there was uncertainty as to whether a wind farm could meet its design target as set out in planning conditions. Many measurements had been taken, and all met the target, but there was an argument that none of the measurements covered the theoretically worst case for wind speed and direction. Without an absolute guarantee, the lending banks were unwilling to sign off on the loan agreement. The promoters argued that it was inappropriate to put yet more effort into the off-chance that the development could be shown to have failed its targets. Instead, it was agreed that the control system would be enhanced, so in the proven to be unlikely event of excessive noise, the responsible turbines could be temporarily powered down. This does not eliminate the uncertainty, but it provides a productive (and relatively inexpensive) way of dealing with it.

Principles of Noise Prediction Methods

Noise Indexes

Environmental noise predictions are made in order to assess the noise impact of an existing or – more usually – proposed project on people and, to a lesser extent, on the natural world. Unfortunately, the effect of noise on people and nature, despite decades of research, is still difficult to quantify. This is discussed in detail in Chapter X, although we cannot avoid the subject here. Whilst we are not sure of the best way to assess noise impact, acousticians have settled on the A-weighted 'equivalent continuous sound level' $L_{Aeq,T}$ as a universal index for assessing noise impact, although a number of other indexes are used for specialist applications.

The prediction techniques discussed here are focussed on the $L_{Aeq,T}$ index or a closely correlated one such as the earlier L_{A10} index (still used in some cases). The prediction process may divide the sound into octave bands for a more accurate calculation of frequency-dependent factors, but this is usually integrated into an A-weighted figure for final presentation. There is a more recent trend that includes prediction of the maximum noise level $L_{Amax,T}$ in some form, but this index is often subject to a higher level of prediction uncertainty than the $L_{Aeq,T}$. The $L_{Aeq,T}$ values thus calculated may then be weighted for the time of day and summed to give a single-figure index such as the day-evening-night L_{den} of the European Environmental Noise Directive or the day-night average sound level (DNL) airport noise metric used in the United States.

Noise Prediction Models

Probst [7] distinguishes between 'scientifically based' noise prediction models and 'engineering' methods. Scientifically based models are analytical models based on numerical solutions of the wave equation using the fast-field programme (FFP) or approximate solutions using parabolic equations (PEs). These may be implemented via a boundary-element (BEM) approach and are commonly used in underwater acoustics and seismology. Probst describes another alternative, the finite-difference time-domain (FDTD) method. He points out that although these methods are used to investigate special problems in detail, they are too complex and slow to be applied in large-scale, complex scenarios. Some of these are capable of giving a full solution, but only in special cases, such as a horizontally stratified medium, or limited to a maximum elevation angle or to non-turbulent atmospheres. These 'numerical methods' may be summarised [2] as having many strengths, principally those of accuracy, and many weaknesses, principally those of practical application and inability to cope with all environmental conditions, frequencies and transmission distances. Moreover, they are not implemented as commercial software, and their deployment requires considerable expertise, with the risk of differing conclusions.

Engineering Models

Engineering methods generally rely on the source-propagation path-receiver model, (Figure 4.1). The basic assumption of these methods is that noise calculations can be divided into three separate stages that are essentially independent and whose characteristics do not interact. This allows each stage to be defined and characterised separately, thereby greatly simplifying the modelling and calculation.

The most complex part of the modelling process is identifying the propagation paths and their characteristics. The usual method is to assume that

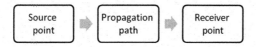

Figure 4.1 Source-propagation path-receiver model for noise prediction.

the sound propagates as many thin 'rays'. There are a number of ways of identifying the path of these rays, usually by starting at the receiver point or receptor – the place where the noise level is to be determined. There are two basic systems: the simplest is to throw out rays at equiangular spacing around the receiver point and to detect whether any intersect a source object (Figure 4.2). Use of equiangular spaced rays can have problems – depending on the angular spacing, they may fail to identify sources or obstructions, particularly at larger distance, or indeed they can overestimate screening. A better solution is to use irregularly spaced rays aligned with the centre or edges of objects (Figure 4.3). However, in complex urban situations, this can lead to a very much larger number of rays with consequently increased computation time. A further issue with ray tracing is the assumption that the ray is a narrow line, whereas sound travels as a wave front. Some implementations take steps to deal with this, but again with consequences of increased computation time.

For each ray, the effect of distance, ground interference and atmospheric absorption is established by means of engineering formulas. This requires the ground topography to be considered (Figure 4.4). Next, it is necessary to determine whether the ray intersects with any obstructions, such as purpose-built barriers or buildings. If so, the potential screening of the object is calculated (Figure 4.5). If the ray intersects with other obstructions, any interaction between these diffracting edges must be considered, as must any interaction with the ground effect (Figure 4.6). The effects of directivity at

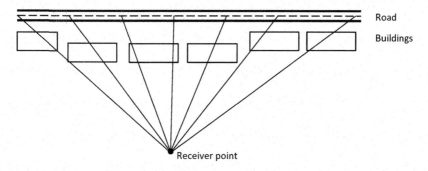

Figure 4.2 Equiangular rays.

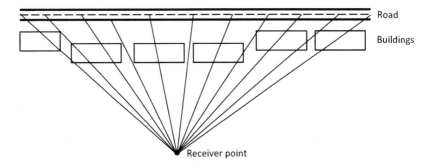

Figure 4.3 Irregular rays.

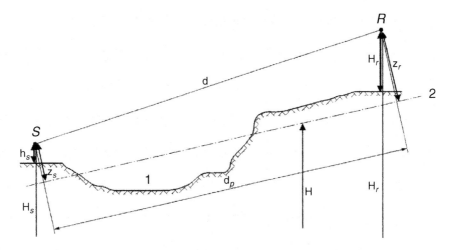

Figure 4.4 Consideration of ground topography.

the source and reflections from buildings and noise barriers are also deter-mined. The effects are summed to obtain the contribution of that source to that receiver. The process is continued until all rays have been investigated. Reflections are a particular problem: each source could appear as an image source in any reflecting surface – typically another building or a noise bar-rier – and there can be multiple reflections between such reflecting surfaces. In reality, diffraction, diffusion and the additional propagation distance mean that the reflected images are much weaker than the main source, and various methods are used to reduce the effort spent on assessing reflected images. Indeed, some methods do not evaluate image sources but add a gen-eralised amount based on the characteristics of the reflector.

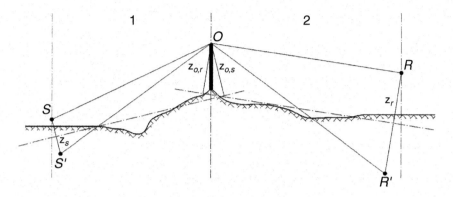

Figure 4.5 Consideration of barrier attenuation.

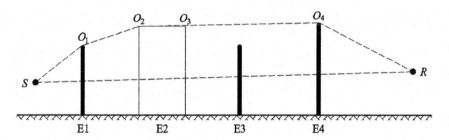

Figure 4.6 Consideration of multiple diffraction.

Thus it can be seen that although engineering implementations are a great simplification of the physical mechanisms and processes, they are nevertheless complex, and their evaluation requires considerable computing power. Although this is becoming less of a problem as computers become more advanced, it is often a limiting factor.

Whilst engineering implementations all use the same general assumptions and approach, they differ in important details. The propagation element of the calculation is the most significant factor in calculation uncertainty, although, as we shall see, not necessarily in the end result. The effects of distance, ground interference, atmospheric absorption and barrier screening are evaluated from engineering formulas. Historically, these formulas were established by measurement studies: measurements made at various distances over 'hard' ground could be used to establish the change in level with distance. Measurements over 'soft' ground could establish the effect of ground interference by reflection and absorption. Similarly, measurements under various meteorological conditions could establish the effect of wind speed and direction and atmospheric stability. But these effects are not in reality independent, and the different methods have distributed the proportion of path attenuation slightly differently between these factors. This could mean, for example, that 'mixing and matching', say, the 'soft ground effect' from *Calculation of Road Traffic Noise* [8] with the atmospheric absorption effect from ISO 9613–1: 1996 [9] could lead to increased uncertainty.

Over the last two decades, much work has been done [10] to establish the numerous propagation effects mathematically using the exact methods described earlier. Although this can theoretically lead to more exact prediction equations that distribute the various propagation factors more precisely, this cannot overcome the reality that they are not entirely independent. This means that the engineering implementation can never be more than an approximation, but they are acceptably reliable in most cases, especially, as we shall see, when such approximations may not be the greatest source of uncertainty.

The situation for which we are predicting is not immutable. Experience indicates that in many situations, source variability is a very large contributor to uncertainty, and even when the source can be stabilised, variability in the weather conditions, ground cover, vegetation and built environment is ever-present. The uncertainty introduced by these environmental factors is at least as important as that arising from approximations in the theoretical equations or the practical implementations of the theory. Moreover, it should be recognised that measurements, which are often taken as the 'true' reference, are subject to similar sources of uncertainty.

It has been suggested that at least a noise prediction represents an estimate of a snapshot of a moment in time and space [2], but this is not necessarily true. No family actually has the average 2.5 children, and no noise prediction can claim to give the 'true' noise level at a receptor. Often it represents an entirely hypothetical situation. Nevertheless, it is a common mistake to believe that when a measurement does not agree with the calculation, the prediction must be in error. Many cases arise where a noise measurement is flawed, and the prediction, though wildly different, could be a better representation of the situation. Moreover, noise predictions are often possible in situations where measurements cannot be made – not simply because the situation is for a past or future scenario but because a measurement is not possible, due to either access restrictions or interference from other noise sources.

An Overview of Noise Prediction Methods

A very large number of prediction methods exist. This section reviews a few of the more common ones, particularly with a view to understanding their approach to uncertainty.

European Union Common Assessment Method for Noise Indicators

Since the European Union's (EU's) Environmental Noise Directive (END) was published in 2002, the EU's Joint Research Centre has been working on a common noise assessment method to be used throughout the EU. In this context, assessment means determination by calculation of the common noise indicators L_{den} and L_{night} at the required assessment positions. Since 2002, a number of calculations methods have resulted, including Harmonoise [10], IMAGINE [11], NMBP2008 [12] and CNOSSOS-EU [13]. These have now been refined into a new method set out in EU Directive 2015/996 and its 820-page annex, 'Assessment Methods for the Noise Indicators' [14]. This defines the calculation methods for road, rail, industrial and aircraft noise in exhaustive detail, with the latter occupying the bulk of the document because it contains data needed for noise calculations of a large range of aircraft, including their variants. From 2019, these methods are mandatory for strategic noise mapping made under the END. It is only possible to outline its broad approach in this chapter.

The first section of this directive describes the derivation of source levels for road, rail and industrial sources in great detail and then goes on to describe the analysis of propagation effects through highly detailed analysis of the intervening ground. ISO 9613–1: 1996 [8] is to be used for determining atmospheric effects.

Disappointingly, there appears to be no indication of the accuracy or uncertainty to be expected from this highly complex methodology. Under the heading of 'Quality', it states that source emission levels shall be determined to an accuracy of ±2 dBA or better and that input values shall reflect the actual usage, implicitly recognising that many test regimes involve contrived standardised situations, which can be very different from real use. However, many of the prediction formulas require information that cannot be easily obtained, and therefore, default input values and assumptions are accepted if the cost of collecting real data is disproportionately high. There is a series of guidance notes [15] produced by European Commission working groups giving guidance on good practice, which includes advice on these default input values. Needless to say, the use of default values can largely negate the sophistication of the new method. No doubt it is hoped that improved remote sensing and other techniques will allow such data to be gathered in the future. Software used for the calculations is to be certified by compliance of results with test cases.

The END requires the evaluation of yearly averages, and it is unclear how differences in vehicle speed, composition and flow rate throughout the year are to be accommodated. The uncertainty resulting in practical application of the method, including the use of default values and yearly average source levels in particular, is not mentioned. This methodology may result in improved uniformity across the EU, but given that calculation errors would seem to be a small component of uncertainty, it remains to be seen whether there will be a meaningful reduction of uncertainty in the assessment of population exposure.

ISO 9613-2:1996: Acoustics: Attenuation of Sound during Propagation Outdoors, Part 2: General Method of Calculation, and ISO 9613-1: 1996: Calculation of the Absorption of Sound by the Atmosphere

ISO 9613–2 [16] is an international standard on the attenuation of sound during propagation outdoors. ISO 9613-1 [8] is an analytical method for calculation of the absorption of sound by the atmosphere and is used within ISO 9613–2. ISO 9613–2 is regarded as an 'engineering method' for general-purpose prediction of environmental sound from a variety of sources. It predicts the equivalent continuous A-weighted sound pressure level under meteorological conditions favourable to propagation from sources of known sound emission. This means downwind propagation or under a well-developed moderate ground-based temperature inversion, such as commonly occurs at night. The method predicts a long-term average A-weighted sound pressure level encompassing a wide variety of meteorological conditions.

The source-propagation-receiver model is used. This assumes that the characteristics of these three elements of the model are essentially independent and do not interact, which allows them to be defined and characterised separately. The method assumes that the noise originates from a point source or an 'assembly' of point sources and travels as a thin ray to the receiver. 'Extended' point sources can be represented by a series of point sources covering the line or area that emits sound, each having a certain sound power and directivity. In addition to the real source, there is an image source in each reflecting surface. The method of tracing these rays is not defined but is left to the individual user.

This is a very analytical approach that lends itself to general-purpose application over almost any source, but this also discards much of what is known about the characteristics of specific sources, such as road vehicles and trains, which can lead to differences in interpretation and to complexity and inefficiency in implementation.

It is clear that ISO 9613–2 is of such complexity that it would have to be implemented in a computer, although the document does not address methods of implementation. There are some charts and diagrams, but they are not sufficiently detailed to be used for calculations. They are likely to have been provided so that users can check the description of the procedures, as there are no other examples of use.

Accuracy and Limitations

The document addresses the accuracy and limitations of the procedure, stating that there is information to support the method for broadband sources. It provides an estimate of accuracy for downwind propagation of broadband noise as follows:

- Average height of propagation up to 5 m; accuracy is ±3 dB up to 1,000 m.
- Average height of propagation between 5 and 30 m; accuracy is ±1 dB up to 100 m and
- ±3 dB up to 1,000 m.

This is in the absence of screening or reflections and is thus only the unobstructed propagation accuracy. Moreover, the soft ground effect is limited to approximately flat terrain. These are significant provisos, as few real applications will be this simple.

Calculation of Road Traffic Noise, 1988

The *Calculation of Road Traffic Noise* (CRTN) [8] was based on research by the United Kingdom's then National Physical Laboratory and the Building Research Station. It is essentially an empirical method based on a large number of measurements made alongside roads in the late 1960s and early 1970s. It was originally published in 1975 and was intended to be used for the determination of noise from new motorways so that provision could be made for the protection or compensation of residential properties exposed to excessive noise. In 1988, a number of significant improvements were made. The source level calculation procedure was revised to introduce a 'road surface correction', as it was found that the type and texture of the road surface had a great influence on source levels, and this was speed dependent. This was introduced after problems had arisen from certain heavily textured concrete surfaces. It was intended to discourage their use and also to encourage the adoption of noise-reducing road surfaces. Whilst the noise-reducing surfaces available at the time were somewhat expensive and unsatisfactory, it has led to the development of more satisfactory reduced-noise surfaces, including those known as 'thin wearing courses'. The CRTN does not require reflected image sources to be assessed; instead, a correction is made for the observed effect of reflecting surfaces close to the source and receiver, including reflections from roads in retained cutting (popular at the time). Further changes were made in 2008, issued as an update within the *Design Manual for Roads and Bridges* [3]. This was done to avoid the legislative process that would be necessary for changes to the CRTN. These changes allow a measured 'road surface influence' to be used in place of the formulaic values for road surface correction in the 1988 procedure. Also, the method of dealing with reflections was changed to reduce the influence of distant surfaces and to allow absorbent barriers to be evaluated. Finally, the soft ground effect was capped at the value achieved at 600 m. This was to permit the formulas to be used for long-distance noise propagation.

A procedure to measure 'road surface influence' (i.e. surface noise) was devised and is much easier to implement than the measurement 'of 'surface texture depth', which required closure of a road in order for a 'sand patch test' to be made. This test was not only risky and expensive but also unreliable because the surface texture is variable both across and along the road. Given the great changes in vehicle construction and noise-control measures since that date and the increasing application of the method to road improvement schemes of a more local nature, the method still gives impressive results, which are discussed in the case studies later.

Prediction Uncertainty

The CRTN is a procedural document that does not discuss uncertainty, but a technical paper [17] published to accompany the first version of the document has considerable detail. Over 2,000 field measurements were obtained. Comparison of 768 points of unobstructed propagation showed a mean error of +0.2 dBA (i.e. predicted level is higher than the measured level). The root-mean-square (RMS) error was 2.4 dBA. Limiting the comparison to the 328 best-quality measurements gave a mean error of +0.1 dBA and an RMS error of 2.0 dBA. When all 2,064 data points, which included screened sites, were included, the mean error was –0.6 dBA, with an RMS error of 2.5 dBA. This implies that 87% of predictions should be within ±3 dBA.

Sources of variance were analysed. Uncertainty was higher on sites with low flows and with low speeds. Somewhat unexpectedly, uncertainty was greater at sites very close to the road. As regards screening, there was some indication that barrier screening was slightly overestimated in deeply screened areas, but the overall mean error is +0.2 with an RMS error of 2.6 dBA, which is not significantly different from the unscreened result. Overall, there was a slight tendency to underpredict the higher noise levels and to overpredict the lower noise levels, with mean errors of –1.2 and +1.6 dBA, respectively. The investigators could not identify any reason for this. Of course, these tests only show how well the prediction method reproduces the measured data on which it is based. It is not a test of the overall reliability of the method.

A paper by Hood in 1987 [18] considered whether the CRTN had maintained its accuracy in the light of changes to vehicle design and regulatory limits on noise. Hood undertook measurements over 'an extended period' at 17 locations during which detailed traffic flow, speed and composition data were recorded. The CRTN gave a mean overprediction error of +0.7 dB L_{A10} with an RMS error of 2.1 dB L_{A10}, compared with an underprediction of –0.2 dB L_{A10} for similar sites in the original study. During the intervening period, permitted noise levels for new vehicles had dropped by 3 dBA for heavy vehicles and by 4 dBA for cars. Clearly, the reduction in test levels had not been translated into corresponding reductions in roadside levels, leading to concerns that the test regimes do not adequately reflect real driving conditions. Hood concluded that the two main factors affecting prediction accuracy were wind direction and road surface texture. He found that beyond 500-m propagation distance, good predictions can only be obtained when there is a strong wind (i.e. 4 m/s or more) blowing from the road to the receiver, which only occurred for 15% of the time at his locations. Lower noise levels would occur under different wind conditions. However, he concluded that uncertainty due to road surface texture is likely to account for the major part of the measured error in the CRTN.

Calculation of Railway Noise, 1995

The *Calculation of Railway Noise* (CRN) [19] calculates noise from railways and other tracked transit systems. In essence, it is an adaptation of the CRTN for use on railways. Railway noise has a number of complications not associated with roads, the main one being that the wheel-rail noise (the main source of noise from trains at moderate speeds) is directional, being strongest at low elevations at right angles to the track.

A further complication is that each type of train has markedly different sound emission characteristics: tread-braked trains (now falling out of use) are much noisier due to the wheel roughness caused by the brake shoes, the many different types of propulsion units have different noise characteristics, and for high-speed trains, aerodynamic noise from the overhead power pickups (pantographs) and associated fittings can be significant. The track type and track support system also have a large influence on noise levels. The basic assumption is that the track is supported on ballast, which is the quietest system, and if other supports are used (e.g. on bridges or on-street installations), a correction is made for this. However, the procedure is not intended for elevated railways carried on concrete or steel structures, which can radiate significant amounts of noise. The CRN has been updated to deal with certain types of high-speed trains but has been found to be too prescriptive and inflexible to apply to recent high-speed proposals in the United Kingdom, as it does not have adequate provision for sources at different heights and with different speed profiles. The methodology does not include any guidance on uncertainty.

Train Noise Prediction Model, 1994

The Train Noise Prediction Model (TNPM) [20] is a more flexible method that was devised for the UK high-speed Channel Tunnel Rail Link, subsequently called HS1. This model uses the source-propagation path-receiver concept and is very similar to the CRN described earlier. However, it allows for any number of sources at different heights and with different speed-varying characteristics to be assigned to a train and has consequently been adopted in the assessment of proposals for a new high-speed railway in the United Kingdom known as HS2.

Validation and accuracy of the TNPM

An appendix to the HS2 Environmental Assessment [21] describes the assessment methodology, assumptions, limitations and tests used in the assessment of the direct effects due to airborne sound arising from the operation of a proposed high-speed railway. The HS1 is a high-speed railway approximately 110 km long connecting from the English end of the Channel Tunnel in Folkestone to St Pancras in London. A regression analysis between predicted and measured levels gave a standard error, for the goodness of fit, of approximately 3 dBA both for the sound exposure level (SEL) and for L_{pAFmax}. This means that the difference between predicted and measured levels is typically within ±3 dBA [22]. Measurements taken since HS1 came into operation show that the prediction method tends to overestimate in-service noise levels.

Sensitivity tests have been made to identify which input parameters have the greater impact on noise forecasts and to identify what input changes could lead to more than a 3-dBA change in predicted levels. These tests indicated that a 10% change in train speed at 330 km/h would have an effect of less than 2 dBA but that HS1 trains would be about 3 dBA noisier than the proposed HS2 specification. The sound-reduction performance of a 3-m absorbent noise barrier depends on the train specification and speed, as the sound emission of the various sources (which are at different heights) varies with speed. For example, HS1 trains have greater noise from the power-collection system on top of the train, and as a result, the barrier is 1.6–2.0 dBA less effective.

The model assumes downwind or 'positive wind' propagation. This is similar to the effect of a temperature inversion (a positive temperature gradient). Measurements were taken in France in relation to the *Train à Grande Vitesse* (TGV) high-speed railway and show that even close to the track (at 25 m), there is a spread of 16 dBA in the pass-by noise of different trains, mainly due to differences in 'source terms'. At 200 m distance, the spread increases to 29 dBA, with downwind having a spread of 12 dBA and upwind having a spread of 15 dBA, all lower than the downwind measurements. The measured mean difference between upwind and downwind measurements is over 10 dBA.

The large differences in the source-noise levels of individual trains are of concern, as it is larger than the mean difference between upwind and downwind measurements. My own observations show that much of the difference is attributable to wheel-rail noise. Driven bogies are noisier than 'trailer' bogies. But some wheel sets are very much noisier than others. Even when the wheels themselves do not suffer from 'corrugations' or flats, differences can arise from the position of the wheels on the rail: the wheels ideally ride on the centre of the tread but can slide from side to side. The wheel flange can make contact with the railhead, the bogie may not be perfectly in line with the track, and differences in loading may accentuate certain defects. Other differences can affect the aerodynamic noise from the power pickup and the train body. This means that the source terms are often a greater source of uncertainty than the rest of the prediction method.

Aircraft Noise

Aircraft noise is a huge topic, and we will consider the issues through one prediction method, the Aviation Environmental Design Tool (AEDT) [23], which will enable the issues to be explored. There are many other tools in use, including ANCON in the UK, but all work on similar principles.

Aviation Environmental Design Tool (AEDT)

This software was developed from the widely used integrated noise model (INM), which it replaced in 2015. It is the mandatory tool for assessing aircraft noise, fuel burn and emissions modelling on new Federal Aviation Administration (FAA) projects in the United States. It is applicable to all aircraft types, including helicopters, and for individual aircraft as well as whole airports. It requires detailed information on aircraft types, flight tracks and track dispersion, fleet mix and flight frequencies. It can create noise footprints and noise contours and generate population exposures. It uses the source-propagation path-receiver model, as in most engineering methods of noise prediction. It uses a database of aircraft information known as the European Organization for the Safety of Air Navigation's (EUROCONTROL's) Base of Aircraft Data (BADA).

This is sophisticated software based on improvements to the science in its predecessor, INM, and other legacy software. The authors' 'Uncertainty Quantification Report' [24] aims to identify how much variability is associated with uncertainties within the model and to identify the input parameters that can cause most variability of the output. This is a lengthy and detailed report that assesses many aspects of the model's performance. It describes how the performance data of each type of aircraft has been validated, including one-third octave and tonal data for different operational modes, both for effective tone-corrected perceived noise level (EPNL) and SEL. Aircraft flight performance data are also validated, as this is an essential aspect of modelling flight paths. A final validation replicates real-world flight test conditions in the model to give a combined validation of noise and flight performance data. If the modelled levels are within 3 dB of the measured data, this is regarded as acceptable; otherwise, the *performance* data (not the modelled result) are queried. This sometimes leads to the performance data being revised, but if the data provider is satisfied that the performance data are correct, they are accepted for inclusion in the database despite the deviation. This approach suggests that the software authors generally consider that the modelled results are more reliable than the measurements. It may be noted that during validation of the AEDT, the authors demonstrated an ability of the system to generate comparable results to legacy software, taking into account that differences were expected owing to the intentional evolution of the algorithms. Nevertheless, it may be noted that although most differences were in the range 0–3 dB, a few differences of up to 5 dB were observed.

Sensitivity tests were run using Monte Carlo simulation on a simplified model. Among the tests, the day-night average sound level airport noise metric was assessed at five example airports, such as John F. Kennedy International Airport (JFK). No single input parameter had the biggest influence on the area of the 65-dB DNL contour at all airports, but headwind, air temperature, takeoff speed and jet thrust coefficient were significant at various airports and taken in combination, and meteorological factors had a large effect on variability. Overall, the coefficient of variation, which is a measure of the standard deviation relative to the mean, was less than 5% across all five airports.

It may be noted that the validation studies, whilst extensive, did not compare AEDT output for airports against measured data but only the output for individual aircraft. This is an implied admission of the difficulty of making validation measurements extended in time and location.

Construction Noise

BS 5228–1: 2009: Code of Practice for Noise and Vibration Control on Construction and Open Sites, Part 1: Noise

General-purpose calculation methods such as ISO 9613–2 can be used for construction noise evaluation. However, this is a highly involved procedure. In order encourage an awareness of the need to consider construction noise and to provide a method that could be readily used by construction engineers, a simple method of estimating noise levels was published in BS

5228: 1975. The method has been developed through further study and experience since that time, with the latest version being BS 5228-1: 2009 [25]. It is based on the usual source-propagation path-receiver engineering model, although with its own formulas for propagation and barrier corrections. Charts are given so that manual implementation would be possible in simple cases. The importance of the method is not in the familiar (and rather basic) propagation equations but in the provision of a database of noise from construction plants. The earliest edition of BS 5228 recognised that good source data would be the foundation of reliable construction noise assessments, and it drew on a large survey of many construction, demolition and minerals extraction operations undertaken on behalf of the Construction Industry Research and Information Association. The original sound level data were based on the concept of 'activity sound level' in terms of the equivalent continuous sound level L_{Aeq} taken over the whole cycle of an operation, measured at 10 m from the centre of the activity. These data were also presented as a sound power level with an 'on time' such that if applied correctly, it would give the same value as the activity L_{Aeq}.

The advantage of this approach is that the noise levels during an activity (such as a loading shovel moving material from a stockpile to a lorry) will vary as the plant works. Thus, the sound power level of a plant taken during a certification test is not representative of real use. Moreover, manufacturers usually do not issue the results of actual sound tests – they rely on issuing a certificate to show that the plant meets the permitted sound levels. Use of 'permitted levels' can greatly overestimate the sound output in real use.

More recent additions to the source data have taken a slightly different approach. The sound power level and on time are no longer presented, but octave-band sound levels are given, particularly for use in the barrier screening calculation, along with the L_{Aeq} at 10 m. For mobile plants, the value presented is of L_{Amax} measured at 10 m during a drive-by. The uncertainty of the source levels or resulting predictions is not mentioned but can be inferred from the case studies presented next.

Case Studies

Commercial Noise Prediction Software Systems

It is beyond the scope of this chapter to review or critique the many commercial noise prediction applications that are available. Each software system will differ in its range of procedures, model sizes, speed of calculation and ease of use, but an evaluation of these differences depends on the intended use, the size and nature of the project and the skill of the operator. It can be assumed that all established commercial noise prediction systems with a reasonable number of current users have been validated against their stated methodologies and found to give acceptable results. All manufacturers will be able to produce validation information, if required.

Most commercial noise prediction systems will use similar methodologies and will require similar input information which is subject to the same uncertainties. Accordingly, this section presents some example studies in order to provide some insight into the issues.

Comparison of Traffic Noise Predictions of Arterial Roads Using Cadna-A and SoundPLAN Prediction Models

This Australian study [26] suggests that an uncertainty of ±1 dBA can result in the costs of implementing noise mitigation to vary dramatically. The CRTN (1988) and Federal Highway Administration (FHWA) algorithms are used in Australia. Previous Australian studies had shown that the CRTN predicted 85% of results within ±2.7 dBA and 95% within ±5.0 dBA of the 'true' (presumably the measured) level. CRTN 1988 is the recommended algorithm and was implemented by both software systems under comparison, namely SoundPLAN and Cadna-A.

Two situations were compared. The first was an idealised straight and level 'proposed road'. A flow of 1,729 vehicles per hour (approximately 20,000 vehicles per 18 hours) at 3.1% heavy and 70 km/h was used. The receivers were spread over distances from 33 to 75 m from the nearest kerb, shielded by noise barriers 1.8 to 2 m high at the boundary. A three-height source model was used, 0.5 m for light vehicle wheel-road noise sources, 1.5 m for light and heavy vehicle engines and 3.6 m for heavy vehicle exhausts. These source heights are due to the larger size of heavy vehicles in Australia. SoundPLAN predicted, on average, 0.28 dB higher than Cadna-A, with the range being 0–0.6 dB.

In the second situation, a model was set up for an existing slightly curving and undulating road with housing irregularly arranged alongside it. The traffic flow was 1,230 vehicles per hour at about 4% heavy and 65 km/h. The four receivers were between 19 and 66 m from the kerbside. The study showed that SoundPLAN predicted, on average, 0.7 dB higher than Cadna-A, with a range of –1.3 to +2.5 dB too high. Cadna-A predicted, on average, 0.25 dB too high, with a range of –2.5 to +2.1 dB. The largest under-prediction by both methods was for the same receiver, but over-predictions were for different receivers in the two cases. No investigation into these differences was reported.

Noise Mapping of London Using NoiseMap SE Software: Quality Checks

The London Road Traffic Noise Map (Figure 4.7) was produced by Atkins in 2004 using the NoiseMap SE software system [27], which implements

CRTN 1988. Independent quality checking was undertaken by R. C. Hill of AIRO. The report considered the accuracy of the modelling process in representing the fabric of London and the accuracy of the final result in relation to a sample of measurements made by AIRO over the years 2002–2003. The London Noise Map covers a huge area of over 1,600 km^2, covering 33 London boroughs with a population of 7.1 million people and containing 2.67 million acoustically significant buildings.

There are over 5,200 km of roads carrying acoustically significant traffic flows, represented by 120,578 road segments with more than 21,100 different traffic flows distributed over the network. The traffic flows vary from virtually nothing up to 171,800 vehicles per 18 hours at speeds varying from 2 up to 106 km/h.

The report noted that there were inconsistencies between the observed and modelled traffic flows on about 2% of the road 'links'. These inconsistencies were randomly distributed and traced back to a replacement of the original traffic flow data set (the London Atmospheric Emissions Inventory, or LAEI) by an update of the same data set. The report states that the updated data set was set out slightly differently, causing misallocated flows, but the effects would be localised. After correction of the misallocation errors, AIRO checked the LAEI flow against the observed traffic flow and found that in some cases there were considerable differences. On one road, a flow of 4,431 vehicles per 18 hours had been assigned, whereas the measured flow was 12,000 per 18 hours. The percentage of heavy vehicles was correctly assigned at 9.5%. Nevertheless, this would amount to a level difference of around 4 dBA. It was further noted that a few road segments (~0.4% of the total) had excessively steep gradients. This was found to be due to rounding errors in the end heights of a few excessively short segments on tile boundaries, again with a very localised effect.

AIRO checked predictions against measurements in 18 cases and, after correcting for differences in traffic flow assumptions, concluded that the NoiseMap predictions were 0.5 dB lower, on average, than the measured results. Manual calculations were made at five locations and were each within 0.4 dB of the noise levels shown on the map. Some example results are shown in the table below.

It will be noted that these are all in areas of high noise levels without screening. The general conclusion of this report was that errors in traffic flows are the main source of discrepancies rather than the modelling or prediction process itself.

Post-environmental Impact Statement: Evaluations of National Road Schemes in Ireland

The environmental impact of national road schemes in Ireland must be evaluated and accepted prior to the scheme being built. The National Road

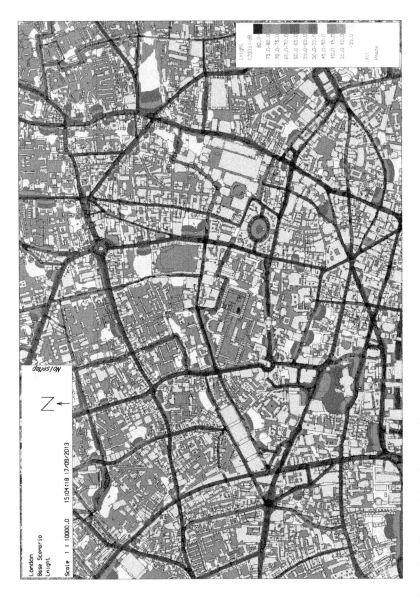

Figure 4.7 Part of the Road traffic noise map of London.

Location	Predicted level, dB	Measured level, dB	Difference, dB
Westway	76	77	−1
Tredegar Road	69	69	0
A406 Barnet	66	67	−1

Authority (NRA) of Ireland issued guidelines on the noise aspects of this process in October 2004 [28]. An important aspect of the guidelines was that noise arising from new road schemes should aim to meet a design goal of 60 dB L_{den} (free field) at residential facades. The NRA undertook a wide-ranging review [5] of the effectiveness of these guidelines in achieving design goals. The study identified one problem in comparing measured and predicted levels – that measurement locations were often not reported with sufficient precision. Moreover, input traffic data for the calculations were not adequately reported. There was also a wide variation in the number of measurement and prediction sites used to represent a scheme. Predictions were often validated by comparing the output of a prediction package against a spreadsheet, which is a validation of the prediction method and not the result. The authors reported an Australian study [29] that speculated that post-project monitoring and auditing were the weakest areas of the environmental impact assessment process. They pointed out that, in practice, noise mitigation in Ireland was synonymous with the use of timber noise barriers, which was not the intention of the original NRA guidelines: these state that a structured approach should be undertaken in order to ameliorate road traffic noise through the consideration of alignment changes, barriers, earth mounds and low-noise road surfaces. They suggest that mitigation measures should be considered in a sequence in which barriers should rank below other options rather than being the preferred choice. Many barriers produced a noise reduction of 3 dBA or less, widely acknowledged to be 'barely perceptible'. They note that barriers several hundred metres long and up to 4 m high are not unusual to protect one or two properties. It is reasonable to question whether the benefit is proportionate to the cost and visual intrusion. They consider that the term 'sustainable' needs to be clearly defined in terms of the design goal.

As a result of this review, the NRA of Ireland issued *Good Practice Guidance for the Treatment of Noise during the Planning of National Road Schemes* [4]. Among much good advice, these now recommend that the route corridor selection process, formerly based on counting the number of properties with certain distances of the proposed road alignment, should take advantage of the ability of noise prediction packages to produce a noise footprint quickly and efficiently.

Channel Tunnel Rail Link: Case Study

HS1 was built to allow eight high-speed trains per hour in each direction through from London to the Channel Tunnel. In 2014, Eurostar was running two to three trains per hour in each direction, with up to eight domestic trains per hour between London and Ebbsfleet. In 2016, ten years since the full line was first opened, 75,000 trains a year (two-way flow) were using the London terminal, St Pancras, although nearly three-quarters of these were domestic services [30]. The current passenger numbers are two-thirds of the original forecast, and the line is running at about 50% capacity.

The original noise study was undertaken using a procedure specially developed for the purpose by Ashdown Environmental Limited, which they termed the 'Train Noise Prediction Model' (TNPM) [20]. The accuracy of the source noise terms of 'the two train types' (domestic and international) was identified as critical and 'a limitation to the accuracy of the modelling'.

In practice, there are now at least four train types using the line, excluding freight trains. Although freight traffic was envisaged, no freight traffic used the line until 2001 owing to the need for locomotives to be modified to use the signalling system and for wagons to meet a particular specification. Speed has a major effect on railway noise, which rises in proportion to the square of the speed. This means that it is crucial to have accurate speed profiles as trains accelerate and decelerate, but rates of acceleration and braking are one of the distinguishing features of the various trains.

It is difficult to know what effect these matters have on the accuracy of the predictions, as it is notoriously difficult to obtain source terms. However, the 50% less traffic in itself would mean that predictions are 3 dBA too high, and it would appear that at the tunnel itself, the flow rate may be only 14% of the presumed flow, leading to an overprediction of about 8.5 dBA. It is clear that these discrepancies in assumptions dwarf the likely errors in the calculation procedure.

Comparing Scenarios

The impact of a scheme such as the HS1 is often assessed by considering a change from the present or existing condition. The process will include a comparison of the range of options. In this situation, uncertainty can be reduced by comparing differences between options, as errors in source terms and other assumptions will be similar in each case, thus tending to cancel out rather than accumulate.

Reducing Uncertainty

Uncertainty in Input Information

This review has shown that computer prediction software using the standard methodologies is capable of giving results of acceptable accuracy. This does not mean that they will always give reliable results, as the method must be used correctly, the model must be constructed accurately, the source data must be correct and the results must be collated and interpreted correctly in order to be fit for purpose. This section discusses how these issues can be controlled.

Source Characterisation

The issue of source characterisation has been mentioned previously. For road traffic, we need the correct traffic flow rate, the correct proportion of each class of vehicle (motorcycle, car, van, lorry, etc.) and the correct speed. For other types of transportation noise, equivalent variations occur in train or aircraft types and pass-by numbers. Since these vary throughout the time period being assessed, whether by time of day, day of the week, or month of the year, some means of typifying the flow must be adopted. In the United Kingdom, for road traffic, the annual average weekday 18-hour traffic flow has been standardised as having a good correlation with subjective response to road traffic noise. This is on the basis that the diurnal variation of traffic flow is very similar on all roads. However, there is concern that increasing congestion is forcing more traffic to travel outside working hours and especially on long-distance roads such as motorways. For railway noise, the 18-hour daytime and 6-hour night-time traffic flows have been adopted for assessment because night-time traffic is under control of the railway operator, and there can be considerable freight traffic at night. For aircraft noise, it is necessary to identify the fleet mix, flight paths, operating mode, operational hours, noise-reduction strategies and other factors.

The preceding case studies help to place quantitative values on the effect of such errors. It is fortunate, perhaps, that the decibel is a forgiving unit in this respect – a doubling of flow increases noise levels by only 3 dB, which is taken to be the smallest change of environmental noise normally perceptible. Consequently, significant errors in a flow may have little effect overall. Furthermore, except in the simplest cases, the noise at a point will arise from many sources near and far, and it is unusual for all the uncertainties to err in the same direction: they are more likely to cancel each other out. A simple statistical sum of all the uncertainties could lead to a misleading result.

Mapping

The first part of the process must be to construct an adequate representation of the physical situation. This has become a great deal easier over the last decade with ready availability of digital mapping, aided by remote surveying techniques. It is now possible to purchase detailed digital maps of conurbations showing both the natural and built environment. The digital maps of a few years ago were two-dimensional digital copies of paper maps made by manual surveys, and currently, these are the basis of much mapping. Digital maps are available in a variety of formats, most commonly bit maps – essentially images of the paper maps – or vector maps. Vector maps represent the outlines of objects: each object can be given a unique code to identify the object, along with many other attributes, such as its name and a code to identify the type of object. More useful still, remote survey techniques such as light detection and ranging (LIDAR) are being used to assign a height to an object. These results can be obtained as 'Shapefiles' or other standardised formats. Shapefiles originated as a proprietary form of vector mapping for use in the ArcGIS Geographical Information System (GIS) [31]. This format permits a map object to be given an arbitrary number of additional attributes such as the traffic flow or surface texture of a road. Depending on the capabilities of the software, such maps can be imported directly into the mapping software. This greatly increases the amount of detail in the noise model whilst speeding up the process and reducing the scope for mistakes in the modelling process. Remote surveying has also made topography much easier to obtain and enter into the model in a similar way to the built environment.

Nevertheless, this does not eliminate error. A significant problem is that maps quickly become out of date and should always be verified on the ground or from recent satellite photography. A second problem can be the issue of object duplication – digital mapping comes in 'sheets', and there can be duplication at the edges or indeed the same sheet could be entered twice. If the duplication is exact (which it should be), then it can be impossible to detect from visual inspection of the model, even from cross sections or three-dimensional representations. The software must have checking tools for this type of problem.

A further significant problem is that of fence-type noise barriers. These are not identified as such on most digital maps – at most, there will be a code showing a fence or wall but no other information. Very often it is not possible to identify them from satellite photography, as they are too thin to show, which also means that their height will not normally be identified from aerial surveys. It takes a keen eye and fortuitous lighting to spot them from their shadows, and a pass-by inspection may be needed.

Indeed, small sheds, outhouses and general 'ground clutter' are also omitted from most maps. At one time, the inclusion of such objects would simply overload the noise calculation software, and it was essential to exclude them. With the automation of mapping, it is increasingly possible for such objects to appear, particularly in the case of topographical information, where maps with a resolution of better than 10 cm can be obtained. There is a temptation to include this detail on the grounds of 'more is better', but as we have seen, it is completely beyond the ability of any current engineering method to use this information. Including it will cause potentially large increases in computation time for no benefit. It is to exclude the uncertainty arising from ground clutter that the END requires maps to be made at 4 m above ground level.

Going from the Modelled Situation to Reality

Noise prediction studies are an attempt to forecast the effect of a proposal. Great effort may be made to ensure that errors in the modelling are minimised, but a problem may occur if the proposal is not implemented in the expected way.

On major schemes, it is usual for the 'reference design' assessed during the planning stage to be revised by the contractor for many reasons – such as practicality, problems overlooked in the design, misinterpretation of the design, cost saving and many other reasons. On large projects, there may be a contractual requirement for the new scheme not to be any worse than the reference design, and this may require recalculation. But when boots get on the ground, difficulties can still arise. One area of uncertainty is in regard to the height of noise barriers. When a barrier is shown on an environmental plan to be 3 m high, does this mean 3 m above the local ground, 3 m above the road or 3 m above the top of a noise bund? All these possibilities can and do exist. There are cases of noise barriers being built at the foot of an embankment, for example. Admittedly, good digital drawings and accurate surveying should reduce this type of blunder in the future.

When a part of the M25 London Orbital Motorway (a major UK strategic round encircling London) was opened, an unexpected level of complaints about noise was received from residents. The environmental design of the road had been particularly thorough and included some very high noise barriers. The contract had specified the minimum surface texture depth to be achieved, but investigation showed that to be sure of compliance, the contractor had applied a particularly deep texture to the brushed concrete road surface. Fortunately, after some months of use, much of the excessive roughness had worn off the surface and noise levels had reduced somewhat. Nevertheless, the only real solution to the problem would have been extensive re-surfacing.

On a minerals extraction project, it was found that the site manager was unaware of the environmental assessment and its assumptions over the method of working and the deployment of noise bunds. He was working from an entirely different set of assumptions based on his experience.

Knowledge of the Receiver Characteristics

In central London, the exact location of residential property is not always easy to ascertain, especially in predominantly commercial developments. Many of these projects have contractual terms requiring the noise exposure of residential properties to be limited to certain trigger levels. Exceeding such a level could require the contractor to protect the residents, even to the extent of providing alternative accommodations. There was particular concern in one case that limit values would be breached, and because of the proximity and elevation of the property, it would not be possible to provide screening. Fortunately, investigation showed that the designated windows were not noise sensitive: the sensitive windows overlooked a screened courtyard. Similar problems can occur on any large-scale assessment: it is impracticable to investigate every building in detail, and this can often lead to inaccuracies.

In most environmental assessments, no account is taken of the quality of the sound insulation of noise-sensitive buildings, nor the location of sensitive rooms, unless they are of exceptional sensitivity, such as a theatre or concert hall. For residential property, it is generally assumed that all facades are equally noise sensitive, even though this is often incorrect. Residents may have chosen to use a quieter facade for their bedroom or to have provided extra sound insulation on a noisy facade, leaving the quieter facade more sensitive. There can be an additional problem to ascertain the noise sensitivity of large sites such as schools and hospitals.

One technique used on UK buildings is to use the post code to identify the occupiable buildings. The Post Office's AddressPoint database contains the location of each post code and also shows the type of building – such as residential, school or hospital. This can be used to classify buildings for environmental assessment, but often the post code identifies the postal delivery point of the site, which may be non-sensitive offices – and in one case a boiler room – rather than classrooms or wards.

Why Do Noise Predictions Usually Turn Out to Be Too High on Average?

There is a concern from scheme promoters, planners and the general public that the noise impact of a scheme will be worse than predicted, despite the fact that it is usual for assessments to be made for a 'worst case' scenario. In reality, noise levels usually turn out to be less than predicted for a number of reasons.

Most importantly, perhaps, is that the intensity of use is often less than assumed. In order to justify the economics of a scheme – be it a transportation scheme, a commercial development or an industrial installation – the tendency is to assume that it will run at full capacity the whole time. In reality, typical use is somewhat lower. In addition, predictions assume that meteorological conditions will be favourable for noise propagation, but in most locations, the wind blows for a relatively small proportion of time from any one direction, even the prevailing direction. Figure 4.8 shows a typical wind rose indicating the strength and frequency of occurrence of the wind from each direction. Calculations also presume that a receiver point is directly downwind of every source, whereas this would rarely occur if there are several sources. Also, small obstructions are usually omitted from a model, whereas these can add considerably to the effects of dispersion and screening. Of course, propagation effects are fairly small at locations close to the source but are increasingly significant at 200 m and beyond. Hood [18] suggests that at 1 km, between the most and least favourable wind direction there is a variation in sound level of +3.5 to –7.0 dBA relative to calm conditions. He states that at one site positioned in the direction of the prevailing wind, calm conditions occurred for 30% of the time, lower noise levels for 30% of the time and high levels for only 15% of the time.

In the case of construction and other open sites applying for consent under environmental noise legislation, it is difficult for a contractor to know exactly

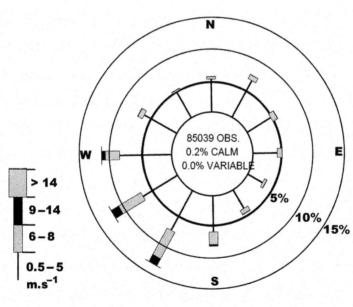

Wind direction and speed (10-year average)

London Heathrow Airport, UK

Figure 4.8 Wind rose showing typical distribution of wind speed and direction around the compass.

when an item of plant will be used or a particular phase of development will take place. Indeed, it may only be during the work that an unexpected difficulty needs the use of a particularly noisy item of plant. Accordingly, it tends to be in the contractor's interest to get permission for a situation in which the noisiest plant is all operating at the same time. This can often be much easier to get at the application stage than to deal with retrospectively. However, there are limiting cases where this assumption will produce excessive noise levels requiring mitigation. Research [32] has shown that, particularly for construction sites, it is useful to reduce the worst case by applying a 'diversity factor'. This is commonly used in situations such as the design of electricity distribution systems, where it is assumed that only a proportion of the appliances connected to the network will be used simultaneously.

Low-Speed and Congested Traffic

There is a popular perception that traffic in low-speed and congested situations is noisier than free-flowing traffic. This probably arises from the fact that people most often come closest to moving traffic when in busy residential or shopping streets, where there is congestion, traffic moves slowly and heavy vehicles in particular are noisy as they accelerate. Furthermore, CRTN gives credence to this by belief because Chart 4 (Figure 4.9) shows that below about 40 km/h, noise levels increase significantly as speed decreases. A study made on behalf of Transport for London by Atkins in 2001 [33] showed that this is incorrect. Noise levels continue to decrease as speeds drop, with the result that the CRTN overestimates noise levels by as much as 6 dB in low-speed cases (Figure 4.10). In essence, in congested traffic, queuing vehicles act as a line of sources of low sound power.

Managed Motorways

In many places, traffic growth has been such that high-speed roads are running above design capacity, with overloading and severe traffic delays at times. Attempts to mitigate this include the use of variable speed limits and hard-shoulder running. Just as in the case of low-speed and congested traffic described earlier, varying the speed limit downwards will reduce the noise from a road. CRTN requires motorways to be modelled as a single line source 3.5 m in from the nearside edge of the running lane. However, when the hard shoulder is used as a running lane, this could bring some of the traffic closer to buildings just beyond the highway boundary. It has been found that in such cases, the effect of hard-shoulder running can be ascertained by modelling each lane of the road as its own line source. It should also be noted that conventionally in the United Kingdom, 18-hour

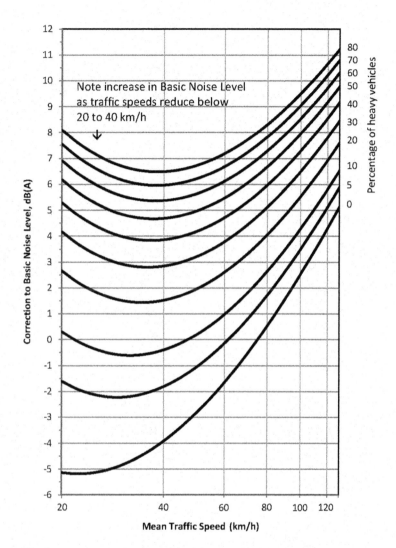

Figure 4.9 Chart 4 of CRTN showing higher noise at lowest speed.

weekday traffic flows are used in noise assessments. However, the effect of congestion can change the usual day-night proportion of traffic such that, in particular, a greater proportion of heavy vehicles travel at night. In such cases, a better estimate of night-time noise levels in particular can be obtained through the use of hourly traffic flows rather than the cumulative 18-hour flow.

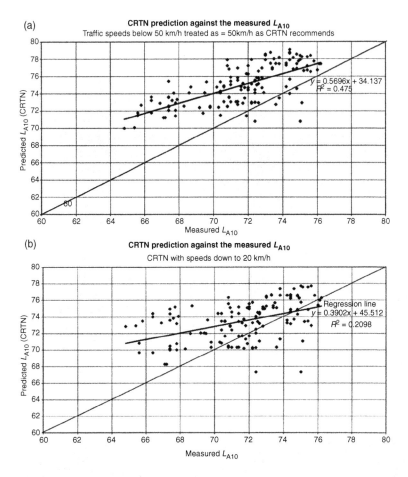

Figure 4.10 Inaccuracy in CRTN at low speed.

Conclusions

'Noise prediction', as a term, carries with it the connotations of uncertainty arising from looking into the future, but practitioners tend to be wary of admitting such uncertainty. Perhaps surprisingly, the literature does not try to address what level of uncertainty is acceptable for use in the assessment of noise impact. For example, the recent EU directive on common noise assessment methods does not directly address the subject, even though its researchers felt that it was necessary to introduce some extremely complex computations, presumably to make the system more accurate. The literature states that numerical methods of acoustic modelling can give 'exact' answers in specific cases, but these are too difficult to implement for everyday use. Almost all noise predictions are done using 'engineering' methods implemented in commercially available software, and these methods are approved and widely used for the assessment of the noise impact of all types of developments. There is little to suggest that these result in unsatisfactory outcomes, which would suggest that current prediction methods are sufficiently accurate for practical purposes.

Some engineering methods, such as ISO 9613–2, are general purpose and can be applied to a wide variety of noise sources. Others, such as CRTN 1988, are specific to a particular type of noise. Comparison of the various commercial implementations shows no significant overall difference in accuracy between them. Taking all prediction methods into account, it appears that a theoretical mean error of around 0 dB and a standard error of 3 dB are typical. However, the uncertainty in real use can be larger owing to uncertainty in the input data. For modelling of transport schemes of all types, actual flow rates can be significantly different from input assumptions. Prediction methods are also susceptible to assumptions about weather conditions. Engineering methods usually assume downwind propagation, but especially in urban situations, receiver points are generally not downwind of all sources simultaneously, leading to an overestimate of noise levels. Aircraft noise predictions are equally sensitive to assumptions about wind and temperature.

Validation of predictions against measurements is not straightforward. Measurements are also susceptible to weather conditions: care must be taken that all the assumptions are comparable, including the assessment locations and averaging periods. Where noise sources vary in a random way, as is often found on construction and similar open sites, it may be desirable to make a correction for 'diversity' (as is done in the electricity supply industry) rather than assuming that all the noise sources are simply additive.

Noise prediction requires a large amount of detailed input data, including source location, source intensity, topography, ground cover, buildings, barriers and receptor locations. Errors are inevitable when gathering and entering all this information into a computer system, and it is essential to use the functionality of computer software to eliminate such modelling errors.

Noise predictions are now sufficiently quick and reliable to be used as the basis for detailed design decisions: they can be used at early design stages, which removes the need for simplistic surrogates such as separation distances which are inaccurate wasteful and costly. This would enable prediction processes to add value to projects.

The availability of quick and accurate noise predictions should enable the development of better indicators of noise impact. The performance of EU common noise indicators of L_{den} and L_{night} (annual average) are little justified or validated by research and indeed would seem to be designed to be difficult to verify by measurement. It should be possible to adapt prediction methods to more sophisticated indicators. It is now clear that predictions are often superior to measurements, even to the extent that they are preferred over measurements in many cases.

References

1. Isukapalli SS, Georgopoulos PG, *Computational Methods for Sensitivity and Uncertainty Analysis for Environmental and Biological Models*, EPA/600/R-01-068. National Exposure Research Laboratory, US Environmental Protection Agency, Washington, DC, 2001.
2. National Physical Laboratory Acoustics Group, *Guide to Predictive Modelling for Environmental Noise Assessment*. 2007 (out of print).
3. *Design Manual for Roads and Bridges*, Volume 11, Section 3, Part 7, 'Environmental Assessment Techniques'. Highways England, London, 2011.
4. National Roads Authority, *Good Practice Guidance for the Treatment of Noise during the Planning of National Road Schemes*. NRA, Dublin, Republic of Ireland, 2014.
5. King EA, O'Malley VP, 'Lessons learnt from post-EIS evaluations of national road schemes in Ireland', *Environmental Impact Assessment Review* 32 (2012), 123–132.
6. Tompsett KR, *West Coast Wind Farms*, London, 2000, unpublished report.
7. Probst W., *Methods for the Calculation of Sound Propagation*, ACCON GmbH und DataKustik GmbH, Greifenberg, Germany, 2008, https://pdfs.semanticscholar.org/f187/e361a37fceb609e22d5c04b943b1d0bb79bf.pdf.
8. Department of Transport, Welsh Office, *Calculation of Road Traffic Noise*. HMSO, London, 1988. ISBN 0 11 550847 3.
9. International Organization for Standardization, *Attenuation of Sound During Propagation Outdoors – Part 1: Calculation of the Absorption of Sound by the Atmosphere*, ISO 9613-1. ISO, Geneva, 1993.
10. HARMONOISE, *Harmonised, Accurate and Reliable Prediction Methods for the EU Directive on the Assessment and Management of Environmental Noise*. EU Report, Brussels, 2005, https://cordis.europa.eu/project/rcn/57829/factsheet/en.
11. IMAGINE (Improved Methods for the Assessment of Generic Impact of Noise in the Environment), 'Final report summary', EU Report, Brussels, 2006, https://cordis.europa.eu/project/rcn/73857/reporting/en.
12. Dutilleux G, Defrance J, Ecotière D, et al., 'NMPB routes 2008: The revision of the French method for road traffic noise prediction', *Acta Acust United Acust* 96 (2010), 452–462. 10.3813/AAA.918298.
13. *Common Noise Assessment Methods in Europe (CNOSSOS-EU)*, https://ec.europa.eu/jrc/en/publication/reference-reports/common-noise-assessment-methods-europe-cnossos-eu.
14. *Common Noise Assessment Methods According to Directive 2002/49/EEC of the European Parliament and of the Council*, Commission Directive (EU) 2015/996.

15. European Commission Working Group Assessment of Exposure to Noise (WG-AEN), 'Position paper on good practice guide for strategic noise mapping and production of associated data on noise exposure', Version 2, 2007.

16. International Organization for Standardization, *Attenuation of Sound during Propagation Outdoors: General Method of Calculation*, International Standard, ISO 9613-2. ISO, Geneva, 1996.

17. Delaney ME, Harland, DG, Hood RA, Scholes WE, 'The prediction of noise levels L_{10} due to road traffic', *J Sound Vibrat* 48(3) (1976), 305–325.

18. Hood RA, 'Accuracy of "Calculation of Road Traffic Noise"', *Appl Acoust* 21 (1987), 139–146.

19. Department of Transport, *Calculation of Railway Noise*. HMSO, London, 1995. ISBN 0 11 551754 5.

20. Ashdown Environmental, Ltd., on behalf of Union Railways, Ltd., *Channel Tunnel Rail Link: Assessment of Airborne Noise Effects*, Final Report, Volume 1 of 4, November 1994.

21. 'Sound, noise and vibration', London–West Midlands Environmental Statement, Volume 5, 'Technical appendices: Methodology, assumptions and assessment (route-wide)(SV-001-000)', November 2013.

22. Hood RA, et al., 'Calculation of railway noise', *Proc Inst Acoust* 13(8) (1991).

23. *Aviation Environmental Design Tool (AEDT)*. 2014.

24. https://aedt.faa.gov/Documents/AEDT%202a%20Uncertainty%20Quantification%20Report.pdf.

25. Committee B/564, *Code of Practice for Noise and Vibration Control on Construction and Open Sites*, BS 5228–1: 2009+A1: 2014. British Standards Institute, London, 2008.

26. Chung M, Karantonis P, Gonzaga D, Robertson T, 'Comparison of traffic noise predictions of arterial roads using Cadna-A and SoundPLAN prediction models', in *Acoustics 2008*, Geelong, Australia, November 2008.

27. Hill RC, 'Noise mapping of London: Quality checks. AIRO, unpublished report, January 2004.

28. National Roads Authority, *Guidelines for the Treatment of Noise and Vibration in National Road Schemes*, Revision 1. NRA, Dublin, Republic of Ireland, 2004.

29. Ahammed, Nixon, 'Environmental impact monitoring in the EIA process of South Australia', *Environl Impact Assess Rev* 26 (2006), 426–447.

30. HS1 Annual Report 2015–16, orr.gov.uk/__data/assets/pdf_file/0020/22547/hs1-annual-report-2015-16.pdf.

31. *ArcGIS Geographical Information System*. ESRI (UK), Ltd., London, www.esri.com/en-us/arcgis/products/arcgis-pro/overview.

32. Tompsett KR, 'Uncertainty and diversity in construction noise assessment', *Acoust Bull* 39(1) (January–February 2014).

33. Tompsett KR, 'Calculation of Road Traffic Noise in low- speed and congested situations', in *Inst of Acoustics London Meeting*, January 2005.

Chapter 5

Uncertainty in Environmental Noise Measurement

David Waddington and Bill Whitfield

Chapter Overview

This chapter brings a qualitative and quantitative assessment of the major factors affecting measurement of environmental noise in the field. Until recently, uncertainty has tended to focus on measurement accuracy and precision and has not considered the stochastic variation inherent in the process, possibly because it is difficult to assess the components individually or separate out the likely dominant contributory factors. If a survey were to be carried out to quantify road noise, using a fixed noise monitor at a specific location, we would expect that there would be differences in the sound pressure levels over the day and night time periods if measurements were carried out over multiple days. The sound pressure levels recorded would be expected to vary on different days in different seasons and in different weather conditions. If the survey were to be carried out by a separate operator with a different instrument, this would introduce additional variability.

It is possible to extract estimates for some of these components of variation using multiple observations and a particular type of linear model, but there are limitations on what can be gleaned from a single observational site; for example, estimates of uncertainty in a rural setting may not be totally representative of a similar survey in an urban environment or one carried out on the coast. Estimates can only be estimates, so the examples and data in this chapter should only serve as a starting point and not be treated as absolute.

It is noted that from the point of obtaining an unbiased estimate of both mean and variation (standard deviation), the data set should comprise a random sample of days. This is unlikely to ever be practical, and the continuous sampling of contiguous days is offered as the best alternative.

This chapter is split into two parts that look at factors that are internal and external to the measurement process. The first part looks at what could be termed 'external factors', which can be defined as factors that are beyond the control of the individual trying to measure the noise, for example, the weather and the individual meteorological factors associated with weather and its effects on the propagation of sound. It considers the noise source, propagation path and receiver together with refraction, atmospheric absorption and scattering.

This approach includes looking at how meteorological data can be correlated with sound pressure level measurements and makes some observations on how measurement data compare to predictions. Finally, it offers guidance by reviewing the requirements of the standards, suggests best practices and offers some 'rules of thumb' to minimise the effects of measurement uncertainty caused by meteorological effects.

The second part of this chapter deals with factors that are generally in the control of the individual or organisation tasked with measuring the environmental noise, for example, location and accuracy of the measurement position, the differences which may occur due to instrumental measurement error, the variation between similar instruments (sound level meters) or between different types of instruments, the variation between operators and the variation of the source (day to day variation). These could be termed 'internal factors', that is, within the control of the organisation undertaking the measurement (rather than the 'external factors'), and with care, these components can be isolated and independently assessed. Initially, a general description of the nature of uncertainty in the context of sound measurement is presented, together with a discussion of the statistical nature of sound measurement and its summary. This is followed by a description of how a particular type of designed experiment can be used to identify specific components of variance in the measurement process. The chapter concludes with some results from an empirical study of the differences in measured levels of environmental noise in a single typical survey setting using road and rail noise sources. The analysis is used to demonstrate the quantification of a budget for measurement uncertainty and the apportioning of uncertainty to variables in the measurement process.

Part I: External Factors Affecting Environmental Noise Measurements

Uncertainty and Variability in Environmental Noise Measurement

An important concept is the distinction between uncertainty and variability. When making practical environmental noise measurements, varying sound levels represent the actual sound field changes occurring at the microphone during the time of recording. Uncertainty arises when particular measurements are used to represent sound levels occurring in other time periods or at other locations.

Environmental sound variability can be considered in three components: source, path and receiver variability. An example of source variability is the traffic sound due to hourly, daily and seasonal changes. Receiver variability mainly concerns the measuring instrument and its competent use. Environmental factors affecting the propagation of sound are most often the largest cause of path variability. It is therefore important to understand these factors in order to best manage the risk that this induced variability does not introduce such uncertainty that the measurement data are unfit for purpose.

Path Variability Due to Weather

Assessing the influence of weather on a particular measurement depends on determination of the prevailing meteorological conditions over the propagation path, as well as an understanding of how the various meteorological factors influence noise propagation. Measured sound pressure levels owe as much to near-surface weather as to ground shape and impedance and factors such as source and receiver heights and locations. Wind and temperature gradients in the atmosphere cause refraction that can increase or decrease sound pressure levels significantly. However, the variability in meteorological parameters over the propagation path and measurement duration creates uncertainties in the measurement of temperature, wind speed and direction in practical situations. Often it is recommended that measurements be performed under stable downwind conditions described in terms of the vector wind. However, a physical interpretation of outdoor noise propagation can be appreciated better through the three principal mechanisms of refraction, atmospheric absorption and scattering.

Refraction, Atmospheric Absorption and Scattering

'Refraction' is the term used to describe the change in propagation path from a straight line to a curve due to temperature and/or wind speed gradients. Sound may be refracted upwards, creating shadow zones of low noise at the ground, or refracted downwards, focusing the sound towards the ground. Atmospheric absorption is the attenuation of principally high-frequency noise due to classical and molecular absorption and is determined by the relative humidity and temperature. Scattering of sound by atmospheric turbulence allows noise to enter 'shadow zones', reducing the strength of interference patterns. Because of the variability and interdependence of these meteorological variables, the combined effect of the three mechanisms is often complex.

Generally, wind speed increases with altitude. On the downwind side, this will cause the path of sound to refract down, focusing the sound and increasing the noise level. On the upwind side, the path of the sound will bend away from the ground, creating a shadow zone and a corresponding decrease in noise level. Similar effects to those of wind gradients can be created by temperature gradients. Effects of temperature gradients differ from those of wind gradients in that they are uniform in all directions from the source. Under normal lapse conditions such as may occur on a sunny day with little or no wind, temperature decreases with altitude, resulting in a shadow effect. Under a temperature inversion such as may occur on a clear night, temperature will increase with altitude, resulting in the focusing of sound back towards the ground.

Standards and Guidance

Noise measurements should be performed under weather conditions representative of the situation under investigation. However, practical environmental noise measurements are usually made at times dictated by other factors, such as the planned progressive shutdown of a factory plant. In this case, it is often recommended that measurements be performed under stable downwind conditions. These are often described in terms of the vector wind, which is the component of wind velocity in the direction from the source to the receiver. Planning Policy Guidance 24 (PPG24, since withdrawn), for example, suggested a light vector wind less than 2 m/s. British Standard (BS) 4142: 2014: *Methods for Rating and Assessing Industrial and Commercial Sound*, by contrast, suggests that noise level measurements should not be made with winds greater than 5 m/s or temperatures less than 3°C. More comprehensive classifications of weather conditions have been described, for example, in the CONCAWE model for calculating the propagation of noise from open-air industrial plants, and these are useful when determining average weather patterns for long-term measurements. Another method is discussed in the International Organization for Standardization (ISO) 1996–2 (2017): *Acoustics: Description, Measurement and Assessment of Environmental Noise*. This standard defines meteorological windows determined from vector wind speed components giving propagation conditions described as 'unfavourable, neutral, favourable and very favourable'. ISO 1996–2 states that if measurements are carried out during favourable or very favourable propagation conditions, unless better information is available, the standard uncertainty is 2 dB for source-receiver distances up to 400 m.

Experiments to Investigate the Variation of L_{Aeq} with Wind Speed and Temperature

Setup

To illustrate the effects of meteorology on outdoor noise propagation, experiments were performed using an omnidirectional point source over flat grassland. These measurements were performed for source-receiver distances typical of community noise problems, specifically up to 1 km. Measurements were performed at a disused airfield along flat grassland parallel to the runway. A high-power omnidirectional electroacoustic source with centre height 2 m was used to provide a sound power of 130 dB. Four reference positions were set up around the source at 5-m distances in order to monitor any source power fluctuations. Ten acoustical monitoring units with a microphone height of 1.5 m were installed at approximately 112-m intervals to the east of the source, and an additional four units were set up at selected positions to the west of the source, as shown in Figure 5.1.

Each station was used as a stand-alone data logger and audio recorder logging L_{Aeq}, L_{Afast} and one-third octave band spectra each second. The source emitted pink noise in five-minute sections separated by one-minute sections of silence to enable background levels to be monitored. The measured noise data were synchronized with meteorological measurements of wind profiles from a SODAR (SOnic Detection And Ranging) system. Automatic weather stations (AWSs) were also set up on 2- and 10-m masts at the source and at the receiver 1 km east of the source. The measurements detailed here were performed between 1900 and 0500 hours on the shortest night of the year.

Categorisation of Acoustical and Meteorological Data

The acoustical data presented here were calculated for source on times only, with allowance made for background noise. Measurements of less than 3 dB above background noise were discarded. Acoustical and automatic weather station meteorological data were measured each one second, and these data were averaged over 150-s time periods to correlate with the SODAR measurements. The 150-s average automatic weather station meteorological data were categorized, with each sample identified by category in terms of temperature gradient, wind speed and vector wind, as summarized in Table 5.1.

Qualitatively, the negative temperature gradient became a positive as the sun set and the ground cooled. The converse was seen as the sun

Figure 5.1 The line of the receiving array and positions of the automatic weather stations and SODAR.

Table 5.1 Summary of meteorological categories

Category	Investigative purpose	Temperature gradient dT between 2 and 10 m	Vector wind at 10 m	Wind velocity at 10 m	Symbol	Number of samples
1	No met.	$-0.01 < dT < 0.01$ K/m	<1 m/s	<1 m/s	Δ (triangle)	2
2	Just turbulence	$-0.01 < dT < 0.01$ K/m	<1 m/s	>1 m/s	◊ (diamond)	2
3	Sonic gradient due to wind only	$-0.01 < dT < 0.01$ K/m	>1 m/s	>1 m/s	* (star)	16
4	Just decreasing temperature with height	$dT < -0.01$ K/m	< 1m/s	< 1 m/s	+ (plus)	19
5	Decreasing temperature with height and turbulence	$dT < -0.01$ K/m	<1 m/s	>1 m/s	□ (square)	9
6	Decreasing temperature with height, turbulence and wind	$dT < -0.01$ K/m	>1 m/s	>1 m/s	**X** (ex mark)	118
7	Just increasing temperature with height	$dT > 0.01$ K/m	<1 m/s	<1 m/s	< (triangle, left)	0
8	Increasing temperature with height and turbulence	$dT > 0.01$ K/m	<1 m/s	>1 m/s	> (triangle, right)	0
9	Increasing temperature with height, turbulence and wind	$dT > 0.01$ K/m	>1 m/s	>1 m/s	○ (circle)	26

rose and the ground warmed. The temperature gradient in between was complex, perhaps due to the effects of winds and a nearby thunderstorm. The wind speed fell during the night and became still between around 0100 and 0300 before picking up around sunrise. The most commonly occurring conditions during this measurement are those of a temperature lapse with some crosswind and some vector wind. Categorisation of the data in this way allows the effects of turbulence, refraction due to wind gradients only and refraction due to temperature gradients only to be identified and analysed.

Correlation with Vector Wind

The correlation of the L_{Aeq} with vector wind speed is illustrated in Figure 5.2 for a receiver distance of 112 m. With a slight vector wind of 1 m/s downwind, enhancement is seen, and it increases slightly with vector wind speeds up to around 7 m/s. Upwind, however, a sharp fall-off of noise level with vector wind is seen as the shadow zone is formed. The shadow increases sharply for vector winds up to around –4 m/s and is then seen to plateau with this data set.

As receiver distance increases through 224 to 1,008 m, as shown in Figure 5.3, the downwind enhancement is found to change comparatively little with vector wind speed. Two counteracting mechanisms are occurring, so refraction due to the positive sound speed gradient is offset by absorption by the ground. Conversely, the sharp attenuation with negative vector wind speeds is seen to deteriorate. This is due to the scattering of sound into the shadow region by turbulence and is discussed further later.

Influence of Range-Dependent Meteorology

Meteorological data are not routinely collected at 10 m close to the source during environmental noise surveys. Consequently, the correlation between

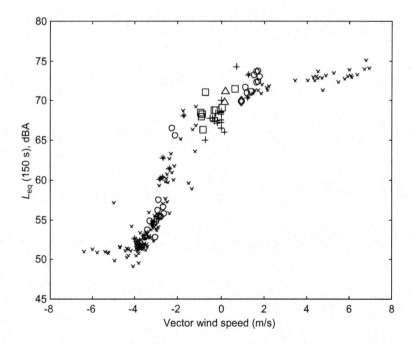

Figure 5.2 Variation of $L_{Aeq(150\ s)}$ with vector wind speed measured at 10-m height for a receiver 112 m east of the source for each meteorological category.

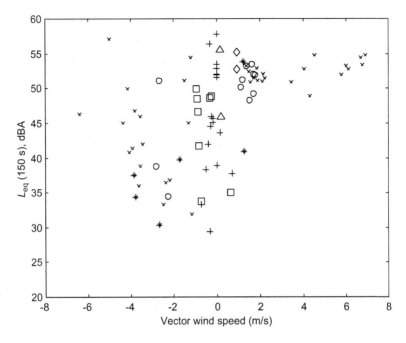

Figure 5.3 Variation of L$_{Aeq(150 s)}$ with vector wind speed measured at 10-m height for a receiver 1,008 m east of the source for each meteorological category.

L_{Aeq} and vector wind as measured at various distances from the source at 2 m is of interest. The correlation of L_{Aeq} with 2-m vector wind speed measured both at the source and at 1,008 m east is shown in Figure 5.4 for a receiver 112 m east. Since the acoustical data are the same as those shown in Figure 5.2 for the correlation with 10-m vector wind speed, the same acoustic enhancement downwind and shadow upwind are seen. However, since wind speed increases with height, the spread of vector winds at 2 m in Figure 5.4 is less than that in Figure 5.2 at 10 m.

This means in practice that the onset of the marked attenuation of the shadow zone occurs at lower vector wind speeds as detected at the ground close to the source and, furthermore, that the L_{Aeq} can vary greatly with only small changes in wind vector in the 0 to 2 m/s range. The range of vector wind speeds measured at 1,008 m from the source is seen to be greater than at the source. The correlation with automated weather station data from 1,008 m from the source is not as tight a fit as with AWS data near the source. This is due to differences in vector wind over the 1 km. These descriptions can be applied to the correlation of L_{Aeq} for a receiver at 1,008 m east of the source with vector winds measured at the source and at 1,008 m. The wider range of vector winds at 1,008 m broadens the spread of the correlation. In

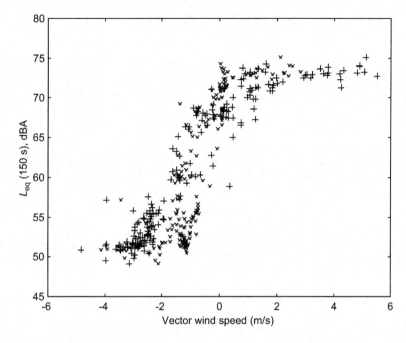

Figure 5.4 Variation of $L_{Aeq(150\ s)}$ for a receiver 112 m east of the source with vector wind speed measured at 2 m at source (X) and at 2 m at 1 km from source (+).

particular, upwind scatter is highly developed because of diffraction into the shadow zone.

Correlation with Temperature Gradient

Figure 5.5 illustrates the correlation between L_{Aeq} and temperature gradient for receivers at distances of 10, 112 and 672 m. These data are selected to have no vector wind component and therefore are influenced only by temperature gradient. The correlation is seen to be weaker than the correlation between L_{Aeq} and vector wind. For conditions of no vector wind, where the temperature gradient and crosswinds influence propagation, a significant spread of L_{Aeq} is seen. This spread is seen to increase markedly with distance. This spread in the data is thought to be due to the variations in the sound speed profile with distance and the influence of turbulence.

Comparison with Standard Prediction Methods

Figure 5.6 shows a comparison between measured and predicted sound pressure levels. Calculations using ISO 9613–2 are shown.

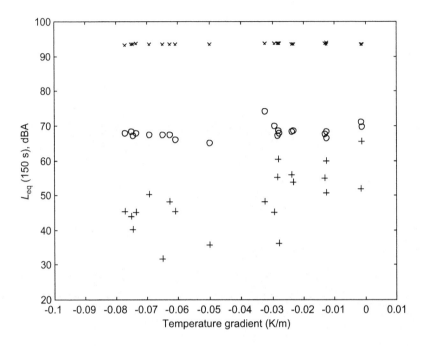

Figure 5.5 Variation of $L_{Aeq(150\ s)}$ with temperature gradient for receivers east of the source at distances of 10 m (X), 112 m (O) and 672 m (+).

The fast-field programme (FFP), a computational technique originally developed for predicting acoustic wave propagation in the sea, has proved useful for calculating sound propagation in the air above the ground. FFP predictions are used here to illustrate the effects of temperature gradients. The FFP predictions assume a logarithmic sound-speed profile based on a linearised temperature gradient between 2 and 10 m of –0.05, 0 and +0.05 K/m. These are compared with experimental measurements selected to best match these conditions, together with a minimum wind speed and minimum positive wind vector. Both FFP predictions and experimental measurements show the enhancement due to refraction by the positive temperature gradient and the shadow due to refraction by the negative temperature gradient.

The ISO 9613–2 method predicts the sound pressure level under meteorological conditions favourable to propagation; such as propagation under a well-developed moderate ground-based temperature inversion, as commonly occurs on clear nights. These predictions are consequently seen to agree best with the FFP prediction using the positive temperature gradient. The effects of turbulence on propagation in a refracting sonic gradient are illustrated by the parabolic equation (PE) predictions of Figure 5.7.

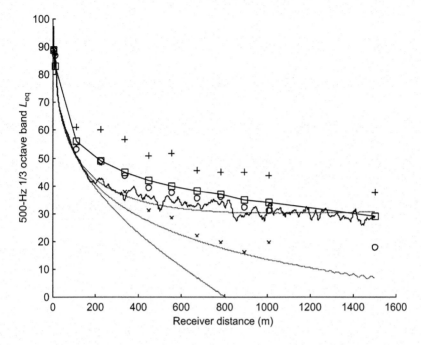

Figure 5.6 Comparing measurement with predictions at 500 Hz. ISO 9613-2 (□), parabolic equation (downwind), measurement and fast-field programme temperature gradients: 0.05 K/m (X), 0 K/m (O), +0.05 K/m (+).

In contrast to the FFP method, the PE method, which is based on an approximate form of the wave equation, is not restricted to systems with a layered atmosphere and a homogeneous ground surface. The PE predictions show the same positive, neutral and negative refraction as the FFP due to –0.05, 0 and +0.05 K/m temperature gradients. The addition of turbulence shows that all cases converge to that of a positive refracting atmosphere due to scattering into the shadow zone. The propagation patterns observed in these predictions illustrate the principal mechanisms of refraction and scattering.

Elevated Sources and Longer Distances

It is usually considered that the influence of meteorology may be regarded as negligible when considering propagation distances of less than 100 m unless frequencies greater than 2 kHz are of particular interest. However, meteorological changes may exert significant influence especially with elevated sources where the effects can be observed over long distances. Depending on wind speed, noise levels may increase downwind by a few decibels. However, when measuring with an upwind or sidewind, level decreases in

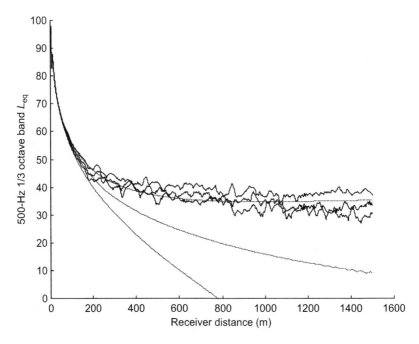

Figure 5.7 Comparing PE predictions with and without turbulence at 500-Hz temperature gradients −0.05, 0 and +0.05 K/m.

excess of 20 dB may result. PE predictions were used to illustrate that turbulent scattering allows noise to enter shadow zones, thereby reducing the strength of interference patterns. For an analytical description of sound propagation in a moving inhomogeneous atmosphere, extensive study is required, and the treatise presented by Ostashev [8] is highly recommended.

Rules of Thumb

Good agreement between measurement and ISO 9613–2 predictions was observed under favourable conditions for propagation as described by the standard. The meteorological corrections given by ISO 9613–2 are in practice limited to a range from 0 to 5 dB, with values in excess of 2 dB said to be exceptional. Meteorological conditions can change rapidly during measurements, and occasionally, measurements are made under unrepresentative conditions for a variety of pragmatic reasons. These are sources of uncertainty in the practical measurement of environmental noise. Unless specific meteorological conditions are required, the results presented here illustrate that measurements should be made under conditions that are favourable for propagation, as described by ISO 9613–2. These and other sources of uncertainty arising in the practical measurement of environmental noise are discussed more fully in the DTI good practice guide [9]. Downwind measurement is normally preferred because the end result is more conservative since the deviation is smaller. As a rule of thumb, for downwind propagation, the wind direction will be blowing from the source to the receiver within an angle of ±45 degrees with a wind speed between 1 and 5 m/s measured at a height of 3–11 m above the ground.

Part I References

1. Department of the Environment, *Planning Policy Guidance: Planning and Noise*, Planning Policy Guidance 24. HMSO, London, 1994.
2. British Standards Institution, *Methods for Rating and Assessing Industrial and Commercial Sound*, BS 4142: 2014. BSI, London, 2014.
3. Manning CJ, *The Propagation of Noise from Petroleum and Petrochemical Complexes to Neighbouring Communities*. CONCAWE, Brussels, 1991.
4. International Organization for Standardization, *Acoustics: Description and Measurement of Environmental Noise*, ISO 1996–1: 2016 (draft revision 2001). ISO, Geneva, 2016.
5. International Organization for Standardization, *Acoustics: Description, Measurement and Assessment of Environmental Noise*, ISO 1996–2: 2017. ISO, Geneva, 2017.
6. Salomons EM, *Computational Atmospheric Acoustics*. Kluwer Academic Publishers, Dordrecht, Netherlands, 2001.
7. Gilbert KE, Raspet R, Di, X, 'Calculation of turbulence effects in an upward-refracting atmosphere', *J Acoust Soc Am* 87 (1990), 2428–2437.
8. Ostashev VE, *Acoustics in Moving Inhomogeneous Media*. E. & F.N. Spon, San Francisco, CA, 1997 (the second edition of this text by V. Ostashev and D. K. Wilson was published in 2010).
9. Craven, N.J., Kerry G., *A Good Practice Guide on the Sources and Magnitude of Uncertainty Arising in the Practical Measurement of Environmental Noise*, DTI Project 2.2.1, National Measurement System Programme for Acoustical Metrology. University of Salford, Salford, UK, 2001.

Part 2: Uncertainty from Factors Internal to the Measurement Process

Introduction

Part 2 of this chapter looks at uncertainty from factors that are internal to the measurement process, for example, uncertainty that can be influenced by the person carrying out the measurements on a different day of the week. Uncertainty in environmental noise measurement is, as discussed in other chapters in this book, a combination of uncertainties that may not always be easily disentangled. Until recently, measurement uncertainty has tended to focus on instrument accuracy and precision and has rarely considered the stochastic variation inherent in the process. Suppose, for example, that a fixed monitor is set up at a specific location close to some environmental noise source (e.g. a road) and that the noise levels are measured every day. One would anticipate that no two days' profiles would be the same, and the means ($L_{Aeq,T}$) would be expected to vary from day to day, with possibly different patterns on different days, in different seasons and in different weather conditions. In fact, day to day variation is compounded of several of these various factors which are not easily separable. Suppose that the same exercise were repeated, but in a different way: sending a different engineer out each day to a specified location to make the same measurements. Then we would introduce a whole new tranche of variation including positional variation, operator variation and instrument variation. These matters will be discussed in this part of the chapter. But first, some general observations are necessary about the empirical assessment of environmental noise and an illustration of these ideas with some examples taken from extended surveys of two common environmental noise sources: road and rail. In the process, it is demonstrated how certain factors can be isolated and inferences made using simple design of experiment (DoE) methods. In this context, DoE is a systematic method to determine the relationship between factors affecting a process and the output of that process. Often used in engineering, science and manufacturing, this information is needed to manage process inputs in order to optimize output.

The ideas presented in this chapter arose from an attempt to interpret the ISO 17025 standard [1] requiring an uncertainty evaluation for all measurements; the standard states that laboratories should attempt to identify all components of variation and make a reasonable estimate of their uncertainty to ensure that results can be assessed fairly. A particular difficulty with this requirement is that some replication (i.e. repetition in a broad sampling sense) is necessary. The implications and practical implementation of this are considered in some detail.

Measurement

For the most part, the measure $L_{Aeq,T}$, appropriate for assessing environmental noise, is used. Conventionally, 'daytime' is defined as the 16-hour period between 0700 and 2300 hours, and 'night-time' is the remaining eight-hour period between 2300 and 0700 hours. Modern instruments 'average' the sound level over a designated period, usually five minutes for road traffic sources (which are usually continuous but variable) and one minute for rail sources. In the latter case, the one-minute interval further allows the number of rail events to be counted with reasonable accuracy during a given period. Since daytime data offer the higher sound pressure levels above the prevailing background noise, the period results are log-averaged to give a daytime mean value using

$$y_d = 10 \log_{10} \left(\frac{1}{16} \sum_{i=7}^{22} 10^{x_i/10} \right)$$

where x_i corresponds to the hourly L_{Aeq} values (based on the start of an hour), so y_d is the log-transformed exponential average of the 16 hourly noise levels from 0700 to 2300 hours (the summation from $i = 7$ to 22 implies the inclusion of both hours, ending at 2300). The hourly noise levels are themselves determined using an analogous formula to integrate the five- or one-minute integrated measures for a location over a one-hour period. Note that the logarithmic average y_d is always greater than the arithmetic mean $\bar{x}_d \left(= 1/16 \sum_{i=7}^{22} x_i \right)$ unless all the x_i values are equal, and that the value y_d is dominated by high sound pressure levels that make the statistic prone to influence by spurious unidentified sources with high sound pressure levels much more so than if simple arithmetic averaging were used.

Potential Sources of Variation

In preparing an uncertainty budget, one should consider *all* the potential sources of variation. The following can be considered as stochastic uncertainties, that is, of a fairly random nature:

- day to day variation
- Variation between operators
- Variation between types of noise meters
- Variation between noise meters of similar type
- Instrument measurement error

The first of these takes account of the fact that sound measurement taken on any two days will differ, so whilst a simple random sample may be representative, assuming that it is taken on a normal day, we do not know by how much it may vary from day to day. The day to day variation represents not only the difference in traffic between days but also differences in climatic conditions, temperature, humidity, and so on. The other factors reflect types of measurement error, which in other contexts might be considered as reproducible. Some of these factors are often not considered: there is a view that once a monitor has been calibrated, then it should record with a very limited measurement error; that all instruments (when calibrated) should effectively record the same signal; and that operators should not make much difference. Empirical statistical studies suggest that this is far from the truth and that these factors can be very important. There are further factors that are sometimes regarded as deterministic, although there will surely be stochastic components associated with them:

- Meteorological conditions
- Choice of monitor position on site
- Distance from noise source
- Terrain (between monitor position and noise source)
- Seasons or periods (e.g. school term or vacation, summer or winter)

Empirical adjustments using the inverse square law can be made for distance, for example, if the monitoring position is not at the specified position, although such adjustments often assume exact measurements and no interactions with other factors such as meteorological conditions and ground cover. Clearly, wind direction is an important factor, and the predominant prevailing wind direction at a site will affect the sound measurement. Although there are guidelines on positioning of monitors and so on, access to sites is not always straightforward, and exact positioning can be difficult, although recent advances in GPS location (particularly on phone apps) should mean that location uncertainty can be minimised. Not all of these characteristics are easily measured, and for some of them (e.g. distance, weather conditions, terrain), deterministic adjustments can be made using formulas taken from academic studies.

Sampling Considerations

The measurement of $L_{Aeq,T}$, where T represents either 16 hours (daytime), 8 hours (night-time) or 24 hours, affords a single measure of the average sound level integrated over the corresponding period, but no standard deviation, because the 'mean' is not arithmetic. To obtain an unbiased estimate of both mean and variation (standard deviation), the data set should ideally comprise a random sample of days. This is unlikely to ever be practicable, and the continuous sampling of contiguous days is offered as the best alternative. What of the possibility of sub-sampling the measurement period? Empirical measurement of dozens of sites for traffic noise suggests that there is no simple rule for sub-sampling, as the time profile for different sites is quite variable and not easily predictable. If data are collected over a shorter time period (e.g. 0900 to 1700), this may incorporate the noisier part of the day, thereby increasing the bias (i.e. the difference between the actual and true measurements) and, consequently, the uncertainty. A potential exception to this occurs when synchronised paired samples are taken (see the section 'Designed Experiments' below), as this affords a measure of uncertainty.

A simple example of 24-hour traffic noise is given in Figure 5.8, illustrating a single sample for Friday, July 20, 2012 on the A428, midway

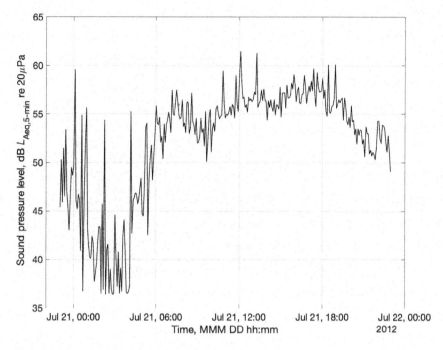

Figure 5.8 $L_{AEq,5min}$ observations on the A428 midway between Coventry and Rugby (Warwickshire) on Friday, July 20, 2012.

between Coventry and Rugby (Warwickshire). The daytime data suggest a bimodal profile with a particular elevation during the evening 'rush hour'. Note also the marked oscillations during the night-time period when occasional vehicles override the background noise level.

Figure 5.9 shows a typical plot of a 24-hour profile of rail sound ($L_{Aeq,24hr}$). The day is actually a Sunday, but the profiles are similar for most days inasmuch as there is a background noise level (ambient, when no trains are passing) and peaks (where a train passes). Clearly, there is very limited traffic between midnight and 06.30, and the background (or base) noise is slightly lower during the night (~35 dB) relative to approximately 40 dB in the daytime. There are clearly troughs in the afternoon where the sound level is lower. If one looks at the peaks, there appear to be three different levels which could be for a number of combined factors. Although there are different train types (e.g. intercity and commuter/local trains), their sound profiles are not hugely different. The one-third octave sound profiles were compared for several samples of background noise, commuter trains and intercity trains, and the latter two were found to be very similar. It has therefore been inferred that the differences in sound levels are most likely to be due to (1) two trains

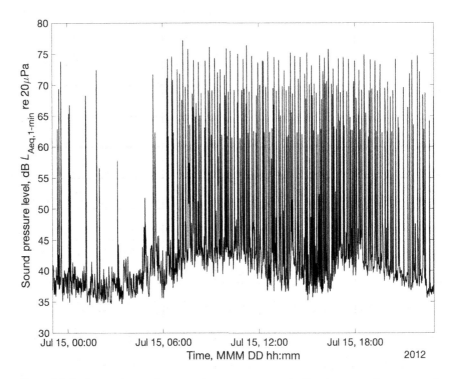

Figure 5.9 $L_{Aeq,1min}$ observations on the west-coast main line between Coventry and Rugby (Warwickshire) on Sunday, July 15, 2012.

crossing and (2) a train passage intersecting two measurement periods. However, a third, more subtle difference emerged later, which appears to be due to a difference in noise levels from the 'up' and 'down' tracks. Figure 5.9 clearly appears to have some periodicity in it, and to illustrate this, Figure 5.10 shows four consecutive two-hour periods from the same profile, which demonstrate the repetition in the timetable on an approximately hourly basis.

A histogram of the one-minute measurements for 24 hours (July 15, 2012) is shown in Figure 5.11. The plot is clearly bimodal – one might claim multimodal – although the twin-peak model will suffice. The distribution centred around 50 dB represents the background noise and comprises about 80% of the measurements, whilst the shallower peak (close to 70 dB) with a much wider distribution represents train noise. Two approximate normal distributions are superimposed on the figure to demonstrate the distinct nature of the two noise components. This will be explored in the 'Results' section, where the estimates of the train noise means for different days can be combined to give an uncertainty estimate for observations on different days.

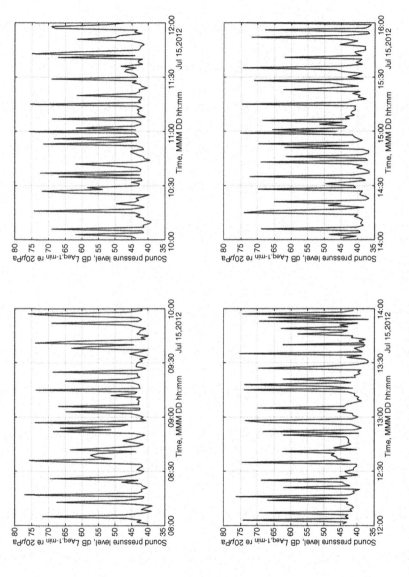

Figure 5.10 Four consecutive two-hour profiles from Figure 5.9.

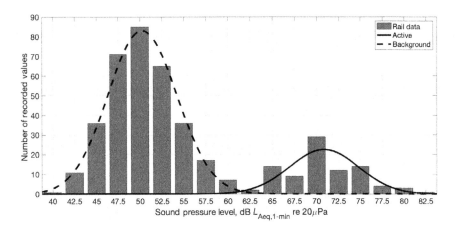

Figure 5.11 Histogram of one-minute data for July 15, 2012, at 10 m.

Outliers

Because log averaging tends to apply a significant weight to higher levels than normal averaging, it is particularly important to identify statistical outliers in the data in order to avoid significant bias in the results [2]. Calculation of measurement uncertainty can be adversely affected by spurious data spikes. There are statistical tests that are often used, such as Cochran's and Grubb's tests, which are both covered in the British standards for determining repeatability and reproducibility: BS ISO 5725–2: 1994 [3] tests for outliers. These tests may prove effective on large data samples or as a first-pass attempt at filtering, but it is equally important to view data graphically, as 'outliers' may pass the statistical test but still look visually out of place in a data cluster. Visual examination is obviously subjective and may be more relevant for the experienced user. Familiarity with the subject being monitored usually offers insights into the data that would not be apparent if the situation being measured were to be viewed for the first time. If doubt exists as to the provenance of the outlier, it can, of course, be left in, as the resulting measurement uncertainty will represent a 'worst-case condition'. Where replication occurs, it is useful to cross-refer data sets to identify unusual values or patterns. For example, where a regular train service exists, it is fairly straightforward to check that train events occur at approximately the same time on successive days, and while there will occasionally be unscheduled traffic (e.g. freight trains at night), a multiplicity of outliers in a single sample might suggest an alternative sound source.

Sampling

Roads

Figure 5.1 illustrates the full range of sound over a 24-hour period. This is generally only achievable when a measuring instrument is left at a site continuously, which is expensive regardless of whether the instrument is attended or not. There can be occasions when an instrument is left at a monitoring site for several days. This will provide some replication (and therefore a direct measure of uncertainty) for $L_{Aeq,day}$ and $L_{Aeq,24h}$ measurements. More commonly, an instrument only records sound during the working day (say, 0800 to 1700 hours) when it can be attended. The integrated measure $L_{AEq,0800-1700}$ may approximate the 16-hour-day measurement. In the Figure 5.8 example, the daytime $L_{Aeq,16h}$ is 55.9 dB, compared to 56.3 dB $L_{Aeq,0800-1700}$ for the shorter working day. But, as commented earlier, there is no simple rule for sub-sampling,

Rail

For environmental sound recording for rail, analysis of several consecutive one- or two-hour samples taken on different days suggests that the difference between days is relatively small compared with the difference between shorter samples on the same day. This does imply that for sound measurement close to busy rail lines, where a regular, repeating timetable is in place, sub-sampling may be appropriate as a proxy for full 16-hour sampling. For example, extracting the values above 60 dBA in Figure 5.10 (essentially the peaks) and integrating them (i.e. forming the log-exponential mean) for the four two-hour sequences gives L_{AEq} values of 73.2, 72.7, 73.1 and 72.9 dB (mean = 73.0 dB, standard deviation [SD] = 0.22 dB), demonstrating the potential effectiveness of sub-sampling. Such a sampling method may be appropriate for the hours between 0800 and 1800, but it needs to be acknowledged that there may be some abatement for the full 16-hour measurement as train frequency diminishes into the evenings. Clearly, this 'neat' separation might not be expected in suburban or urban locations where other noise interference might be anticipated.

Designed Experiments

To assess the uncertainty associated with sound measurement, some estimate of its variability (the variance or standard deviation) is essential. This can be achieved in two major ways: one is to generate a data set comprising a sample of days, and the other is to set up some form of paired-measurement experiment. Generating 24-hour measurements on a random set of days is unlikely to ever be practical, and continuous sampling of contiguous days is suggested as the best alternative (e.g. by leaving a monitor *in situ* for several days); a good starting point for road traffic noise that would give a relatively robust data set would be to take a continuous five-day sample from Monday to Friday. The use of multiple samples (i.e. several 24-hour samples) allows an unbiased estimate of the standard deviation of $L_{Aeq,16h}$ or $L_{Aeq,24h}$ which gives a measure of uncertainty that incorporates primarily the day to day variation, which may well include some assessment of meteorological variability as well as general environmental variability. Our experiments, for example, which were in a semirural location, included birds and other wildlife, farm animals and occasional agricultural machinery noise. Continuous measurement will not usually contain any measure of positional, operator or instrumental variation, as measurement on contiguous days would generally involve the setting of one instrument in one position by a single operator.

The simplest type of experiment involves two instruments arranged side by side (i.e. at the same position) by the same operator on the same day. The most important aspect of such experiments is the synchronicity of the measurements: it is possible to compare different sources of variation/uncertainty, some simple (e.g. two instruments of the same type at the same position positioned by a single operator) and others more complex (i.e. incorporating more than one definable source of uncertainty). Thus, for example, if a single operator sets up two instruments at distances d_1 and d_2 from a noise source (where $d_1 \neq d_2$), then the difference between the integrated measurements will contain a component of variation due to the relative distance of the two instruments from the sound source, but it will also contain a measure of the variability of the instruments. Other sources of uncertainty such as operators can be built into the experimental schedule.

Clearly, observations at the same distance should be identical, assuming proper instrument calibration. So we can consider this type of experiment as giving an estimate of repeatability. The uncertainty for quantitative measurements is given by the standard deviation (the square root of the variance) to ensure that the measure and its uncertainty have the same dimension.

Statistical Methods

Most of the statistical methods used in the analyses presented here are fairly elementary, for example, summary statistics (e.g. mean, median, mode, minimum value, maximum value, range, standard deviation, etc.), regression analysis and analysis of variance, with particular analyses driven by visual interpretation of the data sets. Basic statistical textbooks such as that by Snedecor and Cochran [4] or more specialised books on measurement and analysis such as that by Ellison et al. [5] give typical examples.

There are two types of primary data. The first comprises independent sets to be summarised, such as day to day records, for which some measure of variability is required, and series of data (usually paired data, though sometimes parallel sets with more than two series), which can be subjected to analysis-of-variance methods. The second set comprises correlated data, that is, data that are commensurate, such as observations taken on more than one monitor at the same time, monitors being co-positioned or displaced. More statistical details of these models are given by Fenlon and Whitfield [2]. Where independent daily sets of data are available, the principal summary statistic for 16- or 24-hour data is a single measure $L_{Aeq,16h}$ or $L_{Aeq,24h}$; uncertainty has to be estimated from the sample standard deviation of these separate daily values. If several sets of data are available for shorter periods (e.g. the working day), then a reasonable estimate of uncertainty should be determined by calculating values of L_{Aeq} for each day over the common temporal range and determining their standard deviations. For rail noise, there may be a case for considering estimates based on shorter periods, as earlier, when a repeating train passage timetable is available.

In the case of paired experiments (i.e. instruments at the same position or displaced instruments), we can produce estimates of differences between synchronous samples or estimates of means at distances from the noise source. An important summarising method is analysis of variance (ANOVA); see, for example, Dean, Voss and Draguljić [6], where differences between sets of estimates can be combined to produce overall means corresponding to different sets and a measure of the variability of those means, which is essentially our measure of uncertainty, reflecting the different sources of variation.

Although the sample standard deviation of independent measures of $L_{Aeq,T}$ should be regarded as the 'gold standard' for estimating uncertainty, it is also possible to examine the paired time-series plots of one-minute (rail) or five-minute (road traffic) samples. For example, Figure 5.12 shows simultaneous plots of the five-minute output from two monitors placed side by side close to the A428. Note how one meter reads slightly higher than the other in the first half of the recording period but then trails it later in the day. Summing the squares of the $L_{AEq,5min}$ difference between the monitors provides an estimate of the variance (square of the standard deviation) and therefore of the uncertainty associated with a pair of monitors [3,7]. Alternatively, one can simply integrate the sample over the full measurement period: for the data presented here, the integrated $L_{Aeq,T}$ (0925–1700) values are 58.8 dB for instrument 1 and 59.3 dB for instrument 2, a difference of 0.5 dB. This compares quite favourably with a difference of 0.4 for the arithmetic mean or a similar value for analysis of differences. Although this estimate is based on the set of individual five-minute samples, the fact that the two profiles move almost in unison implies that the uncertainty measure will be very close to that for the integrated measure.

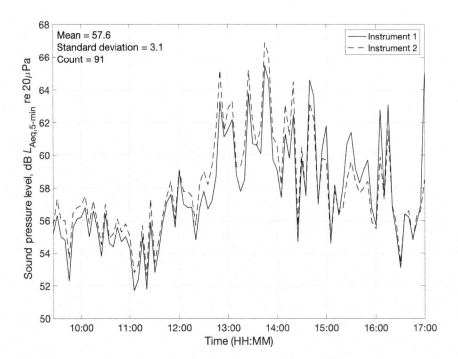

Figure 5.12 L$_{AEq,5min}$ observations for two identical monitors situated side by side close to the A428 midway between Coventry and Rugby (Warwickshire) on July 25, 2013.

Examples

The examples here relate to an extended empirical study carried out to quantify the major components of variance in measuring environmental noise in the field. The measurand chosen in this case was $L_{Aeq,T}$, which is appropriate for assessing environmental noise, and the study had to employ an experimental design that allowed the uncertainties to be identified, isolated and calculated robustly. The field testing took place under controlled conditions over a three-year period. An historic survey database of some 50 road samples with multiple-day samples provided a secondary source.

For the experimental work, the road noise featured a standard mix of cars, lorries and motorbikes on a relatively busy single carriageway road (A428), which, at peak times, formed a continual line source of vehicles passing the monitoring stations. The rail noise featured the West Coast Main Line between Coventry and Rugby, with approximately six high-speed rail (Virgin Trains) and six commuter rail (London Midland) events per hour passing the noise monitoring stations. In both cases, the daytime data provided the main source of information to calculate the uncertainty relating to the measurement process.

Two sampling methods were employed:

1. Controlled extended sampling, where monitors were left in position over several days adjacent to a road or railway source. Instruments were generally paired, and their internal clocks were synchronised at different set distances from the road or rail track source usually in multiples of 10-m distance, for example, 10, 20, 40 and 80 m when measured from a single 'control' monitor set as close as practically possible to the road or rail source. The sampling periods were five-minute averaging for road sources and one-minute averaging for rail sources; note that one-minute averaging was selected for trains to enable the number of rail events to be counted with reasonable accuracy during the day- and night-time periods.
2. On other single days, monitors were deployed in pairs or larger multiples during the working day (approximately 0900 to 1700 or 1730 hours)

Historical survey data provided some 50 samples, where samples included two or more days of noise monitoring. The sample sites were predominantly urban or suburban, where meters had been covertly positioned and left for several days, often for reasons of logistical convenience.

The primary interest was to estimate the magnitude of the variance components in relation to day to day variation, differences between instruments (at the same position), differences due to operators and differences between measurements at different distances from the source. The data that proved most useful in determining uncertainty in these experiments were those associated with the daytime period when the road and rail sources were most active and the 'signal' (source) to 'noise' (background) ratio was greatest.

Results

Historical Road Traffic Data

Taking all measurement data, it was noted that there were three sites out of the 50 measured that appeared as outliers to the general data set. The three sites featured relatively low average sound pressure levels of approximately 55 dB $L_{Aeq,T}$. It is usually the case that environmental noise surveys are requested for sites likely to be affected by high levels of noise, so obtaining an $L_{Aeq,T}$ daytime result of 55 dB is uncommon but not impossible. Including the three low mean sites meant that the pooled standard deviation for all 50 sites was 3.16 dB; removing these three sites as outliers reduced this considerably to 1.98 dB (47 sites). This is clearly a good example of how outliers can affect the overall uncertainty budget if they are not identified. Using the 50 historical sites survey data and filtering out the outliers provides a relatively robust estimate of the standard uncertainty of measurement as it is replicated over a relatively large number of sites. The findings also showed that as the mean sound pressure levels increased, there was a clear decline in the standard deviation of the sample – the mean sound pressure level increases as traffic noise increases and becomes more continuous, thereby reducing the variation in recorded sound. If, additionally, we further differentiate between weekends and weekdays, the standard deviation falls to 1.86 dB, approximately ±2 dB for road traffic data. Such an uncertainty measure is essentially a measure of day to day variation only, as operator and instrument effects are completely confounded with site effects, because the combined measure of standard variation is obtained from amalgamating 'within site' variance (standard deviations) only.

Road Data (A428) Results

The A428 between Coventry and Rugby was measured over 24-hour periods on 19 separate days at 10 m from the control position (road edge). The pooled daytime mean was 59.5 dB $L_{Aeq,16h}$, not especially high but in line with the fact that the A428 at this location is a single-carriageway road. The standard deviation of 2.46 dB is slightly higher than that given earlier for the historical road database. The data are represented graphically in Figure 5.13.

Comparison of instrument measurement data is a key feature of the DoE. In this series of experiments, pairs of meters were installed at 10, 20, 40 and 80 m from the control point, and the data were compared between pairs on the same day generally between 0930 and 1700 hours. Visualisation is essential to eliminate incompatible data, for example, where a meter malfunction suggests that data do not match. Such visualisation can be achieved with either a simple xy scatterplot or overlaid time series. An example of the latter for the four distances is shown in Figure 5.14.

Observing the time histories for several sets of data, the generic estimate of measurement error due to the instrument was 0.5 dB irrespective of instrument make and model. The estimate is based on the statistical methods alluded to earlier in the section 'Statistical methods' and set out in more detail in references [7–10].

For positional data, the experimental setup was similar to that described in the preceding paragraph but with only one instrument at each position. It has been possible to harvest 52 sets of data for which the means (log-average) for daytime (0700–2300 hours) have been calculated. In nearly all cases, the measured differences are lower than expected except for the 20- to 40-m pair. A visual representation of a set of data

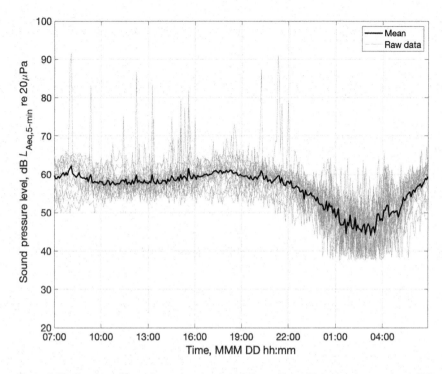

Figure 5.13 Combined sound pressure level data from 19- × 24-h samples on the A428 during July and August of 2012 and 2013. The variability of each sample around the overall mean is shown.

from three monitors at 20, 40 and 80 m from the sound source data is shown in Figure 5.15. The data show the 40- and 80-m data plotted against the 20-m data, which is simply represented by the $y = x$ line. In this way, the decline in sound level over distance can be assessed for different distances, but also any interaction between distance and sound level can be seen. In this particular example, the 80-m sample appears to be parallel to the 20-m sample but some 5 dB lower. Note that the 40-m sample appears closer to the 80-m sample below about 50 dB but migrates towards the 20-m sample at higher sound levels. At lower sound pressure levels, the background noise is more likely to dominate at 40 and 80 m because they are closer to a road that is clearly audible at these positions.

Given that the dominant noise sources in the experiments were line sources, the distance correction can normally be predicted using the distance correction $10\log_{10}(d_1/d_2)$, where d_1 is the distance of the monitor position from the source, and d_2 is the distance from the source that the sound pressure level prediction is required. In this case, a nominal doubling

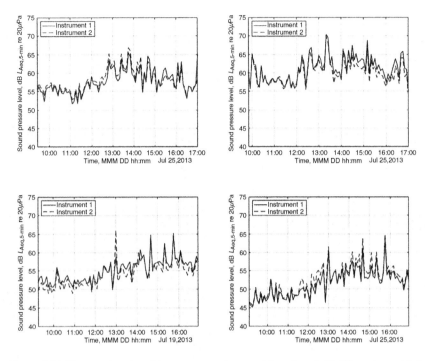

Figure 5.14 Meter pairs time history comparison at 10 m (top left), 20 m (top right), 40 m (bottom left) and 80 m (bottom right).

of distance from the source will result in a reduction of 3 dB. Clearly, the empirical rule of a 3-dB decrease per doubling of the distance does not appear to hold in Figure 5.15. The distances in the various experiments were specifically chosen to represent doubling from the control point, that is, 10, 20, 40 and 80 m, to make visual representations relatively straightforward. The data showed a clearer relationship over the shorter distances, becoming more diffuse as distance increased. In this case, the average differences and their standard deviations are shown in Table 5.2. The standard deviation suggests a departure from the normal distance correction likely caused by diffusion error due to factors other than distance (as well as potential bias) as one moves away from the line source.

Operator uncertainty was assessed using 11 pairs of data associated with six different engineers. In this experiment, the engineers were asked to place a sound level meter at a given ('set') distance from a source identified on a layout plan. A between-instrument error of 0.5 dB is contained within the total operator uncertainty of the operators, as two instruments were used. The resulting operator uncertainty is on the order of 1.0 dB.

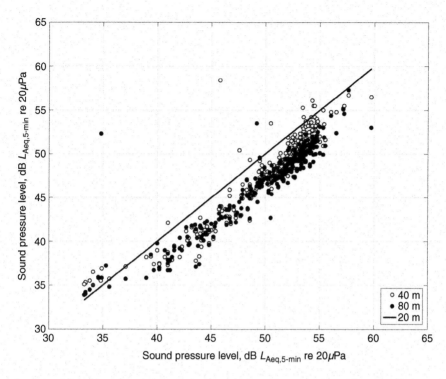

Figure 5.15 An xy scatterplot of sound pressure level data: 40 and 80 m plotted against the 20-m sample for road traffic noise data on the A428, Saturday, August 11, 2012.

Table 5.2 Predicted and actual (measured) change in sound levels (dB) at doubling from the control point (A428, Warwickshire)

Distance	Predicted, dB	Average difference, dB	Standard deviation, dB
Doubling (×2)	3	1.75	±1.52
Quadrupling (×4)	6	4.18	±1.77
Eight times (×8)	9	5.26	±1.91

Rail Data Results

The rail movements occur on a fixed timetable for weekdays, with the train schedules at the same time each hour through much of the day. The weekend schedule was again repeatable through much of the day but with a lower train frequency. The one-minute measurement period identifies each rail passing event with a reasonable accuracy and results in a bimodal plot of sound pressure levels over the period that contains the majority of train events (0700–2300 hours). A graph of a typical 24-hour event sample is shown in Figure 5.9, whilst Figure 5.11 shows a frequency plot of the separation of 'active' and 'background' events based on integrated one-minute measurements. The distribution centred around 50 dB represents the background noise and comprises about 80% of the measurements, whilst the shallower peak (close to 70 dB) with a much wider distribution represents train noise. Excluding the night-time data, during which there is very little train movement, the background noise would only comprise 70% of the total.

Multiple sets of 24-hour train data were collected on three separate occasions during the summers of 2012 and 2013. In all cases, several instruments were set up at different distances from the rail noise. For the purposes of comparing L_{Aeq} values across days, instruments at different distances have been treated as independent, as the uncertainty depends only on day to day variation. Altogether 18 sets of data were combined in groups (between weekdays and between Saturdays and Sundays) to provide a pooled standard deviation – the standard deviations for the different groups were similar and so could be combined. For daytime, this standard deviation was 0.60 dB, and it was somewhat larger (1.04 dB) for the full 24-hour estimate.

Sets of paired instruments were synchronised and put on site at 10, 20 and 40 m from the control position adjacent to the track. Inasmuch as the background sound level will effectively be the same at different distances, we have compared the sound levels for rail event data (i.e. when a train passes and the sound pressure level is greater than 60 dB $L_{Aeq,1min}$), the average difference between all samples and standard deviations were as listed in Table 5.3.

In contrast to the road traffic noise monitoring, sound level meters positioned at 10, 20 and 40 m from the rail line control monitor showed very little difference between the sound pressure levels at these positions. This was confirmed in an ANOVA that showed no effect of group or

Table 5.3 Predicted and actual (measured) change in sound levels (dB) at doubling from the control point (A428, Warwickshire), all data $L_{Aeq,16h}$

Distance, m	Average difference, dB	Standard deviation, dB
10	3.4	±1.13
20	1.9	±1.12
40	2.6	±0.74

significant differences between estimates at different distances. This may have had something to do with the elevated nature of the rail source, which was on a 3-m-high embankment, above the level of the field in which the measurement positions were placed, that is, not a level site.

The difference between the meter pairs at each noise monitoring position is on first viewing surprising because the meters were individually calibrated with synchronised clocks, but when the context of the topography is taken into account, it is easier to understand and one of the problems of following a DoE in the field. This is in contrast to the road traffic noise experiments, where the pairs of instruments had a standard deviation of only 0.5 dB. All standard deviations listed in the tables are taken from summaries of $L_{Aeq,16h}$ measures.

Rail noise monitors at set distances were sited by two engineers, with a control monitor set out by another operator. Analysis of the difference in all measured levels averaged over a seven-day period differed by less than 1 dB.

Conclusion

If a noise survey were to be reproduced on the same site by a different operator using different instrumentation, there would be uncertainties that would be introduced beyond the normal day to day variability of the source. Uncertainty is derived from the standard deviation. When there are several sources of variation, their combined uncertainty is calculated by combining the variances (the square of the standard deviation), so the impact of these additional sources of variation is likely to be small given the magnitude of the day to day variability of the sources. However, the additional components of variance sampled in the two case studies is necessary to understand and allow for this additional measurement uncertainty in practice. The measurement uncertainty calculated from the individual experiments for both a road and a rail affected site for daytime measurement can be summarised as in Table 5.4.

The estimated overall uncertainty associated with the daytime measurement of $L_{Aeq,T}$ (nominally 0700–2300 hours) is 2.9 dB for road and 1.6 dB for rail sources. The main component of uncertainty in long-term measurement for road noise would be the day to day variation in the source. It is possible that the site chosen for rail measurement is not typical because of its exposed position and rural location – nevertheless, the methods proposed here provide a template for further testing.

For the road traffic experiment, instrument-to-instrument variation was low between both meters of the same type and meters of different types; although this was not completely replicated in the rail noise experiment, there were some marked and often unexplained differences in the measured levels of paired instruments (e.g. birds landing on the windshield?).

Table 5.4 Estimated uncertainty for road and rail measurements, taking account of four sources of variation

Uncertainty (standard deviation)	Road noise, dB	Rail noise, dB
Instrument	0.5	0.5[a]
Positional	1.75	1.0
Operator	1.0	1.0
Day to day (source)	2.0	0.6
Total	2.9	1.6

[a] The estimate of instrument uncertainty for rail measurements is taken to be the same as that for road measurement.

The positional effects for the road traffic experiment were similar to but not identical to the predicted results for a line source, but the ones for the rail experiment when tested statistically showed no significant difference between data sets collected. It is noted that the rail line was in an elevated position on top of an embankment above the measurement datum, and this may have influenced the measured values in the relatively near-field positions, that is, shielding some more than others from direct line of sight. It should therefore be noted that this effect was unexpected, and it may not always be easy to predict these components of variance by relying on theoretical equations alone. The operator effects were also relatively small and often can be absorbed within the instrument effects.

As in all cases discussed here, it should be noted that uncertainty is combined in quadrature, so the larger components of variance will dominate, for example, an uncertainty (standard deviation) of 2 contributes four times the variation of an uncertainty of 1. For practitioners wishing to carry out their own assessment of environmental noise measurement uncertainty, a pro forma for the experimental design (DoE) is available on the website www.noise.co.uk/uncertainty together with additional information, further reading and resources to help in the determining measurement uncertainty. We hope you visit the site and welcome your feedback.

Part 2 References

1. British Standards Institution, *General Requirements for the Competence of Testing and Calibration Laboratories*, BS EN ISO/IEC 17025 2017. BSI, London, 2017.
2. Barnett V, Lewis T, *Outliers in Statistical Data.*, Wiley, New York, 1994.
3. British Standards Institution, *Accuracy (Trueness and Precision) of Measurement Methods and Results*, Part 2: *Basic Method for the Determination of Repeatability and Reproducibility of a Standard Measurement Method*, BS ISO 5725-2:1994. BSI, London, 1994.

4. Snedecor GW, Cochran WG, *Statistical Methods*, 7th ed. University of Iowa Press, Ames, IA, 1980.
5. Ellison SLR, Barwick VJ, Farrant TJ, *Practical Statistics for the Analytical Scientist*. RSC Publishing, Cambridge, UK, 2009.
6. Dean A, Voss D, Draguljić D, *Design and Analysis of Experiments*, 2nd ed. Springer International Publishing, New York, 2017.
7. Fenlon JS, Whitfield WA, 'Measurement uncertainty in environmental noise surveys: Comparison of field test data', in *ICSV 24*. 2017.
8. Fenlon JS, Whitfield WA, 'Uncertainty analysis of environmental sound: A statistical analysis of 100 sites with measurement over several days', in *ICSV 26*. Montreal, 2019.
9. Fenlon JS, Whitfield WA, 'Uncertainty analysis of environmental sound: Analysis of a series of experiments', in *ICA 2019*. Aachen, 2019.
10. Fenlon JS, Whitfield WA, 'Design of experiments for determining uncertainty in environmental measurement', in *Internoise 2019*. Madrid, 2019.

Chapter 6

Uncertainty in Room Acoustics Measurements

Adrian James

6.1 Introduction

6.1.1 What Is Covered in This Chapter

Room acoustics is a huge field, and a detailed analysis of its uncertainties could easily fill a whole textbook. The aim of this chapter is therefore to provide an overview for those undertaking room acoustics measurements, calculations and modelling for practical purposes. It is therefore not an academic treatise but a practical guide, with references to acoustic theory and international standards where necessary. I have, of course, referred to the standards current at the time of writing, and for the sake of brevity, I have not included the year of publication every time I refer to one of these; the reader is referred to the list of references at the end of this chapter.

Most room acoustic measurements are reverberation times in ordinary rooms, so no apology is made for an extended discussion of that subject or for the assumption that readers will be familiar with part 2 of ISO 3382. This chapter goes on to discuss other room acoustic parameters used in performance spaces and described in part 1 of ISO 3382. I have also included a discussion of measurements using the "new measurement methods" described in ISO 18233 (2006). Even at the time of publication of that standard, of course, some of these methods were no longer new, but they are still used only by a minority of acousticians, and new measurement systems are still being developed using the principles laid out in that standard.

I have not included in this chapter any discussion of measurements in open-plan offices, as set out in ISO 3382–3. These are effectively sound level measurements and are subject to the same uncertainties as any sound level measurement in a controlled environment. In addition, at the time of this writing, there is an insufficient body of published data for any informed understanding of the uncertainties in such measurements. The use and nature of furnished offices present practical obstacles to the kind of round-robin tests undertaken in other types of rooms (Keränen 2015). Furthermore, the results are strongly dependent on the use of part-height screens, and even small changes to the locations of these screens can significantly affect individual measurements.

Speech transmission index (STI) is another matter. Although more a signal-to-noise ratio than a room acoustic measurement, it is certainly affected by room acoustics, and anyone using room acoustics parameters as part of an STI prediction should take account of the uncertainties described in this chapter. In most cases, however, the overall uncertainly budgets will be dominated by other factors, including uncertainties inherent in the test signal and signal processing as well as fluctuations in the background noise level. Peter Mapp provides an authoritative discussion in Chapter 7.

6.1.2 Why Room Acoustic Measurements Are Difficult

There are some fundamental differences between room acoustic measurements and the measurements described elsewhere in this book. The rest of this book is primarily concerned with measuring sound level in one form or another. In general, when measuring sound levels, we use highly standardised sound level meters for which the measurement uncertainties are well understood, while the sound being measured is subject to temporal and spatial variations over which we have no control.

In room acoustics, however, we have to start by generating a suitable signal over which we have at least some control. There are many different ways of generating and processing the test signal to calculate the desired values, and beyond the statement of fundamental principles, the signal processing is not standardised. These source signals are in one of three forms; interrupted noise,[1] impulsive noise and deterministic signals.

Probably the most commonly used signal for measurement of reverberation time in ordinary rooms is *interrupted random noise* played through a loudspeaker. There are important differences between this and the *impulse response* from which nearly all room acoustic parameters are derived and which can also, in the form of a gunshot or balloon burst, be used to measure reverberation time (RT). This impulse response can also be calculated indirectly by using one of several types of deterministic signals (e.g. sine sweeps or maximum length sequences [MLSs]) which are processed after the measurement.

Whatever type of signal we generate is modulated by the pattern of acoustic reflections in the room, and this modulation will be affected by the positions of our sound sources and microphones. Some parameters,

---

1 "Noise" and "sound" are not quite interchangeable terms, although even in standards their use is not always consistent. I have generally used the terms in common use; it would seem pedantic to refer to "interrupted sound" or to "pink sound".

such as the ubiquitous reverberation time, are relatively independent of source and microphone position, so a spatial average is often (but not always) meaningful. Other parameters, such as clarity, lateral energy and the various forms of inter-aural cross-correlation, are by definition very sensitive to location. In many cases, in fact, we are measuring these parameters to find out how they vary across the room, so a spatial average alone may be unhelpful.

Time variance is another matter altogether. We can normally assume that the acoustics of our room will not change during our measurements. None the less, room acoustics do change over time. In small rooms, opening doors or windows can have a significant effect, as can the presence of furniture, fittings and people, including, of course, those taking the measurements – although it is generally accepted that the acoustics of a completely unoccupied room are of purely academic interest, so most measurement standards now allow the tester to be in the room. In auditoria, of course, the variation in acoustics with audience size is a matter of considerable uncertainty. Another factor, often neglected by musicologists, is that changes in layout, furnishings and decoration also affect room acoustics, so the current acoustics of the Thomaskirche in Leipzig probably bear little relation to those when so many of J. S. Bach's works were first performed there. Hence, in all cases, it is essential to consider and record the conditions in which we take measurements, whether this is for the purposes of commissioning a classroom or for a historic treatise on the acoustics of Roman amphitheatres.

In summary, therefore, any room acoustic measurement will have uncertainties associated with the type of measurement; the source signal; the location, orientation and type of source and receiver; and the signal processing. It is hardly surprising that an authoritative paper concluded, "Doing room acoustic measurements correctly may be more difficult than it appears at first glance" (Christensen et al. 2013).

6.1.3 Why Do We Measure Room Acoustics?

Before we become too depressed about the number of sources of error and uncertainty, it helps if we consider why we are taking measurements in the first instance. The requirements for accuracy and reproducibility depend on the purpose of the measurement, and in many cases, we may find that surprisingly large uncertainties are acceptable. Measurements are normally taken for one of four reasons: research, comparison, commissioning and diagnosis.

Research measurements are usually taken to improve our understanding of how measurable parameters such as reverberation time and clarity affect the user's experience of a room. Obvious examples are those textbooks whose authors have had the pleasurable experience of travelling round the world measuring acoustics in well-known concert halls and then correlating the results with the subjective opinions of performers and audiences (e.g. Barron 1993, Beranek 2003).

In general, these books are more about the auditoria than about the measurement process, and the authors very rarely describe how the measurements were taken or even whether they used the same methods and equipment in each hall. It is very rare indeed to find any discussion of measurement repeatability or uncertainty. This is hardly surprising in cases where the authors were experimenting with measurement methods and acoustic parameters that were relatively new at the time and which therefore did not feature in national or international standards. By describing the measurement techniques and equipment in as much detail as possible, however, they would have allowed subsequent readers to undertake their own uncertainty analysis using knowledge and techniques not necessarily available at the time of measurement.

Many of these books were written when measurement equipment and computers were much bulkier than they are now. It is not easy to travel on commercial airlines with large loudspeakers and other measurement equipment, so it is entirely possible that the equipment used varied from place to place and that the measurements would not have complied with current standards. Hence, we should not be surprised if our own measurements differ significantly from those in the literature; indeed, where several authors refer to the same auditoria in their books, there are sometimes significant differences even in basic reverberation time measurements.

For research measurements, it is generally assumed that the critical requirement for accuracy is the limen or "just noticeable difference" (JND). This is the minimum change in a given parameter that a "normal" listener is expected to perceive. For example, annex A of ISO 3382–1 states that the JND for early decay time is 5% at 500–1,000 Hz. The basis of this statement bears a little thought, as the source of that information is not given. It is not normally possible to adjust the early decay time (EDT) of a room in a controlled manner in a real auditorium, so realistically the JND can only be determined using synthesised signals through headphones or multiple loudspeakers in anechoic chambers. It is actually quite difficult to envisage how we could determine the JND for any acoustic parameter in a real room, as even with electro-acoustic enhancement systems we cannot change any one acoustic parameter without affecting several others, and the subject might in fact be responding to the change in those other parameters. Hence, the JND values quoted in ISO 3382–1 should be regarded with caution, and it seems reasonable to assume that they relate to much more controlled conditions than we will encounter in practice.

Comparison measurements are an extension of the research process. They might, for example, be to compare different types of classrooms or to investigate how changes to a room affect both the measured and subjective acoustics of the room. In these cases, knowledge of the measurement uncertainty is very important so that we can determine whether the differences between our measurement results are significant. For this, we have to understand the repeatability and reproducibility of each set of measurements.

In this category also come the occasional round-robin measurement exercises in which different testers measure the same rooms using different equipment and techniques. The very purpose of these exercises is to assess measurement reproducibility, so a clear description of the measurement equipment and methodology is essential.

Commissioning measurements determine whether the room acoustics meet a predetermined acoustic performance specification. This is a common process in some forms of building contracts, typically where a contractor is made responsible for complying with an acoustic specification provided by the client. Where this is done intelligently and with good communication between the client's and contractor's acousticians,[2] this process can work well. In most cases, however, the acoustic performance specification is often a device to devolve design responsibility from the client's design team onto the contractor.

In principle, there is nothing wrong with this where the acoustic specification is realistically achievable, technically correct and includes a proper understanding of error and uncertainty in the measurement process. This is, sadly, rare outside the most prestigious auditorium design projects, and the acoustician may be tasked with designing a room to a specification that is little more than a list of numbers taken from a textbook, with little or no consideration of measurement uncertainty. When faced with one of these "painting by numbers" specifications, to avoid unpleasant and expensive wrangling at the commissioning stage, we must establish at the outset whether the specification is realistically achievable and ensure that all parties understand the expected uncertainties when it comes to commissioning the completed building. This is often more time-consuming than the acoustic design.

Diagnostic measurements might be undertaken as part of any research, comparison or commissioning exercise but, in contrast to these, may not be concerned with numerical results. A good example is shown later in this chapter in Figure 6.2, in which the most important characteristic of the room being measured is not its reverberation time but a strong flutter echo, which is clearly audible but not easily quantifiable. Much of the time, diagnostic analysis may indeed not be measurement at all, and hence, any uncertainty in the result is likely to be unquantifiable.

---

2 I have used the term "acoustician" to describe anyone working in acoustics. It is not a well-defined term in the United Kingdom, where no specific qualification or registration is required to practice as an acoustician or acoustics consultant.

In summary, therefore,

- When taking research measurements, good practice should include at the very least a description of the measurement equipment and techniques used, including the number of measurements used to achieve the mean values reported. This allows subsequent readers to evaluate the likely reliability and reproducibility of these results.
- When taking comparison measurements, in addition to a full description of equipment and methodology, an assessment of reliability and reproducibility should be included in the report. This need not be a detailed statistical analysis of the data collected, although if you list your individual measurement results in a spreadsheet, it is as easy to calculate standard deviation as it is to calculate the mean.
- Commissioning measurements should include a full description of equipment and methodology and an assessment of uncertainty, even if this is only a statement of the expected uncertainty set out in the relevant standard. This gives the reader confidence that the tester is familiar with the standard or at least knows of its existence, which is by no means always the case.
- Any acoustic specification for a space should make reference to the relevant standards for measuring the acoustic parameters specified and should make due allowance for uncertainties in the design and measurement process. Ideally, the specifier will also draw on extensive practical experience of predicting and measuring acoustic parameters in similar spaces to that specified.

6.1.4 A Note on Standards and Standardisation

Before getting into the details of measurement standards, we should consider the purpose of standardisation. National and international standards are not there to tell us how to take measurements in all circumstances. They exist to tell us how we can take measurements in a way that a committee of experts considered – at the time of writing – to be good practice so that different people taking the same measurements should obtain the same results within known bounds of uncertainty.

We cannot always measure absolutely in accordance with standards, and sometimes we have to take measurements for which standards do not yet exist; it is inevitable that the standardisation of a process lags several years behind the invention of that process. There are also cases in which it may not be necessary to slavishly follow all aspects of a standard. A good example is the measurement of sound insulation between dwellings, for which the latest international standard requires nominally omnidirectional loudspeakers but UK building regulations specifically allow the use of cabinet loudspeakers. This followed on from several practical research projects which determined that the loudspeaker directivity had no significant effect on the results of the measurements in domestic rooms (Critchley 2007), and indeed, the better low-frequency response of some cabinet loudspeakers at the time provided better signal-to-noise ratios and hence more reliable results at low frequencies.

The important principle here is that standardisation (whether in national or international standards, specifications or other guidelines) should be necessary as well as sufficient. It is tempting for standards committees to "raise the bar" at every review of a standard, with a consequent increase in the cost to the tester and, ultimately, to the client. At the other end of the reality scale, the acoustics consultant's client does not want to pay for anything more than the bare minimum, so commissioning measurements in particular are often pared down to the very minimum required by the standards.

It is even, sometimes, justifiable to not comply with a standard, particularly in diagnostic measurements. An example of this is the measurement of impact noise through floors using a tapping machine, which is a very expensive and precisely engineered device. The international standard for tapping machines requires the hammers to be very slightly concave, and the tolerances for this curvature are very tightly specified. Over time, these hammers wear down so that the curvature is outside the standardised tolerance, and manufacturers like to supply new machines rather than provide replacement hammers for old ones. The result is that otherwise serviceable machines cannot be used in accordance with the relevant standards. In practice, we find that the results of site measurements using compliant and non-compliant hammers are indistinguishable. In this case, it might be considered perfectly good practice to continue to use a "non-compliant" machine, provided that we have a fully documented and scientifically valid justification for this.

Naturally, when we choose not to comply with a standard, or when something has gone wrong on site so that our measurements do not comply in all respects with the standards, we must be able to understand and quantify the effect of that variation on measurement results. One of the consequences of standardisation, however, is that people can be trained to follow a standardised process without really understanding the science behind it, and in such cases, compliance with the standard is, of course, imperative.

6.2 RT Measurements in Ordinary Rooms (ISO 3382–2)

6.2.1 Introduction

The vast majority of room acoustics measurements are in fact routine measurements of reverberation times (RTs), as described in ISO 3382–2: 2008, *Acoustics: Measurement of Room Acoustic Parameters*, Part 2:

Reverberation Time in Ordinary Rooms.[3] The "ordinary rooms" in question are listed in the introduction to the standard as including domestic rooms, stairways, workshops, industrial plants, classrooms, offices, restaurants, exhibition centres, sports halls and railway and airport terminals. The last of these may challenge our usual definition of "ordinary," and for some of these larger spaces, the measurement techniques set out in ISO 18233 (discussed later in this chapter) may be more appropriate.

The vast majority of RT measurements, however, are undertaken in relatively small rooms in dwellings, schools, offices and health centres. These are most often commissioning measurements, a common example in the United Kingdom being acoustic commissioning of classrooms to check compliance with building and school premises regulations.

That these measurements are routine does not mean that they are simple or universally understood even by the testers. Even the best acousticians have found substantial differences between their calculations and measurements of RTs in "ordinary" rooms. These differences are generally due to errors rather than to measurement uncertainty. Although this book is about uncertainty rather than errors, to differentiate between them, we have to know the possible sources of error both in our calculations and in our understanding of measurements.

6.2.2 Overview of ISO 3382–2: A Genuinely Useful Standard

ISO 3382–2: 2008 is a relatively practical and comprehensive document and a great improvement over its predecessors, although there are still a few contradictions and ambiguities. The standard covers a wide range of measurement equipment and techniques (including, rather quaintly, a "visual best-fit line" to a decay plotted on a paper chart recorder, a piece of equipment most likely to be found in a museum). It discusses the measurement of RT in terms of both T_{20} and T_{30} and discusses the benefits of each. Most importantly, it addresses and explains the differences between the two most common types of noise source, which are the *interrupted noise method* using a loudspeaker and the *integrated impulse response method* using impulsive noises such as gunshots or (and generally more practicable in most buildings) balloon bursts.[4]

---

3 There is a separate standard, ISO 354, for measuring absorption coefficients in specially constructed reverberation rooms. While ISO 354 provides useful background reading, the measurement of RT outside these laboratories is covered by ISO 3382.

4 The standard mentions other means of generating an impulse response indirectly using sine sweeps and MLS signals, but this is discussed in more depth in ISO 3382–1 and ISO 18233.

For both methods, the standard defines three methods giving different levels of measurement uncertainty:

- The *survey method* is intended for measurements – in octave bands only – of room absorption for noise control purposes and survey measurements of sound insulation. Readers in the United Kingdom should note that sound insulation tests in dwellings and schools for compliance with building regulations require a modified version of the engineering method (and <u>not</u> the survey method).
- The *engineering method* is intended to verify RTs against specifications, that is, commissioning measurements. It is also widely used to measure RT as part of sound insulation measurements.
- The *precision method* is defined merely as "appropriate where high measurement accuracy is required," which definition is undeniably correct, if not very informative.

In practice, the engineering method is used for the vast majority of measurements. The standard specifically and helpfully states that for the survey and engineering methods, there is no specific requirement for the directivity of the sound source, although precision measurements require a nominally omnidirectional source, as defined in appendix A of ISO 3382-1. More importance is attached to the ability of the source to generate a sufficient sound pressure level to provide decay curves unaffected by background noise. This is sensible, as repeated measurement exercises have shown that an inadequate signal-to-noise ratio is a much more frequent source of error and uncertainty than the directivity of the loudspeaker.

The main difference between the three methods is the number of source and microphone positions and the number of decay measurements required. The minima listed in Table 6.1 are specified.

Table 6.1 Minimum numbers of source and microphone positions required for each method outlined in ISO 3382–2

Minimum numbers	Survey	Engineering	Precision
Source-microphone combinations	2	6	12
Source positions	1	2	2
Microphone positions	2	2	3
Number of decays in each position	1	2	3

The last line of this table is ambiguous: does it refer to the number of decays for each source position, for each microphone position or for each source-microphone combination? It seems logical to assume the last of these. In practice, given that each decay takes only a few seconds to measure once the source and microphone are in place, there is no reason to limit oneself to the minimum number of measurements anyway.

Hence, for the engineering method, we would take measurements with three loudspeaker positions and two microphone positions, or vice versa, to give a total of six source-microphone combinations, for each of which we would measure two decays, giving a total of 12 individual decays. For the precision method, there are more possible combinations, but as microphones are easier to move than loudspeakers, it is usual to use two source positions and six microphone positions or three source positions and four microphone positions. In principle, however, it is better to use more loudspeaker positions because this reduces the possible effects of having the loudspeaker at the pressure node of a strong room mode. This is also why we tend to locate our sources close to the corners of the rooms.[5]

To further complicate matters, a footnote in the standard states that where the engineering method is used as a correction to other engineering-level measurements – in other words, for sound insulation measurements – only one source position and three microphone positions are required, giving a total of six individual decays. Readers in the United Kingdom will be familiar with this as part of sound insulation tests to comply with building regulations in dwellings. This "reduced engineering method" is justified because the effect of the slightly larger uncertainties on our RT measurements is negligible compared with the other uncertainties in sound insulation measurement.

These decays may be stored individually and the results averaged arithmetically, or they may be "ensemble averaged" within the measuring instrument. The latter is simpler for routine measurements, but by not storing individual measurements, we lose the opportunity of comparing them at a later date. This is not normally a problem, but if our results show larger variations than expected, it is sometimes helpful to see the differences between individual decays at the same source and microphone positions.

It is important to note that Table 6.1 sets out the minimum requirements for rooms with standard geometry. More measurements are required in rooms with complicated geometry, although no further guidance for such rooms is given – presumably this is left to the expertise and experience of the user. Although it is commonly assumed that more measurements are required

---

5 We should not neglect the effect of vertical room modes, which are the most significant modes in many rooms. Our aim is to excite all the room modes, which is easy with semi-dodecahedron loudspeakers and cabinet loudspeakers set on the floor. Dodecahedron loudspeakers have to be mounted on stands and are therefore more likely to be close to a pressure node of a vertical mode. To minimise the effect of this, the loudspeaker height should be changed between measurement positions.

in large rooms (which the standard defines as having a volume greater than 300 m³), this is not actually stated in the standard, and the analysis in the next section of this chapter suggests that this is not necessarily the case.

6.2.3 Evaluation of Uncertainty: Interrupted Noise Method

6.2.3.1 Theoretical Uncertainty

ISO 3382–2 initially states an assumed "nominal accuracy" in octave and third-octave bands as listed in Table 6.2. This appears to assume that the nominal accuracy (whatever that means, as it is not defined) is independent of frequency. This is not the case, although these nominal accuracies are often approximately correct for mid-frequency measurements in medium-sized rooms such as classrooms.

Table 6.2 Nominal accuracy quoted in ISO 3382–2 for each method

Nominal accuracy for each method	Survey	Engineering	Precision
Measurements in octave bands	10%	5%	2.5%
Measurements in third-octave bands	N/A	10%	5%

Fortunately, later in the standard there is a more comprehensive discussion of measurement uncertainty for both the interrupted noise and integrated impulse response methods. The calculation of uncertainty using this method requires nothing more than basic mathematics and the use of a spreadsheet, so there is really no need to rely on the nominal accuracies shown in Table 6.2.

For the interrupted noise method, ISO 3382–2 states that the standard deviation of T_{20} measurements can be estimated as

$$\sigma(T_{20}) = 0.88T_{20}\sqrt{\frac{1 + (1.90/n)}{NBT_{20}}} \qquad (6.1)$$

The equivalent formula for T_{30} measurements is

$$\sigma(T_{30}) = 0.55T_{30}\sqrt{\frac{1 + (1.52/n)}{NBT_{30}}} \qquad (6.2)$$

where B is the bandwidth in hertz ($0.71fc$ for octave bands and $0.23fc$ for third-octave bands, where fc is the mid-frequency of the band), N is the number of combinations of source and receiver positions, n is the number of decays measured at each position (in fact, at each combination of source and receiver position) and T_{20} and T_{30} are, of course, the RTs that we are measuring, and it seems reasonable to assume that the figure used will be the arithmetic mean of all our measurements.

It is helpful if at this stage to introduce two other parameters:

> $\sigma(T)/T$ is the *coefficient of variation*, that is, the standard deviation divided by the mean. It is also known as the *relative standard deviation*, but annex A of ISO 3382–2 refers to it as the "coefficient of variation", so we will use this term, and for convenience, we will express it as a percentage. We will use it a lot in the rest of this chapter.

> u is the Estimated *uncertainty*. This is not considered in ISO 3382–2 but is defined elsewhere as the standard deviation divided by the square root of the total number of individual measurements, so in this case $u = \sigma/\sqrt{n \cdot N}$(Bell 2001). It is not a ratio or percentage but has the unit of whatever is being measured, so for RT it is expressed in seconds.

Naturally, all the measures of variation and uncertainty (σ, $\sigma(T)/T$ and u) increase and decrease together, so where, in the following discussion, I refer to an increase or decrease in "uncertainty" as a function of some other parameter, this applies to all these mathematical quantities.

Now let us consider an engineering method measurement in octave bands in an unusual but mathematically convenient room with a T_{20} and T_{30} of exactly one second at all frequencies. A little light work with a spreadsheet results in Table 6.3. The same exercise for T_{30} yields Table 6.4.

Table 6.3 Expected uncertainty of T_{20} in a theoretical room using the interrupted noise method in octave bands

fc, Hz	63	125	250	500	1,000	2,000	4,000
Bandwidth B, Hz	44.73	88.75	177.5	355	710	1,420	2,840
T_{20}, s	1	1	1	1	1	1	1
N (source-microphone combinations)	6	6	6	6	6	6	6
n (decays measured at each N)	2	2	2	2	2	2	2
$\sigma(T_{20})$ (standard deviation)	0.075	0.053	0.038	0.027	0.019	0.013	0.009
$\sigma(T_{20})/T_{20}$, %	7.5	5.3	3.8	2.7	1.9	1.3	0.9
u (estimated uncertainly)	0.022	0.015	0.011	0.008	0.005	0.004	0.003

Table 6.4 Expected uncertainty of T_{30} in a theoretical room using the interrupted noise method in octave bands

fc, Hz	63	125	250	500	1,000	2,000	4,000
Bandwidth B, Hz	44.73	88.75	177.5	355	710	1,420	2,840
T_{30}, s	1	1	1	1	1	1	1
N (source-microphone combinations)	6	6	6	6	6	6	6
n (decays measured at each N)	2	2	2	2	2	2	2
$\sigma(T_{30})$ (standard deviation)	0.065	0.046	0.033	0.023	0.016	0.012	0.008
$\sigma(T_{30})/T_{30}$, %	6.5	4.6	3.3	2.3	1.6	1.2	0.8
u (estimated uncertainly)	0.019	0.013	0.009	0.007	0.005	0.003	0.002

The initial reaction of most acoustics consultants to these figures is that the theoretical uncertainties are surprisingly small. I would be very happy indeed to measure the RT of a classroom with a coefficient of variation of less than 3% in the 500-Hz octave band. As discussed later in this chapter, RT measurements using the engineering method generally return estimated uncertainties around two to three times the theoretical values.

Both theoretical and measured uncertainties vary as a function of frequency, the measured parameter (T_{20} or T_{30}), measurement bandwidth (octave or third-octave bands) and, perhaps unexpectedly, the RT itself. These uncertainties are discussed in more depth in the following section; the only generalisation that I can offer is that realistic measurement uncertainties follow the general trends discussed later but tend to be considerably greater than the theoretical values presented in Tables 6.3 and 6.4.

6.2.3.2 Uncertainty as a Function of Frequency

The first and most important trend that we see is that all our measures of uncertainty increase at low frequencies. They are in fact inversely proportional to the square root of the measurement frequency. This agrees with our experience, as we are all familiar with the increasing variation in RT measurements at low frequencies even above the Schroeder cut-off frequency of the room.

6.2.3.3 Uncertainty of T_{20} and T_{30} Measurements

Comparing Tables 6.3 and 6.4, it is clear that, in theory, the uncertainty of a T_{30} measurement should be slightly less than that of an equivalent T_{20} measurement. In fact, at all frequencies, $u(T_{20}) = 0.86u(T_{30})$. Before we decide to measure only T_{30} in the future, however, we must remember that this assumes that all our measurements are unaffected by signal-to-noise issues. In practice, T_{30} is considerably more sensitive to background noise than T_{20} – unsurprisingly, it requires a signal-to-noise ratio at least 10 dB higher at all frequencies of interest. Failure to maintain the necessary signal-to-noise ratio must be treated as an error rather than as a source of calculable uncertainty.

ISO 3382–2 allows us to evaluate RT in terms of both T_{20} and T_{30}, but with a stated preference for the use of T_{20} because

- T_{20} is more relevant to estimation of steady-state noise levels as a function of reverberation time.
- T_{20} is more closely related to the subjective evaluation of reverberation.
- T_{20} is less prone to error as a result of inadequate signal-to-noise ratio.

These are all important issues and normally will more than offset the small theoretical advantage in reduced uncertainty that T_{30} appears to have. In practice, most measurement equipment will measure and store both T_{20} and T_{30} values. A comparison of the two can be a valuable technique to inform our estimate of reliability where the equipment does not store a decay curve but merely provides a set of numbers. Most sound level meters with a room acoustic module will also provide some indication of reliability, whether this is a correlation factor (effectively an indication of how much the smoothed decay curve deviates from a straight line) or some other form of "uncertainty factor," as shown in Figure 6.1. Before relying on any of these, the user should read the manual to

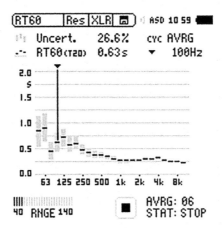

Figure 6.1 Screenshot from typical sound level meter showing T_{20} and nominal uncertainties in third-octave bands.

determine exactly what these mean and how they are calculated. In many cases, when measuring in non-Sabine spaces, these will always show large uncertainties even if they are measuring the RT correctly.

For a detailed investigation, there is no substitute for the ability to examine the individual decay for each measurement at each frequency. Figure 6.2 is a screenshot from an elderly, battery-hungry and heavy (but still portable) real-time analyser which is none the less extremely useful when the user wants to examine the actual decay on site. This particular instrument also allows one to choose the part of the decay over which the numerical result is calculated, which is extremely useful when investigating non-Sabine spaces, where the decay is not a single straight line.

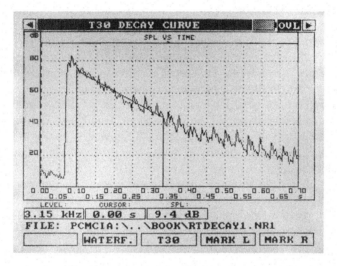

Figure 6.2 Screenshot from another sound level meter showing a decay trace from which T_{30} has been calculated.

Figure 6.2 is in fact the decay in the 3.15-kHz third-octave band, measured in a meeting room, the source being a hand clap. In spite of this low-tech source, we can immediately see that we have a reliable decay over more than 30 dB and that we are well above the background noise. As an added bonus, we also have a very clear indication of a strong flutter echo. Here is another aspect of room acoustic measurement that is frequently neglected; in this room, the most important characteristic was not the RT but the existence of this very distracting flutter echo, which would not necessarily have been detected using the interrupted noise method or a sine sweep. This is a good example of a diagnostic measurement that does not necessarily confirm with any standard but which provides us with a lot of useful information, including a reliable RT measurement at speech frequencies.

6.2.3.4 Uncertainties as a Function of Bandwidth

RT is normally measured in either octave or third-octave bands. Referring to Equations (6.1) and (6.2), we see that the standard deviation is inversely proportional to the square root of the bandwidth B, which itself is related to the centre frequency fc of the relevant octave or third-octave band. For octave bands $B = 0.71fc$ and for third- octave bands $B = 0.23fc$. The net result is that the uncertainty of measurements in individual octave bands is, at all frequencies, 57% of the uncertainty of the same measurement in the corresponding third-octave band.

In some forms of commissioning (notably in schools), it is common practice to use RT for measurements in third-octave bands, which are taken as part of sound insulation measurements, for assessing compliance with RT specifications set out in octave bands. Although various empirical methods exist, there is no intrinsically correct mathematical method to convert between the two; a simple arithmetic average is not mathematically correct because in practice the octave-band RT is dominated by the longest third-octave band RT within that octave. If the RTs in adjacent third-octave bands do not vary excessively, however, an arithmetic average is commonly used, sometimes with some form of weighting towards the highest third-octave value.

It might be thought that this averaging process would compensate for the intrinsically higher uncertainty of measurements in third-octave bands, but because the longest third-octave value has a disproportionate effect, this is not really the case. The uncertainty of the octave band measurement derived from third-octave band measurements will always be greater than that of the octave band measurement measured directly. Whether this is significant depends on the use to which we are putting our data; even a third-octave band RT can be a very blunt instrument when investigating a room with significant room modes. For problems in concert halls and music studios, an investigation in twelfth-octave bands may be necessary.

To reduce uncertainties, it is therefore better to measure RTs directly in octave bands if that is how they are specified, particularly if there are large differences between the values in adjacent third-octave bands.

6.2.3.5 Uncertainty as a Function of Reverberation Time

Now let us see how the uncertainty and coefficient of variation vary with the RT itself. Let us consider the T_{20} of a room with a rather more realistic reverberation characteristic (Table 6.5), although for ease of comparison we will make the T_{20} at 500 Hz exactly twice that of the example in Table 6.3.

Table 6.5 Expected uncertainty of T_{20} in a real room using interrupted noise in octave bands

fc, Hz	63	125	250	500	1,000	2,000	4,000
Bandwidth B, Hz	44.73	88.75	177.5	355	710	1,420	2,840
T_{20}, s	3.2	2.8	2.4	2	1.8	1.6	1.4
N (source-microphone combinations)	6	6	6	6	6	6	6
n (decays measured at each N)	2	2	2	2	2	2	2
$\sigma(T_{20})$ (standard deviation)	0.134	0.089	0.058	0.038	0.025	0.017	0.011
$\sigma(T_{20})/T_{20}$, %	4.2	3.2	2.4	1.9	1.4	1.1	0.8
u (estimated uncertainly)	0.039	0.026	0.017	0.011	0.007	0.005	0.003

Compare the value at 500 Hz with that in Table 6.3; the only change is that the reverberation time has doubled at that frequency. The standard deviation and the uncertainty have, of course, increased, but they have not doubled; the important factor is that the coefficient of variation has decreased. In other words, all other factors being equal, the percentage uncertainty in RT measurements decreases as the RT increases.

We should not be surprised that longer RTs are measurable in a more reproducible way than shorter ones, as the whole process depends on there being a reasonably linear decay above the background noise level, and this is more likely to happen in reverberant rooms. Another way of thinking about this is that reverberant rooms are much more likely to be "Sabine spaces" in which we get a reasonably straight-line decay over the measurement time.

6.2.3.6 Uncertainty as a Function of Number of Measurements

Finally, let us consider how the coefficient of variation relies on different values of *n* and N. What happens if, instead of measuring two decays for each of six source-microphone combinations ($N = 6$, $n = 2$), we reverse the process and measure six decays for each of two source-microphone combinations ($N = 2$, $n = 6$)? For the real room considered in Table 6.5, we get the results shown in Table 6.6.

From this table, it is clear that we achieve greater benefit from measuring a larger number of source and microphone positions than we do from increasing the number of decays measured at each position. This should not be surprising, given what we know about spatial averaging in real rooms. At any single position in a room, the source or microphone may be affected by one or more strong room modes, so we would expect spatial averaging to help average out the effect of multiple room modes. This is particularly important in empty rectangular rooms where there is no

Table 6.6 Expected uncertainty of T_{20} in a real room using the interrupted noise method in octave bands with different values of n and N

fc, Hz	63	125	250	500	1,000	2,000	4,000
Bandwidth B, Hz	44.73	88.75	177.5	355	710	1,420	2,840
T_{20}, s	3.2	2.8	2.4	2	1.8	1.6	1.4
N (source-microphone combinations)	2	2	2	2	2	2	2
n (decays measured at each N)	6	6	6	6	6	6	6
$\sigma(T_{20})$ (standard deviation)	0.191	0.127	0.083	0.054	0.036	0.024	0.016
$\sigma(T_{20})/T_{20}$, %	6.0	4.5	3.5	2.7	2.0	1.5	1.8
u (estimated uncertainly)	0.055	0.037	0.024	0.015	0.010	0.007	0.005

furniture to break up these room modes. This influence of room modes on our results is an inevitable result of using the interrupted noise method and is discussed in more detail later in this chapter.

6.2.3.7 Comparison with Real Results

So how do these theoretical figures compare with our own measurements? Of course, it depends on many factors, but we will look at some typical results next.

Figure 6.3 shows the T_{30} values calculated for the 12 individual decays measured using the engineering method ($N = 6$, $n = 2$) in a small but reverberant village hall. It is clear that the uncertainties are greater at low frequencies than at high frequencies and also that there are only a few outlying values (e.g. the two highest values at 500 Hz), which, if removed, would slightly reduce the uncertainty at that frequency.

As already indicated, it is dangerous to make any sweeping generalisations because the uncertainties rely on so many factors, including the RT itself, but

- We generally find that for the interrupted noise method, the uncertainties are, on average, between two and three times those calculated from Equations (6.1) and (6.2).
- The general trend indicated by the theory seems to be followed, particularly the increase in uncertainty with decreasing frequency.
- The differences between measurements at different loudspeaker and microphone positions tend to be larger than the differences between repeated measurements at the same loudspeaker and microphone position. This is hardly surprising and bears out the distinction between n and N in Equations (6.1) and (6.2).

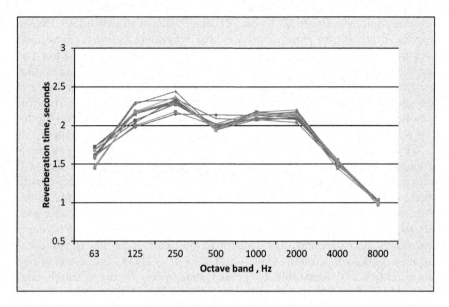

Figure 6.3 Individual RT measurements in a village hall using the engineering method (*N* = 6, *n* = 2) with an interrupted noise source.

- When choosing between omnidirectional and cabinet loudspeaker sources, it is worth noting that variations as a result of changes in loudspeaker directivity (by re-orientating a cabinet loudspeaker at a given position) are generally smaller than the variations between different loudspeaker locations.

So why do we find that the uncertainties in our measurements are so much greater than those expected from the equations set out in ISO 3382–2? The standard informs us that these equations are derived from two papers published in 1979 and 1980 (Davy et al. 1979, Davy 1980) and "based on certain assumptions concerning the averaging device." In fact, these papers include a some fairly sweeping assumptions, including that the loudspeaker and microphone have perfect frequency responses, that the loudspeaker is a point source, that all the room modes have randomly and independently distributed amplitude and phase angles and that the modal frequencies are spaced so closely that they can be integrated rather than summed. These assumptions are obviously necessary to reduce the otherwise impossible mathematical analysis to something that is merely extremely complicated. The rooms in which the subsequent experimental estimates were carried out were all reverberation chambers with nonparallel floors and ceilings and are therefore unlikely to have modal responses in any way similar to the "ordinary" rooms to which ISO 3382–2 applies. Even so, the paper concludes that agreement between theory and measured results in these rooms has been demonstrated "except at low frequencies." Brilliant though the analysis may be, it is therefore not clear that the results apply to measurements in ordinary rooms. As the equations and the subsequent discussion in annex A of the standard are derived from this idealised model, it is perhaps not surprising that we find the uncertainties of our practical measurements to be rather greater than this model predicts.

We also have to consider the essential difference between routine measurements and those undertaken for academic research. A researcher will generally review every individual measurement and discard those with obvious anomalies or outliers. Most routine measurements on site, however, are highly automated; the results are automatically saved and processed, and there is rarely scope for a time-consuming review of every individual measurement. Consequently, the vast majority of practical measurements will contain some anomalies or outliers, and the resulting uncertainties will inevitably be greater than those of the research data from which the calculated uncertainties are derived.

Looking at another common type of measurement, Figure 6.4 shows a series of T_{20} decays measured in a classroom, again using the engineering method with three loudspeaker locations and two microphone locations. In

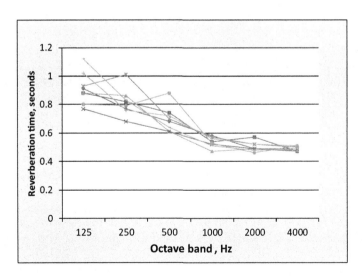

Figure 6.4 Individual RT measurements in a classroom using the engineering method with additional measurement to allow for removal of outliers.

this case, the measurement system automatically measured two separate decays for each individual measurement and averaged these within the meter. Because the time available for the measurements was very short, rather than examine each decay individually, two extra measurements were taken so that when obvious outliers were removed at post-measurement analysis, there would still be enough measurements to comply with the standard.

Table 6.7 shows the mean and variance for the raw data and with the outliers removed. It is obvious that removing the outliers very significantly reduces the uncertainty.

Table 6.7 Effect on variance of oversampling and removing outliers

	Octave band, Hz					
	125	250	500	1,000	2,000	4,000
With outliers						
Mean T_{20}	0.91	0.82	0.70	0.54	0.50	0.49
Coefficient of variance, %	12.4	11.8	12.7	6.7	6.7	3.3
Without outliers						
Mean T_{20}	0.86	0.81	0.66	0.55	0.49	0.49
Coefficient of variance, %	7.3	4.8	7.1	4.3	3.1	2.9

This very basic means of reducing our uncertainty is, of course, open to abuse; there is a fine line between the removal of obviously incorrect values and the deletion of inconvenient data. This is particularly so in the measurement of RT because virtually all obvious outliers are well above the mean, so their removal reduces the mean value, often to the benefit of the contractor trying to demonstrate compliance with a maximal RT criterion. There are, of course, various statistical ways of dealing with this; it is vital to ensure that whatever technique you use is justified and documented.

6.2.4 Evaluation of Uncertainty: Integrated Impulse Response Method

In our discussion of the interrupted noise method, we established that uncertainty can be reduced by repeated measurements using the same combination of loudspeaker and microphone positions. The reason for this is not immediately obvious in what appears to be a non-time-variant system; why is there is any variation at all between decays measured for the same source and microphone positions if nothing in the room has changed? Manfred Schroeder (1965) explained this in a beautifully clear and insightful paper more than 50 years ago:

"Random fluctuations in the decay curve result from the mutual beating of normal modes of different natural frequencies. The exact form of the random fluctuation depends on, among other factors, the initial amplitude and phase angles of the normal mode at the moment the excitation signal is turned off. If the excitation signal is a bandpass filtered noise, the initial amplitude and phase angles are different from trial to trial... . The differences being a result of the randomness of the excitation signal, not of any changes in the characteristics of the enclosure." (Schnoeder, 1965)

Schroeder then goes on to develop the reverse-integrated impulse response technique, which in those analogue days was achieved by playing a tape recording of the decay backwards through an integrating circuit. Figure 6.5 shows the modern equivalent, with the original decay, a smoothed decay and the reverse-integrated decay. Conveniently, this can now be done in one of several programmes that were designed originally for computer modelling and prediction; as these programmes calculate an impulse response and then analyse it, it is a simple matter to import a measured impulse response to be processed in the same way.

There is a fundamental and sometimes imperfectly understood difference between the interrupted noise and integrated impulse response methods of

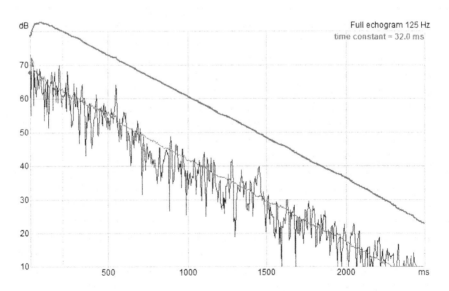

Figure 6.5 Result at 125 Hz showing actual decay, smoothed decay and reverse integrated decay (CATT-TUCT).

measuring RT. The impulse response is, in theory, the room's response to a Dirac delta function, that is, an excitation with infinitely short duration at all frequencies. For a given source position, repeated excitations of this type should excite the room modes in exactly the same way, so there should be no variation between the decays. Hence, for a given source and loudspeaker position, a single backward-integrated impulse response of a true Dirac delta function corresponds mathematically to the averaging of an infinite number of excitations using the interrupted noise method in the same position.

Obviously, if we could generate a true Dirac delta function in practice, we could substitute $n = \infty$ in Equation (6.1), which would result in our standard deviation and coefficient of variance being zero. ISO 3381 correctly points out that in practice it is impossible to create and radiate true Dirac delta functions, although short transient sounds such as gunshots can offer close approximations for practical measurement. The standard goes on to state that for measurements using the integrated impulse response method, only one decay is required at each source-microphone position, but the value $n = 10$ can be substituted into Equations (6.1) and (6.2) and that no additional averaging is necessary to increase the statistical averaging accuracy for each position.

If we take this statement literally, of course, it would mean that there should be no variation – or at least very little variation – between repeated measurements of RT in a given room for given source and receiver positions. Unfortunately, this is not the case. Figure 6.6 shows the results of 12 individual measurements using the integrated impulse response method in the hall referred to in Figure 6.3.

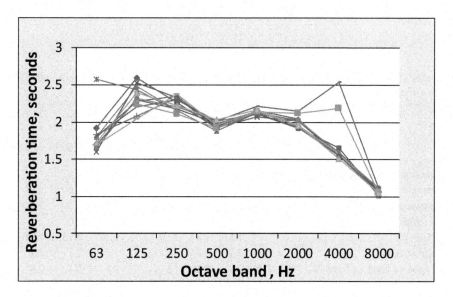

Figure 6.6 Individual RT measurements in a village hall.

Measurements using an impulsive source method (in this case, balloons) initially gave significantly greater uncertainty than those using interrupted noise, although the noise floor was low and reverse integration was used. On reviewing the results, there are only three clear outliers, and once these three points are removed, we obtain the results shown in Table 6.8, which also shows the results for the measurements using the interrupted noise data from Figure 6.3.

From this table, it is clear that the uncertainties associated with the two methods are similar, with neither being consistently better than the other, and in both cases the actual uncertainty is between two and three times that predicted by ISO 3382–2.

These data are typical of those from other comparisons, and it seems that the theoretical benefits of the integrated impulse response technique do not materialise in practice. In fact, we find that the results of repeated measurements with the same source and microphone positions give similar variations to the same process with the interrupted source method. In addition, this technique is more likely to generate anomalous results or outliers, so there is a significant risk associated with taking only a single measurement for each source position. My experience over 20 years of measurements is that both methods are valid and give similar uncertainties, provided that we use the same number of measurements at each source-microphone position for both techniques.

If this is discouraging, it is worth noting that better results can be obtained by generating the impulse response indirectly using an MLS or more commonly these days a sine-sweep technique. These are among the "new" techniques introduced in ISO 18233 and are discussed later in this chapter.

Table 6.8 Comparison of interrupted noise and integrated impulse response techniques in a village hall

	Octave band, Hz						
	63	125	250	500	1,000	2,000	4,000
Integrated impulse response method, outliers removed							
Mean RT	1.75	2.32	2.25	1.95	2.14	2.00	1.57
Variance, %	5.5	7.3	3.9	2.6	1.6	3.3	2.4
Interrupted noise method							
Mean RT	1.61	2.14	2.30	2.00	2.12	2.13	1.52
Variance, %	5.3	4.7	3.4	3.1	1.7	2.1	1.9
Expected variance from ISO 3382–2, %	3.5	2.2	1.5	1.1	0.8	0.5	0.5

6.2.5 RT Measurements in Residential Sound Insulation Tests

Earlier in this chapter we saw that where the engineering method is used as a correction to other engineering-level measurements, ISO 3382–2 allows us to use only one source position and three microphone positions, albeit still with two measurements at each microphone position, giving a total of six individual decays [in Equation (6.1), $N = 3$ and $n = 2$]. This is commonplace in sound insulation measurements, where we measure the RT in the receiver room to determine a standardised sound level difference D_{nT} in each third-octave band, and from this we may obtain a single-figure weighted standardised sound level difference $D_{nT,w}$.

From Equation (6.1), however, it is immediately obvious that this halving the number of source-microphone combinations increases the uncertainty of our RT measurements by $\sqrt{2}$. At first sight, this seems to be a significant increase in uncertainty, and given the ease with which automated systems now measure and average the RTs for these measurements, the savings in time that this relaxation of the standard allows is arguably so small as to not justify this increased uncertainty.

In fact, studies have found no statistically significant correlation between the $D_{nT,w} + Ctr$ results and the methodology used for measuring RT, and the overall reproducibility in $D_{nT,w} + Ctr$ measurements measured in accordance with ISO 140:4 is better than 1.0 dB in most cases (Critchley et al. 2007). This is perhaps not surprising given the relatively small effect of the RT on measurements in any third-octave band and on the averaging across third-octave bands to obtain a single-figure result. This frequency averaging is not, in fact, entirely straightforward because frequency-weighted values such as $D_{nT,w} + Ctr$ are heavily influenced by the values in the 100- and 125-Hz octave bands, where, as we have seen, the uncertainties in RT results are highest. In practice, we find that the room modes that cause variations in low-frequency RT have an even stronger effect on the source and receiver sound levels, and it is these, rather than the RTs, that dominate the uncertainties in sound insulation measurements.

The uncertainties associated with sound insulation testing are discussed in more detail in Chapter 8. In passing, however, it is worth noting that when measuring sound insulation between dwellings, we can obtain significantly different results depending on whether we measure between furnished and unfurnished rooms, even after the correction for RT has been made. It is quite common to find that sound insulation measurements between furnished rooms provide values up to 3 dB higher than between the same rooms without furniture. There are two reasons for this:

1. Unfurnished rooms generally exhibit strong room modes, especially at low frequencies. These room modes tend to increase both the space-averaged sound level and the RT in the receiver room. The presence of

furniture nearly always eliminates these room modes, as well as reducing the RT in the receiving room. The RT correction in ISO 16283–1 does not, of course, correct for the elimination of these room modes.

2. The equations used to calculate sound insulation assume that the source and receiver rooms are Sabine spaces with diffuse sound fields and straight-line decays. This is, in fact, rarely the case in furnished domestic rooms. This is therefore an error in the measurement methodology rather than a quantifiable uncertainty. Approved Document E of British building regulations states that sound insulation measurements should normally be between unfurnished rooms, which provides a worst-case result, although in practice residents will not experience this condition.

The 2015 revision of Building Bulletin 93: *Acoustics of Schools*, recognises this difficulty and specifically states that measurements may be taken between furnished or unfurnished rooms, and the better of the two results can be used to assess compliance with the minimum requirements for sound insulation.

6.3 Measurements in Auditoria: ISO 3382–1

6.3.1 ISO 3382–1

ISO 3382–1 discusses acoustic measurement parameters and techniques for "performance spaces", which implies spaces designed for music and speech performance. It is, however, common to measure these parameters in some other types of spaces, including spaces for worship, studios, listening rooms and even, for research purposes, teaching spaces. As these are all spaces where a good listening environment is desirable, I have referred to these spaces as "auditoria".

In fact, the main body of the standard discusses only the measurement of RT and essentially repeats much of the information in ISO 3382–2, but with less specific guidance on matters such as the number of measurement locations. It repeats the discussion of measurement using the interrupted noise and integrated impulse response methods, including the same equations for estimating measurement uncertainty [Equations (6.1) and (6.2) in Section 2 of this chapter], with the same observation that these are based on certain assumptions. For the reasons discussed in the preceding section, it follows that we can expect the uncertainties from our routine measurements, however carefully undertaken, to be greater than expected from these equations.

Annex A of ISO 3382–1 provides a list of the auditorium measures derived from impulse responses, along with typical ranges for these and an indication of the limen or JND for each of them. Annexes B and

C provide the same information for binaural measures and stage measures, again derived from impulse responses. See Table 6.9.

Of these, only the RT (and possibly the EDT) can be measured using the interrupted noise method already discussed for "ordinary rooms" earlier in this chapter. All the other parameters rely on measurement of the impulse response, and as this method also provides us with the RT and EDT, in practice any detailed acoustic survey of an auditorium requires us to measure the impulse response. The standard acknowledges that these can be measured using "special signals" such as MLSs and sine sweeps but does not provide any more information on these techniques.

There is no further information on the expected uncertainties of any of these measurements, and it is perhaps worth noting that while the standard provides a long list of information to be included in the test report, this does not include an estimate of the uncertainty.

6.3.2 ISO 18233

ISO 18233, *Application of New Measurement Methods in Building and Room Acoustics*, was published in 2006 and was reviewed and confirmed without change in 2015; this is perhaps surprising given the enormous progress made in these new measurement methods in the intervening years.

The standard provides some discussion of how to generate this impulse response, including the use of "special signals" such as MLSs and sine sweeps. It provides some background information on both of these techniques in the annexes, but this is very generic, and there is no attempt to standardise either of these techniques or their application in the measurement process. This is understandable because their implementation is complex and

Table 6.9 Room acoustic parameters listed in ISO 3382–1

Descriptor	Typical frequency range	JND
RT T_{20} and T_{30}, s	100 Hz–5 kHz	5%
Early decay time (EDT), s	500 Hz–1 kHz	5%
Sound strength G, dB	—	1
Music clarity C_{80}, dB	500 Hz–1 kHz	1
Definition D_{50}	500 Hz–1 kHz	0.05
Centre time T_s, s	500 Hz–1 kHz	10 ms
Lateral energy fraction (JLF)	125 Hz–1 kHz	Not known
Inter-aural cross-correlation function (IACF)	125 Hz–4 kHz	0.075
Early stage support ST_{Early}, dB	250 Hz–2 kHz	Not known
Late stage support ST_{Late}, dB	250 Hz–2 kHz	Not known

is generally through commercial software packages; while it is possible to discuss general principles and desirable end results, it is not really possible to standardise these complex signal-processing systems, many of which in any case had not been developed when the standard was written.

As a result, ISO 18233 might be thought of more as a discussion document than a standard, and anyone using these measurement systems will probably find more useful information in their software documentation and references than in the standard itself.

6.3.3 Sources of Uncertainty

Compared with traditional RT measurement methods, the new measurement techniques are subject to a larger number of potential errors and sources of uncertainty. First of all, there are intrinsic differences in the different types of signals used to derive the impulse response. In general, systems using MLSs have been overtaken by systems using exponential swept sine signals, which have generally been found to be more resistant to background noise (particularly time-variant noise) and better resistance to harmonic distortion. In addition, because of the narrow bandwidth, higher levels of output at the frequency of interest can be generated without the risk of damaging the source loudspeaker. Any harmonic distortion caused by nonlinearity in the loudspeaker can be removed by windowing.

The hardware used to generate the signal can introduce unwanted artefacts, and particular care is paid to sound cards, power amplifiers and loudspeakers. The most significant problem here is the non-ideal performance of the loudspeaker, particularly in terms of frequency response and directionality. Some researchers have developed complex three-way loudspeakers to improve on the omnidirectionality of commercially available dodecahedron loudspeakers (Dietrich 2007).

We would not expect small changes in source-loudspeaker location to be very significant in most auditoria, but microphone location is another matter, particularly when we are measuring parameters which we know will vary from seat to seat, such as lateral energy and clarity. Variations in clarity of up to 3 dB in individual octave bands can occur within the space of a single seat in auditoria with significant lateral room modes (Dietrich 2007).

Assuming the use of a good-quality measurement microphone, signal detection is the next issue, with specific processing issues related to windowing and noise. Again, different researchers have developed different approaches to this (e.g. Lundeby et al. 1995).

6.3.4 Background Noise

Of course, we know that background noise will have an effect on all our auditorium measurements, but it is also an acoustic parameter in its own right. It is surprising how often acousticians fail to consider the effect of background noise on the listener's or performer's subjective reaction to a space. By "background noise" in this context, I mean the constant, time-invariant noise from building services – generally the ventilation system. Modern auditoria are generally designed such that time-variant noise from sources such as road traffic, aircraft or railways is inaudible and therefore unlikely to have a measurable impact on our measurements. This is, of course, far from the case in some very famous historic concert halls, and the uncertainties in room acoustic measurements in these halls may very well be affected by time-varying noise.

The level and spectrum of constant background noise will affect the acoustic quality of a room, although few listeners or performers are directly conscious of it except where it contains tonal or time-varying components. The ear very rapidly adapts to constant broadband noise, so we may not directly notice a relatively high level of ventilation noise, but it will reduce our ability to hear very quiet passages of music and hence the dynamic range of the music, which, in turn, affects our perception of the "loudness" of a hall.

There is a surprisingly large variation in design criteria for ventilation noise in auditoria, initially because some designers overestimated the intrinsic noise level generated by the audience (Newton and James 1992). Very low background noise levels in concert halls allow performers to play more quietly and allow listeners to hear more detail. Performers may describe this using subjective terms such as "responsiveness", referring to the ease with which they can make the quietest note heard; singers often refer to the ability to "float" a note to the back of a hall. Audiences may, confusingly, describe this as a form of clarity. In fact, it is, of course, just a signal-to-noise ratio that becomes significant during the quietest passages of a piece of music or during the dramatic silences in a piece of theatre.

If we have to measure in a hall with significant background noise, the obvious way to minimise the effect of that noise on our measurements is to measure with plant and ventilation systems switched off. This reduces the measurement uncertainty both by increasing the signal-to-noise level of our measurements and by eliminating any possible effect of time-varying noise levels as active building management systems automatically adjust airflow speeds as a function of temperature and carbon dioxide levels. These complex building management systems are not often equipped with local controls, and even the most helpful hall manager may not be able to locate someone capable of turning the system off for our measurements.

In any case, our measurement programme should include measurements of noise levels before and after our room acoustics measurements and comments on any audible characteristics of the background noise, using those much-neglected acoustic detectors on either side of our heads. Uncertainties in the measurement of background noise are covered in Chapter 10.

6.3.5 Quantification of Uncertainty in Auditorium Measurements

Various studies have investigated the uncertainty of different acoustic parameters both in models and in actual auditoria. There is, however, no universally agreed estimate of uncertainty for any parameter other than RT. This is not surprising given the very large number of variables, each of which introduces potential sources of uncertainty. Consequently, the only reliable way of assessing uncertainty is by taking repeated measurements of any parameter in an auditorium and undertaking statistical analysis to identify the standard deviations and coefficients of variance between these measurements. This is, in any case, good practice, but it does mean that measurement exercises have to be planned so as to generate enough information (but not more than can reasonably be analysed) to inform us as to the uncertainty and hence the reliability of our measurements.

This is made much more complicated if we wish to measure several types of acoustic parameter. For example, we generally need to know how clarity varies across the audience seating area in an auditorium, so we will need to measure the difference between seats as well as the variation within a single seat. The same applies to stage support parameters, with the added complication that these are much more sensitive to source location and directivity. In practice, we are unlikely to have access to an auditorium for long enough to measure every parameter at every location that we want, and the best way to proceed may be to create a computer model of the auditorium, calibrate this model with such measurements as we can conveniently take in the auditorium and undertake more detailed "measurements" in the model.

6.4 Differences between Calculation and Measurement

Earlier in this chapter I referred to the large systematic differences sometimes encountered between our predictions and our measurements of acoustics, even in ordinary rooms such as classrooms. Many room acoustic measurements are taken to validate the results of predictions, so it is also helpful to understand the main sources of error and uncertainty in room acoustic predictions. These come in three categories: prediction methodology, input data and human error.

6.4.1 Prediction Methodology

6.2.4.1 Statistical Methods

The most basic prediction methodologies in common use are statistical calculations based on Sabine theory or its derivatives (Norris-Eyring, Fitzroy, etc.). These are really only applicable in "Sabine spaces", that is, rooms with good mixing geometries and a reasonably even distribution of acoustic absorption. A Sabine calculation may be valid in a very large space where the acoustic absorption is reasonably evenly distributed about the space and there is a good degree of diffusion or scattering. Similarly, a Norris-Eyring calculation may be quite accurate in a smaller space with a very short reverberation time, such as a well-designed studio control room.

A fundamental problem in both of these cases is the assumption that the RT should be the same throughout the whole room. Some spaces are specifically designed so that this is not the case. Obvious examples are "live-end, dead-end" studios, proscenium theatres where the acoustics on stage are very different from those in the stalls, some lecture theatres and music recital halls where the acoustics on the platform are much more "live" than those around the audience, and irregularly shaped spaces with linked volumes, such as cathedrals. Any calculation of RT using statistical methods is clearly nonsensical in such spaces.

In fact, even very ordinary rooms are seldom Sabine spaces. Let us take the very common case of an empty, unfurnished newly built secondary school classroom with dimensions 7 × 8 × 2.6 m. Typically, the walls will

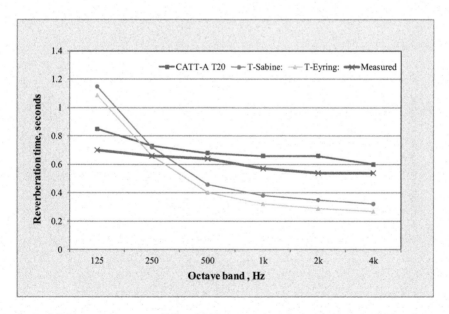

Figure 6.7 Measured versus calculated RT in an unfurnished classroom with absorbent ceiling and little scattering – showing T_{Sabine}, $T_{Norris-Eyring}$, T_{20} (CATT) and measured T_{20}.

be acoustically reflective, there may be some middle- to high-frequency absorption from carpet on the floor, but the vast majority of acoustic absorption will come from a suspended ceiling. It is immediately obvious that the acoustic absorption is not evenly distributed throughout the room, and it is therefore unsurprising that a calculation using the Sabine or Norris-Eyring equations will not yield the correct result, as shown in Figure 6.7.

As Figure 6.8 shows, the same process in the same classroom with furniture in place gives surprisingly different results in the measured RT. The classroom furniture presents very little acoustic absorption but adds significantly to the scattering of the sound field. This tends to break up flutter echoes running horizontally across the room and direct some of this acoustic energy into the acoustically absorbent ceiling. It is therefore not surprising that this results in a reduction in the measured RT, although the extent of this reduction is perhaps surprising.

Note that the scattering effect of furniture can be modelled either by inserting the individual items of furniture, all with edge diffraction, or modelling furniture as a single large surface with a high scattering coefficient. The use of scattering coefficients in geometrical acoustics (GA) models is discussed later in this chapter.

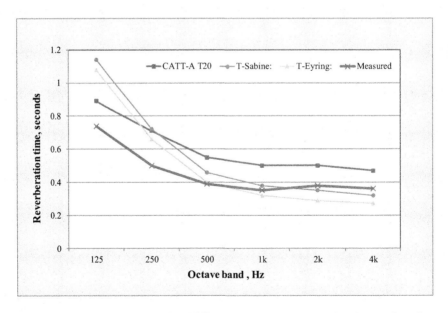

Figure 6.8 Measured versus calculated RT in the same classroom with scattering from furniture – showing T_{Sabine}, $T_{Norris-Eyring}$, T_{20} (CATT) and measured T_{20}.

It is very clear from the preceding that simple statistical calculations just do not work for this type of space. We do not yet have a reliable and accessible method of solving the wave equation in this type of room largely because of the difficulties in characterising the boundary conditions. The overwhelming majority of calculations in such rooms therefore rely on geometrical acoustics.

6.2.4.2 Geometrical Acoustics (GA)

GA modelling programmes were originally developed from ray-tracing techniques and are still sometimes wrongly called *ray-tracing programmes*. They do not solve the wave equation but approximate how sound propagates based on the geometrical features and surface properties of the space. Savioja and Svensson (2015) provide an excellent overview of current techniques, albeit without much discussion of the limitations. Ian Rees (2016) provides an expert and practical discussion of these intrinsic limitations.

At the core of the GA limitation is that it does not inherently model wave behaviour. In particular, the effect of object and detail size in relation to wavelength is handled by modelling the effects of wave behaviour on the resulting impulse response. Over the years, the better modelling systems have developed remarkably effective techniques for modelling scattering and angle-dependent reflections, but to understand the resulting limitations, users must have a good understanding of the calculation algorithms. This, in turn, requires a good practical as well as theoretical understanding of room acoustics. This seems to come as an unpleasant surprise to a minority of users who regard these programmes as a substitute for knowledge and expertise. They are nothing of the sort – they are powerful tools which, with care and experience, will assist acoustics experts in testing their room acoustic designs. As the vast majority of serious room acoustic design is undertaken using GA models, it is essential to understand their limitations.

A technique that is much neglected in GA modelling is use of the auralisation modules provided with some of these programmes. Some users consider auralisation to be a gimmick for impressing clients, while the "real" acoustics consists of generating numbers and impressive graphics. In fact, auralisation is a valuable tool for analysing results, identifying problematic late reflections and flutter echoes and reminding ourselves of what it is that we are trying to create, which is a room that is pleasing to the ear.

6.2.4.3 Finite-Element and Boundary-Element Methods

At the other end of the methodology spectrum are efforts to solve the wave equation either directly or indirectly, most typically using the finite-element method (FEM) and the boundary-element method (BEM). These techniques have been in use since the 1980s and have been markedly successful in modelling complex acoustic fields in small spaces such as vehicle cabs. In the 1990s, it was widely considered that the chief limitation of these methods was the amount of computing power required and that as more powerful computers became available, they would take over from GA methods for predicting room acoustics. In fact, it has transpired that a more fundamental limitation is the complexity of the input data needed to describe the boundary conditions. These input data are very much more difficult to determine by measurement than the simple absorption and scattering coefficients used in GA models, and it appears that for the foreseeable future, we will still be relying on GA models for practical room acoustics design.

6.2.4.4 Physical Scale Modelling

The painstaking creation and testing of physical scale models suffer from some of the difficulties associated with both measurement and computer modelling (James 1997). The chief difficulties are the need to scale everything, including the absorption coefficients of all the finishes and the acoustic absorption of air. In small-scale models, very precise modelling of boundary elements and reflectors is required, while large-scale models can be prohibitively expensive to create, adjust and store – an extreme case being the very impressive scale model of London's Royal Albert Hall, which having fulfilled its original function enjoyed a second career as a bicycle shed.

It is therefore not surprising that over the last 20 years physical scale modelling has largely given way to GA modelling.

6.4.2 Input Data

Whatever prediction methodology is used, it is obviously important to have the right inputs. For both statistical and GA predictions, the most important of these are the absorption coefficients of the room boundaries and contents. For GA predictions, scattering coefficients are also important.

Absorption Coefficients

As acoustic absorption coefficients are the most basic and important of all the inputs to any calculation methodology, it would be nice to think that we have an accurate and reliable source of these. Unfortunately, this is not the case. Many textbooks provide absorption coefficients for various generic materials, building elements and contents, but the variation between these coefficients can be enormous, and they very rarely include a reference to the original source of the data. My own research has revealed that many textbook authors have simply reproduced data from previous textbooks, some of which goes back to measurements taken in the 1950s and whose reliability is most certainly questionable.

Some textbooks and manuals supplied with GA modelling software are more helpful, with some of these even listing the sources of the data, although where different coefficients are quoted for the same or very similar materials, it is not immediately obvious to the user which coefficients to use. This is a particular problem for the absorption of seats and people in auditoria.

In the absence of a generally available and reliable list of acoustic coefficients, it is tempting to rely on data measured in a reverberation room in accordance with ISO 354. Unfortunately, one of the features of the ISO 354 methodology is a tendency to return absorption coefficients which are, in defiance of the laws of physics, greater than one. While this is encouraging for the manufacturers of proprietary acoustic absorbers, it is also one of the most common reasons for finding that our finished room is significantly more reverberant than expected. One reason for this strange anomaly in ISO 354, which has hindered acousticians for decades, is that the absorber does not cover an entire wall or floor of the reverberant room, and edges of the absorber therefore add to the overall absorption. Another reason is that the laboratory test process places all the absorbent material on a single surface, so the room used for the measurements is about as far as it is possible to be from a Sabine space, which assumes that any acoustic absorption is distributed evenly about the room.

As this is the whole basis for the theory on which the calculation of absorption coefficients is based, using uncorrected laboratory-derived absorption coefficients in our predictions is clearly an error. None the less, it would be helpful to quantify the likely uncertainty in our absorption coefficients. Unfortunately, there is no clear mathematical basis for this,[6] but in my experience it is common for absorption coefficients measured in acoustic laboratories to be overestimated by up to 15%. It might be hoped that the "practical absorption coefficient" α_p described in ISO 11654 would overcome this problem, but in fact it is a very blunt instrument, achieved merely by rounding to the nearest 0.05, with values that would otherwise be greater than 1.00 being rounded to 1.00. The resulting absorption coefficients will still often be significantly higher than those which apply in normal conditions.

The use of impedance tubes to determine absorption coefficients (ISO 10534–2) has its own problems, but compared with a reverberant room, an impedance tube is cheap, quick and convenient for testing small samples of material. Practical acousticians have been known to undertake

---

6 ISO 354 states that this is still under consideration, and as this standard was last reviewed and confirmed in 2015, it seems that little progress has been made in this matter since 2003.

their own measurements in an impedance tube in order to "normalise" laboratory-measured absorption coefficients. This does not help us, however, with large elements such as audience seating. Neither do the otherwise useful, and strangely underused, intensity techniques for the estimation of absorption coefficients (Farina 1997).

Often, therefore, the only practical approach is to rely on data measured in real rooms. For example, to model the absorption of books in a library, I have used a relatively reverberant room (in this case, a plastered living room that had been stripped for redecoration) and measured the change in RT as a result of different areas and configurations of books and bookshelves distributed around the room. This yielded a truly practical absorption coefficient for bookshelves in different configurations. As the data were to be used in GA computer models, it was simple enough to create a GA model of the test room and of the individual bookshelves and from this to reverse engineer the absorption coefficients. This method also allowed some investigation of the model's sensitivity to scattering coefficients.

Acousticians should take every opportunity to acquire these sorts of data and to store them for future reference. For example, audience seating is almost always the last element to be installed in an auditorium, so we often have the opportunity to measure RTs before and after seating is installed, with no other change to the room contents. In the case of retractable seating, a few hours spent acquiring data with and without seating retracted can save a great deal of effort and uncertainty in future projects. Likewise, it is often possible to reverse engineer a calculation to estimate the absorption of a surface, which is otherwise not known. Common examples of such materials are plasterboard, timber and plaster ceilings under large roof voids, which can be surprisingly absorbent at low frequencies.

Another complication arises with individual absorbent elements such as the vertical absorbers used in many swimming pools and the horizontal "rafts" used under concrete soffits in many classrooms. These have considerable advantages over suspended ceilings, as both sides of the absorber are in the sound field – but this does not mean that we necessarily achieve twice as much absorption. The acoustic absorption of these elements depends not just on the properties of the materials but also on their dimensions and the distance separating these elements from each other and from the ceiling from which they are suspended. At least one manufacturer supplies comprehensive data for different geometries of these absorbers measured in realistic conditions.

To complicate matters even further, the acoustic absorption of some absorbers is quite strongly dependent on the angle of incidence. At least one GA modelling programme allows for angle-dependent absorption coefficients, but the source data for this can only come from measurements using intensity techniques. Fortunately, this is not normally necessary, but when looking at individual reflections off non-homogeneous materials such as slatted timber finishes, it may be necessary to model different absorption coefficients for different angles of incidence.

Finally, there is a vexing question of how to deal with acoustically transparent absorbers. A heavy curtain 200 mm from an acoustically reflective wall will exhibit more low-frequency absorption than the same curtain close up against the wall. A geometrical model does not "know" this because it does not directly model the wave behaviour. Hence, we have to input the acoustic absorption of the curtain and the wall as a single element, which means that we rely on there being, somewhere, a reliable and realistic measurement of the acoustic absorption of the curtain of the right weight and fullness at the right distance from a wall.

Acoustic Scattering Coefficients

A note on nomenclature: the terms "surface diffusion" and "scattering" are used almost interchangeably, even in the two parts of ISO 17497, which describes different methods for measuring the "scattering coefficient" (ISO17497–1: 2004) or the "diffusion coefficient" (ISO17497–2: 2012). "Scattering" is currently the preferred term among GA modelling specialists because GA models represent the diffusion of waves by the scattering of the rays (or cones or beams) representing those waves. I shall use the term "scattering" throughout this discussion.

Elementary wave theory defines a "specular reflection" as one that obeys Snell's law; that is, the angle of reflection is equal to the angle of incidence. Specular reflection is in fact the exception rather than the norm and only occurs when a wave strikes a plain, rigid surface with dimensions very much larger than the wavelength of the incident sound. It follows from this definition that the scattering coefficient of the surface is a function of its dimensions and "roughness" relative to the wavelength that we are considering. Hence, there is no simple answer to the frequently asked question "What scattering coefficients should I use?" because this depends on the dimensions of the surface in the model.

There are also implications for the degree of detail required in the model itself. A model with many small surfaces will only be valid at high frequencies. In general, better results will be obtained by omitting fine detail in the geometry and compensating for this by increasing the scattering coefficients at high frequencies. The opposite approach (lots of small ornate surfaces with low scattering coefficients at low frequencies) merely increases the computing time while reducing the accuracy of the modelling process. Ian Rees (2016) describes a particularly striking example of this. Figure 6.9 shows a model of a cathedral which is visually impressive but ineffective for GA purposes.

As well as having simply too many surfaces, this model includes a great deal of irrelevant fine detail, particularly as the main purpose of the

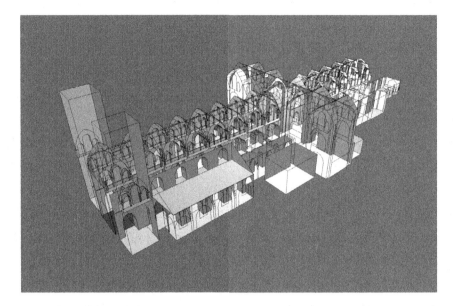

Figure 6.9 Model of a cathedral – visually impressive but acoustically wrong.

model was to assess direct and early reflected sound from digitally steered column loudspeakers in the nave. The detail in the roof zone has no effect on this; effectively what the model needs to reproduce is the early reflections close to the loudspeakers and to the congregation, and the long RT induced by the large volume. A much better representation of the actual acoustics of the building was achieved with Figure 6.10 using many fewer surfaces, all of them being much larger and with much higher scattering coefficients.

Hence, the selection of scattering coefficients in a model is much more complex than that of absorption coefficients because it depends on dimensions, wavelength and the geometry and purpose of the model itself. For this reason, there is no such thing as a database of scattering coefficients, and each GA modelling programme provides its own guidance as to approximate coefficients to use in different circumstances depending on the scattering algorithm used by that programme.

Expertise, common sense and experimentation also help in determining appropriate scattering coefficients. For example, my experience of acoustics tells me that a long, thin, tall room with smooth parallel walls will exhibit strong flutter echoes. If my GA computer model of the room does not indicate those flutter echoes (best identified by listening to auralisations), this suggests that I am using scattering coefficients that are too

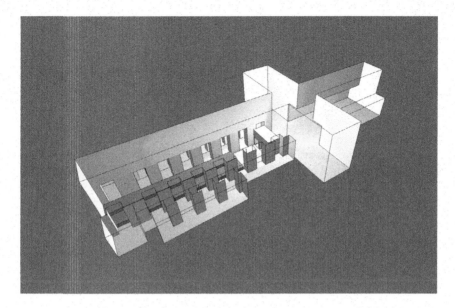

Figure 6.10 Model of a cathedral – acoustically correct.

high. In such cases, there is generally a value of scattering coefficient above or below which the model ceases to become very sensitive to small changes. Anyone wishing to develop real expertise in this field should expect to spend a good deal of time creating and testing models of rooms with known acoustics and assessing the sensitivity of those models to changes in the modelled geometry, absorption and scattering coefficients.

6.5 Further Reading

Of necessity, this chapter can only provide a summary of the main sources of uncertainty in this very complex area of acoustics. For a more detailed discussion, an excellent starting point is the 2013 paper by Christensen et al. (2013). This provides an excellent overview of the sources of uncertainty in measurement and simulation of auditorium acoustics. As well as presenting a practical and realistic assessment of the limitations in accuracy that we can expect, it also provides a very clear summary of the different measurement techniques in general use and covered under ISO 3381 and a comprehensive bibliography for those keen to expand their knowledge of the subject.

References

Barron M, *Auditorium Acoustics and Architectural Design*. E. & F.N. Spon, San Francisco, CA, 1993.

Bell S, "A beginner's guide to uncertainty of measurement", *Measurement Good Practice Guide No. 11* (Issue 2). National Physical Laboratory, Teddington, UK, 1999.

Beranek L, *Concert Halls and Opera Houses*. Springer, New York, 2003.

Christensen CL, Koutsouris G, Rindel JH, "The ISO 3382 parameters: Can we simulate them? Can we measure them?", in *International Symposium on Room Acoustics*. June 2013.

Critchley et al., "Sound insulation measurements in residential buildings", in Proceedings of the Association of Noise Consultants' Conference, Birmingham, AL, 23 October 2007.

Davy JL, "'The variance of impulse decays': Decay rates in reverberation rooms", *Acustica* 44 (1980).

Davy JL, Dunn IP, Dubout P, "The variance of decay rates in reverberation rooms", *Acustica* 43 (1979).

Dietrich P, "Modeling measurement uncertainty in room acoustics", *Acta Polytechnica* 47(4–5) (2007).

Farina A, Torelli A, "Measurement of the sound absorption coefficient of materials with a new sound intensity technique", *Proc AES* (1997).

International Organization for Standardization, *Acoustics: Sound Absorbers for Use in Buildings – Rating of Sound Absorption*, ISO 11654. ISO, Geneva, 1997.

International Organization for Standardization, *Acoustics. Determination of Sound Absorption Coefficient and Impedance in Impedance Tubes: Transfer-Function Method*, ISO 10534-2. ISO, Geneva, 2001.

International Organization for Standardization, *Acoustics: Measurement of Sound Absorption in a Reverberation Room*, ISO 354. ISO, Geneva, 2003 (reviewed and confirmed in 2015).

International Organization for Standardization, *Acoustics: Sound Scattering Properties of Surfaces, Part 1: Measurement of the Random-Incidence Scattering Coefficient in a Reverberation Room*, ISO 17497-1. ISO, Geneva, 2004.

International Organization for Standardization, *Acoustics: Application of New Measurement Methods in Building and Room Acoustics*, ISO 18233. ISO, Geneva, 2006.

International Organization for Standardization, *Acoustics: Measurement of Room Acoustic Parameters, Part 1: Performance Spaces*, ISO 3382-1. ISO, Geneva, 2009.

International Organization for Standardization, *Acoustics: Measurement of Room Acoustic Parameters, Part 2: Reverberation Time in Ordinary Rooms*, ISO 3382-2. ISO, Geneva, 2008.

International Organization for Standardization, *Acoustics: Measurement of Room Acoustic Parameters, Part 3: Open Plan Offices*, ISO 3382-3. ISO, Geneva, 2012.

International Organization for Standardization, *Acoustics: Sound Scattering Properties of Surfaces, Part 2: Measurement of the Directional Diffusion Coefficient in a Free Field*, ISO 17497-2, ISO, Geneva, 2012.

James A, "Practical considerations in acoustic modelling of auditoria", *Proc IoA* (1997).

James A, "Workshop on room acoustics measurements: Analysis of comparative measurements", *Proc IoA* 25(4) (2003).

Keränen JS, "Measurement and prediction of the spatial decay of speech in open-plan offices", *Aalto University Publication Series Doctoral Dissertations* 23 (2015).

Lundeby A, Vigran TE, Bietz H, Vorländer M, "Uncertainties of measurements in room acoustics", *Acustica* 81(1995).

Newton J, James A, "Audience noise: How low can you get?", *Proc IoA* 14(1992), 65–72.

Rees I, "Common pitfalls in computer modelling of room acoustics", *Proc IoA* 38 (Pt. 1) (2016).

Savioja L, Svensson P, "Overview of geometrical room acoustic modeling techniques", *JASA* 138 (2015), 708.

Schroeder MR, "New method of measuring reverberation time", *JASA* (1965).

Chapter 7

Speech Transmission Index (STI): Measurement and Prediction Uncertainty

Peter Mapp

7.1 Introduction

The speech transmission index (STI) is the most widely used method employed for rating the potential speech intelligibility of sound and communication systems and is used internationally. The technique is detailed and standardised in international standard International Electrotechnical Commission (IEC) 60268–16 (BS EN 60268–16) [1]. Many standards and codes of practice relating to sound and emergency communication systems and speech intelligibility refer to and call on the use of STI measurements or predictions, including British, American and international fire/voice alarm and emergency communication system standards.

There are potentially six primary areas of uncertainty associated with STI measurements and predictions. These are

- Limitations imposed by the method itself (e.g. monaural measurement/binaural hearing),
- Uncertainties naturally inherent within the technique itself (use of a pseudorandom noise–based measurement signal),
- Uncertainties produced by errors introduced within the implementation of the technique (e.g. errors in equipment algorithms, signal incompatibility, measurement durations and speech spectrum variations),
- Uncertainties within prediction programs (e.g. algorithmic errors, incorrect or poor application of background noise and interpolation of data),
- Uncertainties associated with normal site measurements (e.g. variations in conditions or number of measurement sample points, equipment calibration errors), and
- Uncertainties introduced by misapplication of the technique (e.g. omission or erroneous use of the speech masking function or background noise level corrections).

7.2 Background to Speech Transmission Index

STI was initially conceived to help evaluate the intelligibility of communications equipment and transmission channels, with the first public reference of the technique being published in 1973 [2]. Since then, the technique has been continually developed so that now it covers a very much wider application, addressing both measurement and prediction of speech intelligibility under a wide variety of conditions and applications [1,3–5]. The method was first standardised by the IEC in 1988 [6]. The original STI technique required very specialised equipment that only a very few laboratories possessed. However, a simplified version of the technique, Rapid STI (RaSTI),[1] was introduced in 1985 [7], and the associated introduction of commercially available portable measurement equipment paved the way for the method to become more widely adopted. This also opened up a new range of potential applications, ranging from the certification of aircraft public address (PA) systems to the measurement of speech intelligibility in rooms and auditoria. The technique continued to be developed with the introduction of separate male and female spectra and associated intelligibility weightings, for example, together with the incorporation of 'adjacent band redundancy contribution' weightings. These latter introductions were standardised in 1998. In 2001, Speech Transmission Index for PA systems (STIPA), a second condensed version of the technique, specifically designed for use with PA systems, was introduced. This overcame many limitations of RaSTI, which has now effectively ceased to be used – though it is still occasionally seen in outdated standards and specifications. The 2003 (revision 3) of the STI standard formalised the incorporation and formulation of STIPA together with an upward spread of masking function, making STI and STIPA sound level dependent. In 2011, revision 4 of the standard was introduced. This clarified a number of aspects and provided greater guidance on the use of STI for a range of applications. Corrections for the use of STI with non-native-language listeners were also introduced. At the time of this writing, revision 5 is going through the IEC standards approval process and will, apart from providing further clarifications relating to implementation and use of the technique, also specify adjustments to the speech reference spectrum. (It is expected that this new version of the standard will be published in 2020).

The STI technique has been adopted by a number of international standards relating to the design and measurement of public address and voice alarm (PAVA) systems and emergency communication systems and has become the primary method of assessing and predicting the potential intelligibility of such systems. In 1981, Schroeder published a paper that

---

1 Sometimes also referred to as 'room acoustic STI'.

showed that STI could also be derived from a room or system impulse response [8]. This led to the introduction of software that not only enabled full STI measurements to be made via a number of computer-based measurement platforms but also to led to STI's eventual direct prediction in computer-based modelling programs. However, owing to the nature of such measurement and prediction programs, the output can sometimes be erroneous due to either to the incorrect use of the program or input data, incorrect adjustment or use of the parameters used to make the predictions or measurements or indeed errors within the computer programs themselves.

The Speech Transmission Index is based on the precept that in speech a significant contribution to intelligibility is based on the preservation of the modulations of the waveform and in particular the relative depth of these modulations. Analysis of the speech waveform shows that the main carrier contains frequency modulations in the range from approximately 0.50–20 Hz. STI mimics this behaviour by modulating a carrier signal centred on the seven octaves from 125 Hz to 8 kHz over the range of 0.63–12.5 Hz. (i.e. with 14 modulation frequencies) and measuring the reduction in modulation in each band (see Table 7.1). This results in a 98-point measurement data matrix. The average of the modulation reductions in each octave band is then computed, producing seven transmission index (TI) values. These are then weighted according to their relative contributions to speech intelligibility and combined to form a single

Table 7.1 STI matrix

Octave band, Hz	125	250	500	1,000	2,000	4,000	8,000
0.63	x	x	x	x	x	x	x
0.80	x	x	x	x	x	x	x
1.0	x	x	x	x	x	x	x
1.25	x	x	x	x	x	x	x
1.6	x	x	x	x	x	x	x
2.0	x	x	x	x	x	x	x
2.5	x	x	x	x	x	x	x
3.15	x	x	x	x	x	x	x
4.0	x	x	x	x	x	x	x
5.0	x	x	x	x	x	x	x
6.3	x	x	x	x	x	x	x
8.0	x	x	x	x	x	x	x
10.0	x	x	x	x	x	x	x
12.5	x	x	x	x	x	x	x

overall STI value. (STIPA employs a sparse matrix of just 14 modulation data points but evenly distributed across the seven octave bands – as shown in Table 7.2. The TI for each octave band is then weighted in the same manner as for the full STI computation.) STIPA is a subset of STI and, apart from the reduced number of modulation frequencies, is otherwise identical. The agreement between STI and STIPA is extremely good, being within 0.02 STI under most circumstances [9,10].

STI employs a rating scale of 0 to 1.0, where 0 relates to completely unintelligible speech and 1.0 to perfect intelligibility speech. To put the scale into context, the minimum target for most emergency sound systems and paging systems is 0.50. The just noticeable difference (JND) is 0.03 STI. Typical measurement errors are also around 0.03.

The Speech Transmission Index is able to account for the effects of both ambient noise and reverberation on intelligibility, as well as a number of other factors that are further discussed later. By examining the modulation matrix, it is possible to determine whether the reduction in modulation is due to the effects of noise, reverberation or echoes.[2] The sparse matrix, of just 14 data points, employed by STIPA, whilst generally

Table 7.2 STIPA matrix

Octave band, Hz	125	250	500	1,000	2,000	4,000	8,000
0.63			x				
0.80						x	
1.0		x					
1.25					x		
1.6	x						
2.0				x			
2.5							x
3.15			x				
4.0						x	
5.0		x					
6.3					x		
8.0	x						
10.0				x			
12.5							x

2 It is not possible to determine the presence of echoes from STIPA due to the sparse nature of its matrix.

enabling extremely good agreement with full STI measurements to be obtained, does result in some limitations, particularly when dealing with strong, discrete reflections and echoes, and is not valid to deal with these effects.

7.3 Limitations and Uncertainties Inherent within the STI Technique

STI was originally conceived to be measured using a modulated speech-shaped pseudorandom noise signal. This is termed the 'direct method' of measurement. However, the mathematical relationship, discovered by Schroder [8], between a system's or communication channel's impulse response and STI led the way to computer-based analysers and software being able to convert a captured impulse response into STI. This approach is termed the 'indirect method' of measurement. The two methods use very different types of test signals and so excite a system quite differently. This can give rise to discrepancies in the measured/computed STI value unless adequate precautions are taken. There are a number of advantages and disadvantages to using each method that are discussed in Section 7.4. This section discusses potential uncertainties associated with the basic technique itself which are independent of the measurement/prediction method.

7.3.1 Binaural Listening and Monaural Measurement Technique

The Speech Transmission Index was originally conceived to assist with evaluating the potential intelligibility of communication channels and so is inherently a monophonic system. Whilst in some communication applications both ears listen to the speech signal, this is not a binaural or stereo signal but double mono, as both ears receive the same signal (i.e. diotic listening). However, normally, when listening to either natural speech in an auditorium or that of a PA system, the speech is generally heard binaurally. This enables a listener to capitalise on small differences detected in the signals received by each ear and so benefit from binaural/spatial release from masking. Such binaural listening can lead to significant improvements in the perceived intelligibility. However, the STI/STIPA technique is based on monophonic measurements, made with an omnidirectional microphone or single-channel analyser, so this potential benefit is lost. This is not to say that binaural measurements cannot be made, but the STI technique itself has no direct way to deal with this information. Instead, it has been found that the data from the 'better ear' can be employed (i.e. the ear either facing directly towards the source of speech or away from interfering noise). Figure 7.1 illustrates the effect. Here standard STI measurements made with an omnidirectional microphone, are compared with those measured binaurally using a calibrated dummy head. The standard method using an

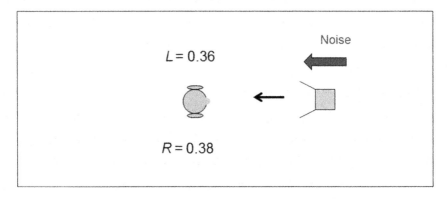

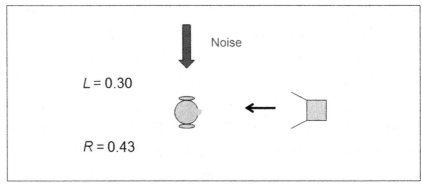

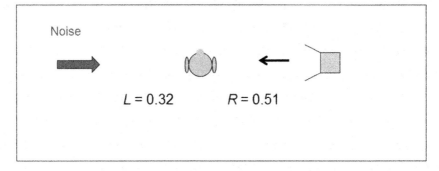

With noise omnidirectional measurement mic STI = 0.41

Figure 7.1 Measurement differences in STI using binaural head and omnidirectional microphone.

omnidirectional microphone produced an STI of 0.41 under all conditions. However, depending on the direction of the noise and orientation of the head, a range of values was found to occur binaurally and is summarised in the following table:

Configuration	Omnidirectional microphone (standard STI)	Left ear	Right ear
1. Noise from in front	0.41	0.36	0.38
2. Noise from left	0.41	0.30	0.43
3. Noise from behind, head at 90 degrees	0.41	0.32	0.51

A similar binaural release from masking effect can be observed with echoes and discrete, late-arriving reflections.

7.3.2 Effect of Echoes and Discrete Reflections

Apart from the potential binaural effect of reducing reflection and reverberation reductions of intelligibility, echoes and strong, discrete late-arriving reflections can also lead directly to an erroneous assessment when using STI. The potential error occurs when the reflection/echo arrives with a delay equivalent to the period of one of the modulation frequencies. For example, an echo or secondary signal received with a delay of 100 ms would cause there to be a cancellation or attenuation of the modulation at 5 Hz and an echo at 80 ms at 6.3 Hz. This can lead to an overall excess reduction in the measured STI. For the two cases cited above, this could theoretically lead to STIs of 0.71 and 0.72 instead of 1.0. The overall effect of an echo affecting a measured STI value, however, is more complex than this, as the losses are not directly additive. Figure 7.2 further illustrates this issue.

It should be noted that discrete echoes and reflections can also increase as well as decrease the apparent modulation reduction. Therefore, if such effects are anticipated or observed, the modulation transfer function (MTF) matrix should be examined, and the normal monotonic decay of the MTF with frequency should be plotted and reviewed. Corrections can then be applied. Figure 7.3 shows a typical set of MTF curves, illustrating some effects of echoes/late reflections. In particular, the modulation reduction is affected at 10 Hz, suggesting a strong reflection occurring at around 50 ms.

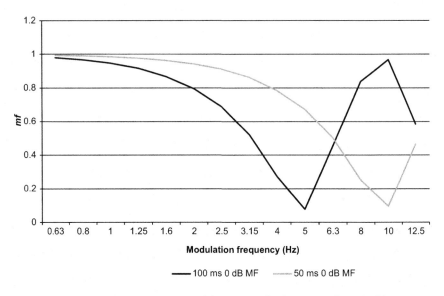

Figure 7.2 Effect of echo or delay on modulation transfer function at 50 and 100 ms.

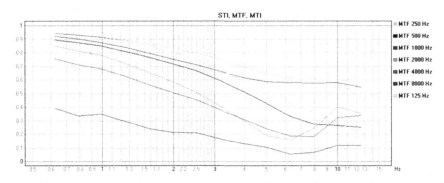

Figure 7.3 Typical set of MTF curves.

Although misunderstood by some, STI does, in general, account for the loss of intelligibility caused by a late arrival or echo. This effect is shown in Figure 7.4. After a steep reduction up to about 50 ms, the loss in STI (intelligibility) levels out. This behaviour, at first sight, may appear to be counterintuitive, with the loss of intelligibility expected to increase as a function of delay time. Casual subjective experience would tend to agree

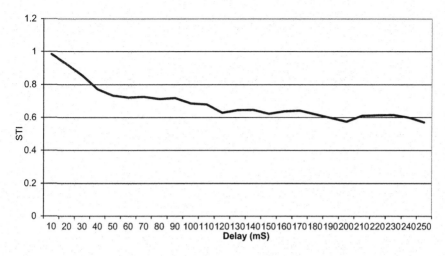

Figure 7.4 Effect of echo or delayed reflection on STI (delayed signal is of equal amplitude to direct signal).

with this, as delayed speech certainly gets harder to listen to [11], but as the delay increases, syllables and parts of words can be resolved, so objectively measured intelligibility (and STI) level out. Given the sparse nature of the STIPA modulation matrix, STIPA is not suitable for use when echoes or late reflections are present.

7.3.3 Effects of Equalisation

Whilst STI can account for the effects of strong spectral distortion, for example, bandpass filtering, it is not able to account for the effects on intelligibility brought about by small changes in system/channel frequency response and equalisation, particularly when auditioned under quiet or relatively quiet conditions and high-signal-to-noise-ratio situations [9,12,13]. However, even relatively minor adjustments to a system's frequency response or equalisation can in practice have a significant effect on perceived intelligibility and ease of listening. Conversely, under noisy conditions, with signal-to-noise ratios of 20 dB or less, the effect of such filtering is generally accounted for.

7.3.4 Effects of Amplitude and Data Compression

Although not extensively researched, it has been shown that some forms of amplitude compression and automatic gain control (AGC) can aid intelligibility, although this may not be indicated by STI. The effect on

STI is heavily dependent of the degree of compression applied and the attack and release times of the compressor [14]. Differences can be expected in the values and effects obtained between the direct and indirect methods and the type of test signal employed. Some forms of digital compression (e.g. MP3) may affect a measured STI value but not affect the perceived intelligibility of the associated speech.

7.3.5 Differences between Actual Speech and STI Standard Speech Spectrum

The STI method employs a standardised speech spectrum, making measurements repeatable and independent of the source. However, this can lead to discrepancies in perception and the measured STI value, as the spectrum of an individual announcer's voice may be different to that of the STI test signal. For comparison purposes and uniformity of approach, the standardised spectrum should always be used, but additional measurements, using a facsimile of the actual announcer's voice spectrum, can also be made and reported. (In practice, many voices have less low-frequency energy and more energy at high frequencies; see Figure 7.5, for example [14,15].)

In previous versions of the STI standard, different spectra and weighting factors were provided for male and female voices. However, this has led to confusion, so for edition 5 (2020), it was decided to just provide

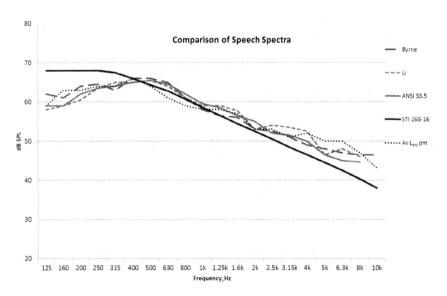

Figure 7.5 Comparison of STI spectrum (edition 4, 2011) with other speech data.

data for the male voice, as this would give the worst-case scenario. The spectrum was also changed slightly to better reflect other speech data and standards. Figure 7.6 shows the new and old male spectra – normalised for 0 dBA. This adjustment to the spectrum also better enables small transducers/mouth simulators to produce the signal with lower distortion at the lower frequencies. The change to the standard spectrum does result in a small difference in computed STI, but this is less typically than 0.02 and so is effectively insignificant in practice [16]. However, measurements should cite which version of the standard was used when making them. (Note that the levels in the 31.5 and 63 Hz and 16 kHz bands should be 20 dB lower than those in the 125 Hz and 8 kHz bands.)

7.3.6 Fluctuating Noise

Although STI is able to nicely account for the effects of interfering noise, this is only the case when the noise is steady state. Where the background or interfering noise fluctuates or varies with time, such as would be the case with traffic noise or spectator noise, each STI reading will give a different result and may flag up a measurement error message. There is no really good way to overcome this, but either multiple STI measurements can be made and the average taken or a statistical and spectral analysis of the noise can be made. A 'noiseless' STI measurement is then

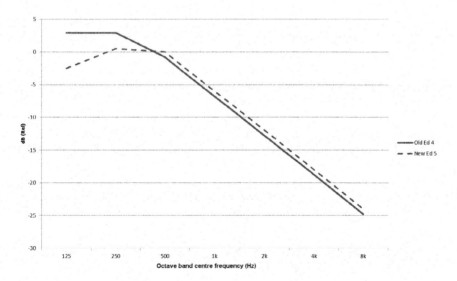

Figure 7.6 Comparison between BS EN 60268-16 edition 4 and 5 spectra.

made, and the noise spectrum L_{eq} (or possibly the L_{10}) is then applied during post-processing of the measurement.

7.4 Uncertainties Associated with the Measurement Method

As noted in Section 7.3, the direct method of STI measurement is based on the use of a modulated speech spectrum–shaped pseudorandom noise signal. The indirect method, however, generally employs an exponential/ log sine sweep or MLS signal. These signals have very different dynamic properties, crest factors and repeatability [17].

7.4.1 Natural Variation in STI (STIPA) Measurements

Due to the pseudorandom nature of the direct method test signal, it differs slightly from test to test. This can give rise to variations in the overall measured STI value. Generally, these variations are within 0.02 STI, but variations of 0.03 are also common. For this reason, it is necessary to take the average of three measurements. Readings that vary by more than 0.02 should be disregarded, and a further measurement should be made. Occasionally, much larger variations are also encountered – often for no apparent reason. These can be purely random spurious readings but are also often due to an acoustic event occurring (e.g. door opening/closing) or external noise that may not be audible to the test operator. If this persistently occurs, then the cause should be investigated and rectified. The indirect method employing an exponential sine sweep has a naturally high immunity to background noise and so is generally more repeatable/consistent. Indeed, by averaging a number of sweeps, a clean or effectively 'noise-free' impulse response can be obtained. This is a useful technique that can be employed in particularly noisy environments, but the effects of the background noise must be taken into account when the STI is computed. When an MLS signal is used, provided that it is appropriately shaped to match the STI speech spectrum, it can directly give the true STI value. Equally, by averaging several sequences, the signal-to-noise ratio can be improved, and noise-free measurements can be obtained. The signal-to-noise ratio improvement is given by $10 \log N$, where N is the number of sweeps or averaged sequences employed. Therefore, two sweeps, for example, produce a 3 dB improvement in SNR, six sweeps give an improvement of 7.8 dB, and 10 sweeps yields an improvement of 10 dB. When using the direct STIPA method, the signal should be run such as to continuously excite the space.

7.4.2 Effects of Signal Processing

It is common in modern PA and VA systems to employ extensive signal processing. This can range from simple input and output equalisation to dynamic processing (e.g. compression) of the signal. Whilst STIPA mimics a speech signal reasonably well in terms of its spectrum and modulation, it does not mimic the dynamic behaviour in the same way, and the crest factor characteristics are different – although in practice this latter factor is of lower importance. Exponential sine sweeps and MLS signals have low crest factors (e.g. a sine wave is just 3 dB) and have notionally constant amplitude characteristics. This means that neither the STIPA signal (i.e. direct method) – and most certainly not the indirect signals – affect dynamic signal processing nor signal limiting and clipping in the same way as real speech. Figure 7.7 shows the effect that an automatic gain control (AGC) circuit can have on STIPA, whilst Figure 7.8 shows the effects that a typical compressor may have. Table 7.3 compares a number of notable speech and test signal characteristics [17].

Apart from compression and limiting (which can be frequency dependent), other forms of nonlinear signal processing include analogue phase and frequency shifting. Full direct STI and STIPA are usually immune to these effects, but impulse response–based measurements cannot operate with such processing. All methods are unsuitable for processes where the digital sample rate is changed. Nonlinear distortion such as clipping can normally be accounted for by the direct methods, but again it is not possible to use indirect techniques in such situations. The testing of smart devices/loudspeakers (e.g. Alexa, etc.) may be possible using a direct STI method – but extreme care needs to be taken, as the processing that such signals undergo may invalidate the STI result. Table 7.4 summarises the suitability of the methods.

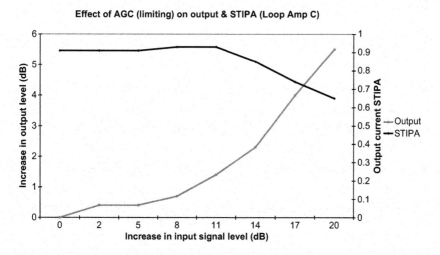

Figure 7.7 Effect of AGC on STIPA (upper curve is STI, lower rising curve shows the input-output level control characteristic).

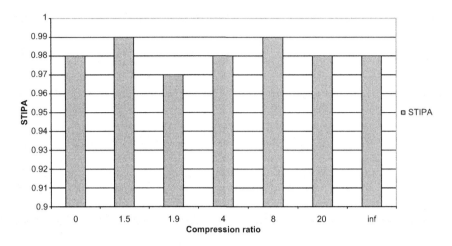

Figure 7.8 Effect of typical amplitude compression on STIPA signal.

Table 7.3 Speech and test signal dynamic parameter comparison

Parameter	Crest Factor, dBAwtd	Crest Factor, dBCwtd	$L_{AI} - L_{Aeq}$ dB	$L_{AI0} - L_{Aeq}$	$L_{AI} - L_{A90}$ dB	$L_{Amax} > L_{Aeq}$ dB	$L_{Cmax} > L_{Aeq}$ dB
Speech (average)	20	16.7	7.1	3.3	27.9	8.9	12.1
Pink noise	12	11.2	1.8	0.1	0.7	0.4	2.6
STIPA	12.4	11.6	1.8	1.0	3.1	1.8	9.2
Sine wave	3	3	0	0	0	–	–

7.5 Uncertainties Produced by Errors Introduced within the Implementation of the Technique

7.5.1 Equipment Algorithms and Test Signal Compatibility

Although the STI technique is standardised, variations in its implementation can and do occur. In particular, the manner in which the signal analysis is truncated to ±15 dB can give rise to noticeable errors. This is illustrated, for example, in Figure 7.9, which shows a laboratory test conducted by the author on four different manufacturers' STIPA meters [14]. Figure 7.10 shows a similar laboratory test but looking at the effects of the signal sound pressure level (SPL) on the indicated STI value.

Table 7.4 Test-method suitability

Type of distortion	Full STI (direct)	STIPA (direct)	Full STI (indirect)	Limitations	Work-arounds
Nonlinear	√	√	X	–	–
Reverb	√	√	√	–	–
Echo delay	√	X	√	–	–
Noise	√	√	?	Depends on test signal	Post-measurement addition of noise to MTF matrix
AGC	√	√	?	Depends on test signal	–
Reverb + noise	√	√	?	Depends on test signal	Post-measurement addition of noise to MTF matrix
Analogue phase or frequency shifting	√	√	X	All methods unsuitable with changes to the digital sample rate of the test signal	–
Channels that do not permit artificial test signals, such as vocoders	?	?	–	Currently standardised methods are inaccurate	Use a speech-based STI test signal or listener tests

It is important that the STIPA test signals from different equipment manufacturers are completely compatible as in practice only a single test signal will be broadcast during PA system commissioning, for example, but meters from different manufacturers may be employed to measure this. A surprising number of errors have occurred in this regard over the years, with one manufacturer's signal, whilst giving correct readings with the manufacturer's own meter, caused all other meters to give much higher values (e.g. a 0.45 condition was read as being 0.60). Conversely, a much lower reading was indicated by the one manufacturer's meter when using any of the other available test signals. Owing to better communication and more detailed guidance being provided in the later versions of 60268-16, this problem, as far as is known, has now been overcome. Variations between microphones and calibration of individual meters, of course, can still give rise to measurement differences between equipment.

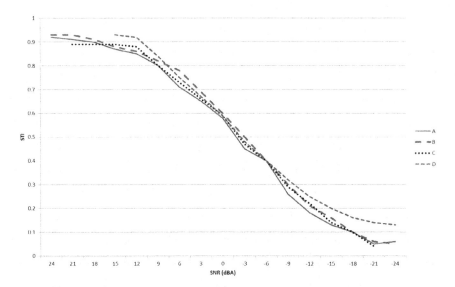

Figure 7.9 Comparison of four STIPA meters operating at various signal-to-noise ratios under controlled laboratory conditions.

At 80 dBA SPL, the upward spread of the masking function begins to operate and introduces a reduction to the measured STI/STIPA value to account for the reduction in intelligibility that is experienced by listeners at higher sound levels. Prior to version 4 of the standard, this was a stepped function but now has been replaced by a continuous function to better mimic the response of a listener. Because this function is well described within the standard, there should be little problem with its interpretation and implementation. However, it does require the meter or analyser to be able to accurately detect an absolute sound level of 80 dBA SPL. The effect of the spl/masking function is shown in Figure 7.11.

It should be noted that the masking function is not applied globally to all STI readings (e.g. reducing them by 0.2 STI, say) but is introduced as another term in the general modulation reduction. This is illustrated in Figure 7.12, which shows the effect of increasing the SPL in a reverberant space (Reverberation time ~ 3 s). This shows, for example, that the STI at 100 dBA is not reduced by 0.20 as it would be in an anechoic space but by 0.06, as the modulation has already been reduced by the reverberation.

The measurement period over which a STIPA measurement is made can also affect the result. For optimal performance, the measurement period

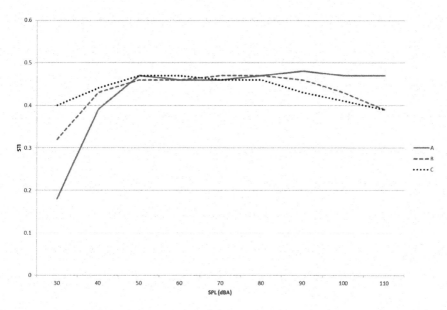

Figure 7.10 Effect of SPL on three STIPA meters.

should be approximately 18 seconds, with a recommended range of 15–25 seconds. There is no 'gold standard' STI calibrator, so there is great reliance on each manufacturer or implementer to ensure that they have correctly applied the technique and the standard.

7.5.2 Indirect Methods

Indirect methods can be extremely useful in determining the STI in noisy or changeable noise conditions, as they can enable a 'noiseless' measurement to be made. Different noise profiles, levels or signal-to-noise ratios can then be evaluated mathematically by post-processing the noiseless STI data. However, skill is required in this process, and the opportunity for errors to occur can be significant unless great care is taken. Whilst the technique has the advantage of enabling different scenarios to be examined, it suffers from the disadvantage that an 'on the spot' result cannot be obtained nor witnessed, which may be necessary as a part of formal commissioning or handover of a system. As noted in Section 7.3, the test signal is unlikely to exercise a sound system in the same manner as a direct STI or STIPA signal or indeed real speech.

A further issue is that the impulse response must be long enough to ensure that the lowest modulation frequencies are adequately accounted

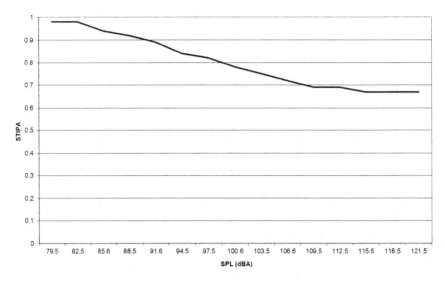

Figure 7.11 SPL masking function.

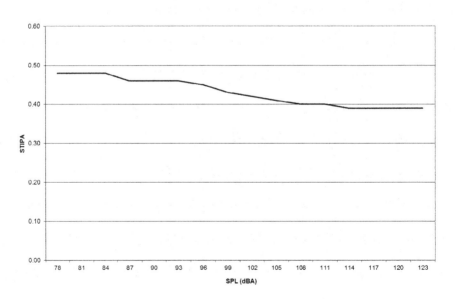

Figure 7.12 Effect of SPL masking in reverberant space with reduced STI.

for. In practice, this means that the impulse response (IR) must be of at least 1.8–2.0 seconds in duration – even in low reverberation time environments. A two second minimum-length IR is recommended. Figure 7.13 demonstrates the effect that the IR length can have on STI by sequentially truncating the impulse length and computing the associated STI.

7.6 Uncertainties with Prediction Programs

Several acoustic prediction programs enable the Speech Transmission Index (full STI or STIPA) for a sound system or individual talker to be predicted. Many national bodies and clients require PAVA system performance to be predicted before a system is installed. Such programs are invaluable in enabling loudspeaker coverage to be optimised, expected or required SPLs to be calculated and potential STI performance to be checked prior to installation. However, such predictions can and all too often are prone to considerable error and uncertainty [18]. There are a number of reasons for this.

7.6.1 Statistical and Impulse Response (Ray-Tracing) Simulations

As noted in Section 7.3, STI can be derived from the impulse response of a system, with this latter parameter being obtained by means of measurement or prediction. The more sophisticated acoustic computer programs are based on 'ray tracing' or similar techniques that enable the impulse

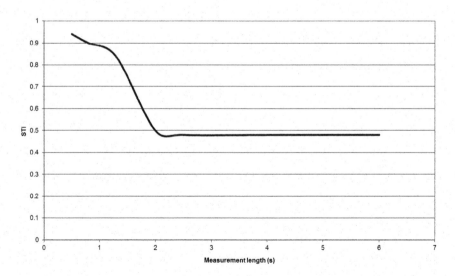

Figure 7.13 Effect of impulse response truncation of computed STI value.

response of a room or system in a space to be predicted. This, in turn, enables the potential STI to be obtained using the relationship discovered by Schroeder[8]. Simpler programs, however, employ a statistical approach and calculate the 'direct-to-reverberant' ratio or assume an exponential sound decay curve. No information can be obtained, using such a program, regarding the strength or arrival of 'early' and 'late' reflections, yet these factors play a significant role in speech intelligibility and within STI. In simple rooms or those exhibiting a truly exponential sound decay, a statistical calculation may be able to predict the STI reasonably well, but in complex spaces or where more than a simple sound system is employed, considerable prediction errors may occur. An example of this is shown in Figure 7.14. Here the predicted STI values for a sound system covering a concourse connecting two reverberant spaces are presented. Statistically, the STI is predicted to be essentially constant and achieves the minimum PAVA goal of 0.50 STI. However, a prediction based on ray tracing shows the STI to vary considerably and to be well below an acceptable minimum. The table in Figure 7.14b presents the results and percentage errors involved. (This assumes that the ray tracing is correct, which later measurement confirmed.) The data illustrate the critical importance of using the correct STI calculation technique, as the system in question formed part of a life safety system, where a minimum of 0.50 STI was required. The statistical technique indicates that the target will be met and the system will be satisfactory. However, in reality, the system was an abject failure, only being marginally intelligible in two of the sample areas.

Another example of inaccurate prediction, based on a statistical calculation, concerned a football stadium stand. The STI prediction for the stand, based on a statistical calculation, was 0.65 to 0.75 – ridiculously high values for an empty stadium. The ray-tracing/impulse–based calculation indicated STI values of 0.45–0.55, which were considerably more likely and in the event agreed well with measurements made on the final installation.

7.6.2 Uncertainties and Errors in Prediction Data

When employing either ray-tracing or statistical prediction techniques, there can be considerable uncertainty regarding the absorption properties of the materials and surface finishes employed in the construction of a space. In the case of the stadium example given earlier, it is not possible to look up the absorption properties of complex structures such as the stadium roof, nor have accurate absorption coefficients for the plastic seats and the configuration in which they are deployed. Even relatively standard materials such as carpet can vary significantly in practice. The absorption properties of building occupants, other than when in typical

(a)   (b)

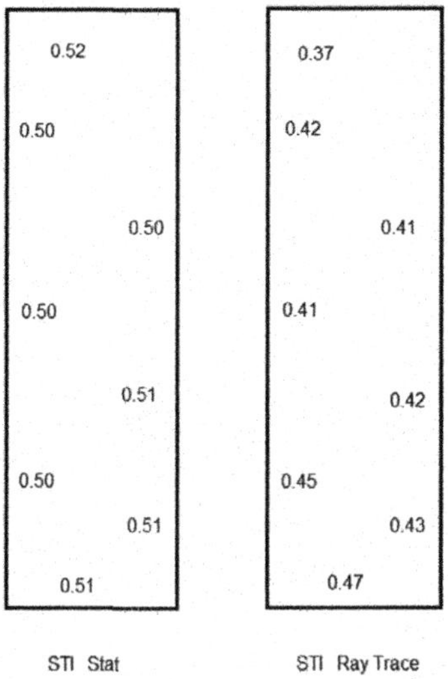 0.52	0.37		
0.50	0.42		

Predicted STI Statistical	Predicted STI Ray Tracing (IR)	Percentage Error
0.52	0.37	40%
0.50	0.42	19%
0.50	0.41	22%
0.50	0.41	22%
0.51	0.42	21%
0.50	0.45	11%
0.51	0.43	18%
0.51	0.47	8%
0.51 mean	0.42 mean	21% Average error

STI Stat   STI Ray Trace

Figure 7.14 (a) Comparison of statistical and ray-tracing-based STI prediction. (b) Comparison of the prediction data and associated percentage error.

auditorium seating, are also open to considerable uncertainty [19]. However, STI is not unduly sensitive to small errors in reverberation time. Conversely, an error of 3 dB in the signal-to-noise ratio can give rise to an error of 0.1 STI, which is a very significant error. When predicting the potential STI of a sound system, great reliance must also be placed on the loudspeaker data in addition to the room acoustic data, as noted earlier. Current loudspeakers are generally provided with three-dimensional polar (balloon) directional data, but these are sometimes interpolated from one-dimensional vertical and horizontal data. Equally, errors can creep into the provision of such information, with files containing missing or corrupted data being released. Furthermore, there have been several instances where the SPL data provided have been up to 15-dB adrift. This can have a very significant knock-on effect when taking background noise into account, particularly when it may be relatively high, so the resulting signal-to-noise ratio (and hence STI) is much lower than predicted – particularly where amplifier power is limited.

Errors can also occur in the prediction programs themselves. At the time of this writing, one well-known program has an inherent error of 2.9 dB when predicting the SPL from loudspeakers and a corresponding error of approximately 0.1 STI. Appendix K of edition 5 of the STI standard provides further details on the uncertainty in the calculation of STI based on the associated uncertainties in predicted interfering noise and reverberation.

7.7 Uncertainties Associated with STI Site Measurements

7.7.1 Equipment Calibration

STI and STIPA measurements are not absolute acoustic measures but effectively comparisons between the measured and expected signals. They are therefore less prone to equipment calibration errors. However, this still means that the sound level meter or microphone and analyser employed to make the measurements should be accurate and calibrated. In particular, it is important that SPLs are correctly read, as this affects the speech masking function and speech reception threshold correction factors. The frequency responses of the measurement system and microphone are of less importance, as any anomaly will equally affect both the reading of the background noise and the test signal and so effectively cancel out.

A 'loop back' measurement can (and should) be performed to ensure that the meter/analyser is reading correctly. For a STIPA meter, this means directly connecting the output of the signal generator/STIPA source to the analyser, and this should yield a reading of between 0.98 and 1.00. Unfortunately, there is no 'gold standard' STI calibrator that can be used to check the performance of a test system.

In the early days of STIPA, the test signal was supplied on a compact disc (CD), and the author found that some CD players can corrupt the signal and so inadvertently produce lower STI readings than should have been obtained [9,14]. The now universal use of .wav files to distribute and play back the STIPA/STI signal has effectively overcome this problem.

7.7.2 Emitting the STI Test Signal

As noted in Section 7.1, STI/STIPA may be used to measure the potential intelligibility of the natural room acoustic environment (e.g. classroom, lecture theatre or auditorium etc.), or it may be used to assess the potential intelligibility of a sound or communication system. In the latter case, the test signal may be produced acoustically at the system microphone, or it may be injected electronically into the system, without an acoustic path being involved. Where the signal is transmitted acoustically, it is important that the test loudspeaker or voice simulator not only has a very accurate frequency response (within ±1 dB over the bands of interest) but also matches the radiation (directivity) characteristics of a human talker. Edition 4 and the previous editions of the STI standards have recommended the use of an artificial mouth or test loudspeaker with a 100-mm driver.

Research by the author [20] has shown that such a driver is too directional, and not only can this give rise to higher STI readings than should be obtained, but such a directional source also is unlikely to provide similar excitation and coverage of a space as compared with a human talker. The author has also seen standard decahedral loudspeakers employed, which are an even more inaccurate source in this respect. The author's research and later research by others [21] show that if a 'talker simulator' loudspeaker is employed, it should not have a drive unit diameter greater than 65 mm. This should also be contained within an enclosure that has dimensions similar to the human head (e.g. ~150 × 250 mm). Artificial mouth simulators or talker (head) simulators can be used when injecting the test signal acoustically via a microphone, but generally speaking, artificial mouths have a lower directivity than a human talker and may therefore produce lower STI values when assessing the acoustic transmission within a room. (A head or head and torso simulator or equivalent should then be employed.) In order to ensure that the effects of background noise are correctly taken into account, a mouth or head simulator should have a frequency response within ±1 dB within the octave bands of 125 Hz to 8 kHz (i.e. from 100 Hz to 10 kHz).

Most artificial mouths have a far from flat frequency response and therefore require significant equalisation and frequency response correction. Figure 7.15 shows the frequency response for a typical artificial mouth with the ±1 dB tolerance overlaid. Small transducers, however, do have the disadvantage that their acoustic output is limited, particularly at low frequencies, so they can be overdriven, leading to undesirable distortion that may affect the STI measurement. (The new edition 5 spectrum, with its reduced levels at 125 and 250 Hz, should help to reduce this problem.)

7.7.3 Effect of Background Noise and Signal Level Variations and Variations in Conditions

As noted previously, the STIPA signal is based on pseudorandom noise and so will vary from sample to sample and from second to second. Figure 7.16, for example, shows the variation in level that occurs within the signal itself over a typical 15-second measurement period. The level variation is based on a consecutive series of 1 second, A-weighted L_{eq} readings measured directly at the output of the test signal generator, set to give an output equivalent to 70 dBA. The variation in output level is 1.5 dB.

Such signal deviations can give rise to variations in complete STIPA measurements. This is illustrated in Figure 7.17, where a series of 27 consecutive STIPA measurements, made at the same position, was made in a quiet (~30 dBA) auditorium. (The STIPA signal level was nominally 70 dBA. The signal-to-noise ratio was therefore nominally 40 dB.)

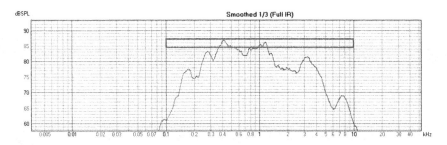

Figure 7.15 Typical artificial mouth uncorrected frequency response.

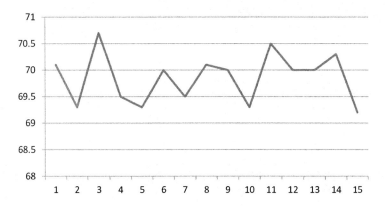

Figure 7.16 Typical STIPA signal level variation (1-sec L_{Aeq} samples).

The monitored sound level variation over the 27 fifteen-second, STIPA measurements was 0.6 dBA. Whilst this may seem to be a very small variation in SPL, the effect was to cause a variation in the measured STIPA value of 0.04. Not only is this subjectively just noticeable, but it is also enough to change the assessment category and requires a number of measurements to be made to achieve a stable result. (Note that the mean STI was 0.55 at an average SPL of 70 dBA.)

7.7.4 Voice and PAVA Announcement Level Matching

When testing PA and communication systems and when assessing the potential intelligibility of the unaided voice in acoustically noisy situations, it is important to accurately match the level of STI test signal to the normal broadcast or spoken SPL. This not only ensures that the

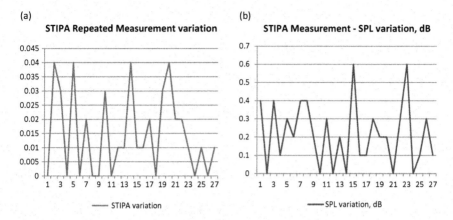

Figure 7.17 (a) Variation in measured STIPA results recorded in an auditorium under low-noise conditions. (b) Variation in the corresponding SPL measurements.

correct signal-to-noise ratio is achieved but also that the SPL masking functions within STI are appropriately activated. In the case of a PA or communications system, the L_{Aeq} of a number of typical (or specific) broadcasts should be measured. A sample length of approximately 20 seconds is ideally required, though many announcements may be less than this. If this is the case, then the level average of a number of shorter announcements should be obtained. A common error is to try to directly match the SPL (L_{Aeq}) readings. However, unlike the STI or STIPA signal, real speech is not continuous but contains many gaps and pauses in the signal. Therefore, for equivalence, the STIPA signal must be broadcast 3 dBA louder than the measured speech level (i.e. if the measured broadcast level is 80 dBA (L_{Aeq}), then the STIPA signal must be broadcast at 83 dBA). This requirement is partially demonstrated in Figure 7.18, where the 1-second L_{Aeq} profile of a STIPA signal is compared to a typical speech announcement – with both, in this instance, being set to 70 dB L_{Aeq}.

7.7.5 Inadvertent Clipping/Overdriving of Sound Systems

It should not be forgotten that an amplifier does not 'see' the A-weighted signal but the full-bandwidth linear signal, and this has considerably more energy and voltage associated with it. Figure 7.19 compares the A- and linear (Z-) weighted signal levels for a STIPA signal. As the figure shows, the linear (Z-weighted) signal is nominally 5 dB greater than the A-weighted signal. The maximum root-mean-square levels (L_{ZFmax}), however, are typically 6.5 dB higher than the L_{Aeq} level. In other words, to

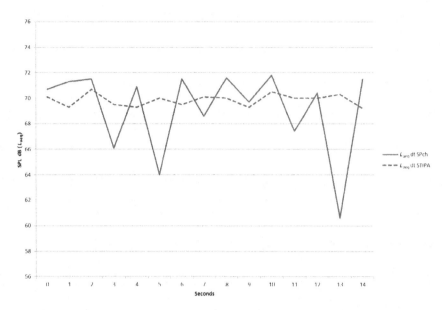

Figure 7.18 Comparison of typical speech announcement (solid line) and STIPA signal both at the same nominal SPL (L_{Aeq}).

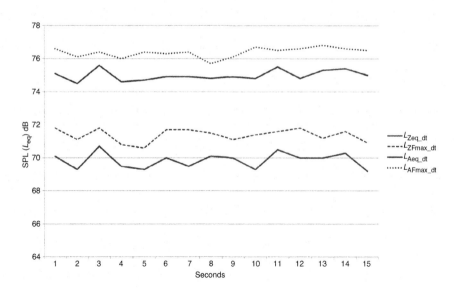

Figure 7.19 Comparison between linear and A-weighted STIPA signal levels (at nominal 70 dBA).

generate the STIPA signal without undue distortion, four times the power or twice the voltage is required than the A-weighted level might suggest. This is particularly important when testing PA systems employing 100-V line distribution, as the maximum voltage can only be 100 V. Therefore, the STIPA signal must be broadcast at a maximum equivalent level of 50 V or less. This effect is illustrated in Figure 7.20, which plots the output voltage and STIPA for a typical modern 100-V line digital amplifier. As the figure clearly shows, as the output voltage reaches approximately 50 V, the STI begins to rapidly decrease as the signal begins to distort.

7.7.6 Location of Sample Points and Number of Measurements

In a similar manner to SPL measurements in general, making a measurement close to a large surface (e.g. wall), room boundary or corner can give rise to a discrepancy or erroneous result. Equally, if the operator stands too close to the measurement microphone, spurious reflections or blocking of the sound can occur. Both of these effects can give rise to a significant error in the measurement. Because the highest frequency of interest is approximately 10 kHz, the orientation of the measurement microphone (assuming a ¼- or ½-inch free-field type) is not particularly critical, but it is appropriate to aim the microphone towards the signal source.

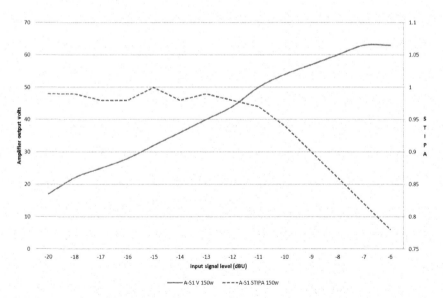

Figure 7.20 A 100-V PA digital amplifier output (solid curve) and corresponding effect on STIPA.

It is important to ensure that a sufficient number of measurements are made to adequately characterise the STI performance of a system or space. ISO 7240 [22] provides some guidance on this for VA systems. Table 7.5 summarises these guidelines. However, in the cases of critical listening spaces (e.g. lecture theatres and auditoria), it may well be necessary to carry out a greater number of measurements to ensure that the space is adequately characterised. Equally, when assessing large venues, such as arenas and stadiums, a more strategic approach may be required based on the seating layout and distribution of the loudspeakers. For example, for two recent projects, a 27,000-seat stadium and 55,000-seat arena, the author found that 193 and 81 measurements, respectively, were the minimum needed to sensibly characterise and assess the PA systems (equivalent, respectively, to 0.72% and 0.15% of the seating capacity). However, in comparison, an 800-seat auditorium required 45 measurements (equivalent to 5.6% of the seats). Some sound system standards use a similar approach to Table 7.5, related to each acoustically distinguishable zone or area.

7.7.1 Calibration of Equipment

As with any acoustic or electronic measurement, it is important to know that the equipment making the measurement is accurate, so therefore it should be calibrated. With respect to STIPA meters, this poses a problem because there is no 'gold standard' against which to compare a meter. A meter's SPL/L_{Aeq} measurement accuracy and microphone response can, of course, be checked and calibrated in a similar manner to any sound level meter. Otherwise, the basic STIPA functionality of a meter can be tested by directly connecting it to the STIPA signal source (as discussed in Section 7.7.1). The functionality of the absolute SPL/STI characteristic/correction factor can be checked by adjusting the input level to simulate a range of equivalent SPLs and monitoring the decrease in STIPA as the SPL increases beyond 80 dBA (see Figure 7.11).

Table 7.5 ISO 7240 recommendations

Area, m^2	Minimum number of STI measurements
<25	2
25–100	3
100–500	6
500–1,500	10
1,500–2,500	15
>2,500	15 per 2,500 m^2

7.8 Uncertainties Introduced by Misapplication of the Technique

Measurement (and prediction) errors can be introduced by misapplying the technique; in particular, this could apply to the omission or erroneous use of the speech masking function or background noise level corrections. Some STIPA meters allow the speech masking function (i.e. reduction of STI with increasing SPL above the 80-dBA threshold) to be disabled or switched off. This is useful and indeed necessary when testing assisted listening systems such as deaf aid loops (AFILS) or infrared transmission systems where an electronic pickup of the signal has to be made [5,23]. (Alternatively, an electronic attenuator can be employed, although this can be cumbersome in the field.) The direct testing of speech communication systems (electronic chain) may also require a similar approach to be adopted. When calculating the potential STI performance of a sound system using an acoustic prediction program, the effect of masking is often neglected, leading to optimistic predictions. (Indeed, some programs inadvertently almost encourage such discrepancies.)

The effect of background noise on the STI is often neglected or not correctly accounted for when employing 'out-of-hours' measurements (i.e. measurements made under quiet or abnormal conditions). Such measurements are regularly required in practice in order not to disturb the public, customers or staff or disrupt normal operation of venues and businesses. Speech level matching, as discussed in Section 7.7.4, is also often neglected or incorrectly applied, again leading to an incorrect measurement or conclusion. STI has also been attempted to be applied to the measurement of speech privacy, but the technique has not been validated for such potentially low levels of speech intelligibility, so privacy or confidentiality should not be assessed in this way. As noted in Section 7.4.2, some signal-processing devices and techniques affect the STI signal and its measurement and so invalidate an assessment. Indirect measurements using an impulse response of less than two seconds will give rise to optimistically inaccurate STI values.

Some Indirect measurement and prediction methods allow the user to select a range of STI weighting factors. The most common are male and female speech weightings, although other, nonstandard weightings also exist. Applying such weightings to the computed STI value can give rise to a range of potential values and considerable confusion. In order to overcome this problem, edition 5 of the STI standard only provides the male speech weighting function because it has been found that this will nearly always produce the worst-case scenario. To put the issue into perspective, Figure 7.21 shows the effect of applying the various weighting factors to six different scenarios. The results are also summarised in the following table, which shows that variations of up to 0.06 STI can typically occur. This is a large range and very significant uncertainly.

	Position					
	1	2	3	4	5	6
STI old ('standard')	0.69	0.61	0.66	0.61	0.67	0.70
STI male	0.70	0.63	0.68	0.63	0.69	0.72
STI female	0.71	0.64	0.69	0.64	0.69	0.74
STIPA male	0.70	0.63	0.68	0.63	0.70	0.72
STIPA female	0.71	0.65	0.69	0.64	0.70	0.74
Variation	0.02	0.04	0.03	0.03	0.03	0.04

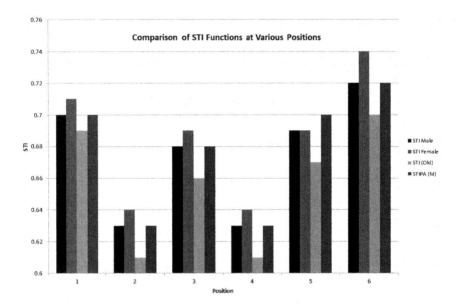

Figure 7.21 Comparison of STI functions at various positions.

References

1 International Electrotechnical Commission, *Objective Rating of Speech Intelligibility by Speech Transmission Index*, IEC 60268–16. IEC, Geneva, 2011.
2 Houtgast T, Steeneken HJM, 'The modulation transfer function in room acoustics as a predictor of speech intelligibility', *Acustica* 28(1973), 66–73.
3 Steeneken HJM, Houtgast T, 'Some applications of the speech transmission index (STI) in auditoria', *Acustica* 51(1982), 229–230.
4 van Wijngaarden. S.J. ed., *Past, Present and Future of the Speech Transmission Index*. TNO Pub Human Factors, Soesterberg, Netherlands, 2002.

5 Mapp P, 'Performance and installation criteria for assistive listening systems', in *Proceedings of the AES 145th Convention*, convention paper 10109. New York, 2018.

6 International Electrotechnical Commission, *The Objective Rating of Speech Intelligibility in Auditoria by the RaSTI Method*, IEC 268–16. IEC, Geneva, 1988.

7 Steeneken HJM, Houtgast T, 'RaSTI: A tool for evaluating auditoria', *B&K Technical Review* (3) (1985), 3–29.

8 Schroeder M, 'Modulation transfer functions: Definition and measurement', *Acustica* 49(1981), 179–182.

9 Mapp P, 'Is STIPA a robust measure of speech intelligibility performance?, in *Proceedings of the AES 118th Convention*, convention paper 6399. Barcelona, 2005.

10 Steeneken HJM, Verhave JA, McManus S, Jacob KD, 'Development of an accurate, handheld, simple-to-use meter for the prediction of speech intelligibility', in *Proceedings of IoA 2001*, reproduced sound 17. Stratford-upon-Avon, UK, 2001.

11 Hammond R, Mapp P, Hill A, 'The influence of discrete arriving reflections on perceived intelligibility and speech transmission index measurements', in *Proceedings of the AES 141st Convention*, convention paper 9629. Los Angeles, 2016.

12 Mapp P, 'Frequency response and systematic errors in STI measurements', *Proc IOA* 27(Pt 8) (2003).

13 Leembruggen G, Hippler M, Mapp P, 'Further investigations into improving STI's recognition of the effects of poor frequency response on subjective intelligibility', in *Proceedings of the AES 128th Convention*, convention paper 8051. London, 2010.

14 Mapp P, 'Some practical aspects of STI measurement and prediction', in *Proceedings of the AES 134th Convention*, convention paper 8864. Rome, 2013.

15 Morales L, Leembruggen G, 'On the importance of the speech spectrum for STI calculations', *Proc IOA* 35(Pt. 4)(2013).

16 Leembruggen G, Verhave J, et al., 'The effect on STI results of changes to the male test-signal spectrum', *Proc IOA* 38(Pt. 2)(2016), 78–87.

17 Mapp P., 'Some effects of speech signal characteristics on PA system performance', in *Proceedings of the AES 139th Convention*, convention paper 9477. New York, 2015.

18 Mapp P, 'Some effects and potential errors in acoustic and electroacoustic systems computer modelling', in *Proceedings of the ASA 171st Meeting*, Vol. 39. Salt Lake City, UT, 2016.

19 Hammond R, Hill AJ, Mapp P, 'Discrepancies between audience modelling methods in performance venues', in *Proc IOA* 40(Pt. 4) (2018).

20 Mapp P, 'Simulating talker directivity for speech intelligibility measurements', in *AES 137th Convention*, convention paper 9156. Los Angeles, 2014.

21 Wijngaarden S, Verhave J, 'Designing an acoustic source of the STIPA signal', *Proc IOA* 38(Pt. 2)(2016), 145–156.

22 International Organization for Standardization, *Fire Detection and Alarm Systems*, Part 19: *Design, Installation, Commissioning and Service of Sound Systems for Emergency Purposes*, ISO 7240. ISO, Brussels, 2007.

23 Mapp P., 'Assessing the potential intelligibility of assistive audio systems for the hard of hearing and other users', in *AES 125th Convention*, convention paper 7626. San Francisco, 2008.

Chapter 8

Uncertainty Associated with the Measurement of Sound Insulation in the Field

Bill Whitfield

8.1 Chapter Overview

The focus of this chapter is to provide a quantitative assessment of the major factors that contribute to the uncertainty in the measurement of sound insulation on development sites using airborne sound insulation as an example. (Of course, a similar approach can be employed to quantify the measurement uncertainty in impact sound insulation, although the published research on this is less comprehensive.) The challenge is to do this without adding to the burden of experimental testing that is essential for data collection, which takes a significant amount of time and effort and forms the major cost component of any uncertainty study.

As this is more of a "how-to guide", the design of experiment and statistical analysis procedure are described in summary for the novice, and a web link is provided to an electronic resource that will assist anyone with a rudimentary understanding of statistics to carry out basic analysis without having to resort to proprietary software. Full referencing and further reading are suggested for the informed reader together with a comprehensive statistical reference section on the link to the electronic web resource.

Understanding uncertainty and the ability to evaluate and understand uncertainty are important parts of any scientific measurement process, and quality-management standards [1] require laboratories to assess measurement uncertainty and attempt to: identify all significant components, make a reasonable estimate of the size of the total uncertainty and its variability and ensure that the reported results do not give a false estimate of uncertainty.

The basic uncertainties relating to the measurement of sound insulation have historically been assessed using the methods in BS 5725 [2,3] and result in several examples of round-robin measurement studies usually between test laboratory facilities, although there are others that look at on-site measurement in existing buildings [4]. However, identifying components that contribute to the total variability beyond "repeatability" and "reproducibility" is outside the scope of the international standard and the specific design of experiment (DoE) and the analysis method it employs.

8.2 Introduction

[There] are many apparent discrepancies on the published data on sound-insulation. These discrepancies may not be real but they are none-the-less responsible for a great deal of unfortunate, and unnecessary, confusion amongst architects, builders and even acoustical engineers. In the absence of satisfactory data, the inquirer may have doubts concerning the reliability of all published data on sound-insulation.

Vern O Knudsen, 1929 [5]

8.2.1 Statement of the Problem

The challenge of understanding and quantifying measurement uncertainty in the field is related to adopting an experimental approach that mimics the most common field-test situations, and in this there may be several scenarios, for example, a situation where a single operator measures the same separating element more than once ("repeatability") and/or more than one operator measures the same separating element at difference times, that is, on separate days with different test kits ("reproducibility"). There may also be a third situation where uncertainty may be higher, where two or more operators measure a different separating element(s) [part(s)] of the same construction on the same residential development site. This could happen if different test operators measure walls between different plots on different days, which is common on larger construction sites where a relatively large test sample is required and multiple visits are needed. In this case, the variability of the separating element or part being measured is added to the total uncertainty budget, and some account for this is needed. There is some evidence that the uncertainty relating to the part influences the total uncertainty of measurement and may have an uncertainty signature of its own [6,7].

In any event, it is key that the testing body develops their own budget for measurement uncertainty based on their own specific work instructions, test equipment and operators and they "understand" the uncertainty associated with each.

8.2.2 Airborne Sound Insulation Field Testing

The sound insulation test procedure used in this chapter's worked example in Section 8.6 follows both the British and international standard test procedures [8,9] and, in order to comply with the building regulations for England and Wales, *Approved Document E* (revised 2004) [10] guidance where the requirements differ. The positions of the loudspeaker and the microphone in the room and in relation to each other are prescribed in the relevant standards British Standard (BS) EN ISO 140 Part 4: 1998 [8] and BS EN ISO 354: 2003. It should be noted that because the building regulations *Approved Document E* (2003, revised 2015) [10] references BS EN ISO 140, Part 4 specifically, the later standard (current) BS EN ISO 16283–1:2014+A1:2017 [11] is not used.

8.3 Calculation of Measurement Uncertainty: Current Guidance

It is worthwhile to briefly look at the limitations of the two contemporary uncertainty assessment methods when the objective is to quantify and understand the components contributing to measuring sound insulation in the field. They are the *Guide to the Expression of Uncertainty in Measurement* (GUM) [12] and BS ISO 5725-2:1994, *Basic Method for the Determination of Repeatability and Reproducibility of a Standard Measurement Method* [3], taking the one used almost exclusively for sound insulation testing first (BS 5725).

8.3.1 BS 5725: Description and Limitations

BS 5725 uses an empirical approach that is easy to follow and is well documented in the literature, and it can be a starting point for any novice practitioner. Ideally, it should involve measuring the sound insulation performance of more than one element more than once with multiple test operators (or laboratories). Most of the relevant literature on measurement uncertainty in sound insulation centres on inter-laboratory studies and their quantification of "within-lab uncertainty" and "between-lab uncertainty", and BS 5725 is specifically designed to assess this "trueness and precision" for a whole range of scientific measurement disciplines. The main reason for assessing the measurement uncertainty in and between laboratories is to ensure that laboratory measurements fall within a reasonable pattern of variability so that there is no competitive advantage favouring one laboratory over another. The normal method of assessing measurement uncertainty is by carrying out inter-laboratory studies that follow an empirical method detailed in BS 5725 [2,3,13–15], and the method detailed in this standard can also be applied to field-testing situations if needed.

The procedure is well documented and well used, and there are several seminal inter-laboratory and round-robin studies that identify the repeatability and reproducibility components [16–24]. They use the BS 5725 assessment process for its intended purpose to determine the repeatability r and reproducibility R and to compare the measurements obtained by each participant laboratory with the reference values in International Organization for Standardization (ISO) 140-2 [25]. It is important to note that the ISO r and R reference values are the product of several inter-laboratory studies themselves, where a chosen element was reconstructed or remounted in each laboratory, measured and the r and R values pooled. These uncertainty reference levels have been updated and in some additional cases redefined (Situations A, B and C) in the latest international standard BS EN ISO 12999–1:2014 [26], which is covered later in this chapter (Table 8.1).

See Situation B, which is closest to our field-testing scenario, although it will not in itself allow the components of variance to be broken down beyond reproducibility. Situations A and C cover other test scenarios that commonly occur, although these are not analogous to our particular investigations.

Although this recent standard states general expected uncertainties for measurements undertaken in the scenarios defined in Table 8.1, it follows the general concept and the procedures in BS EN ISO 5725–1 and BS EN ISO 5725–2, respectively, and includes large inter-laboratory studies.

Most of the participants in inter-laboratory studies are national or commercial testing laboratories, although they can be mobilized to test at any external location, and this may be a requirement of the specific round-robin test method. Each laboratory has their own in-house test, equipment, facilities or consultants who can carry out testing at a specific location or in-house on a single test element. The test samples selected vary from lightweight partitions (e.g. see Farina et al. [16]) to heavyweight walls (e.g. see Luxemburg et al. [24]). The samples are ideally reconstructed from readily available homogeneous materials, or samples are transported between and remounted in each laboratory.

Table 8.1 Measurement Situation B: BS EN ISO 12999–1:1994

Situation A: A building element is characterized by laboratory measurements. In this case, the measurand is defined by the relevant part of ISO 10140, including all additional requirements, for example, for the measurement equipment and especially for the test facilities. Therefore, all measurement results that are obtained in another test facility or building also comply with this definition. The standard uncertainty thus is the standard deviation of reproducibility, as determined by inter-laboratory measurements.

Situation B: Described by the case that different measurement teams come to the same location to carry out measurements. The location may be a usual building or a test facility. The measurand thus is a property of one particular element in one particular test facility or the property of a building. The main difference to Situation A is that many aspects of the airborne and structure-borne sound fields involved remain constant because the physical construction is unchanged. The standard uncertainty obtained for this situation is called *in situ standard deviation*.

Situation C: Applies to the case where the measurement is simply repeated in the same location by the same operator using the same equipment. The location may be a usual building or a test facility. The standard uncertainty is the standard deviation of repeatability, as determined by inter-laboratory measurements.

Field Testing

There are also similar studies on field testing of sound insulation. These focus on existing buildings [4,20,27]. The studies are informative but only comparable if they follow the BS 5725 methodology. Closer inspection reveals deviations and inconsistencies in the test procedures, which can lead to discrepancies in the results. An example is the delta study by Hoffmeyer et al. [20], which undertook field measurements of separating walls between a pair of terraced houses. The reproducibility obtained showed good agreement with the reference values in ISO 140-2, but further scrutiny showed that to reduce the uncertainty caused by differences in test equipment, the five participating test laboratories in some instances used the same test kit. More importantly, given the time constraint, it was not possible to repeat all measurements. This meant that repeatability was not included in the reproducibility value, as required by BS 5725. It therefore underestimated the value of *R* and probably was the prime reason the reproducibility was lower than the reference values. It is also noted that the reproducibility was calculated for each room measured. The single test specimens therefore were *identical*, not only of similar construction or made of the same material. The reproducibility therefore did not incorporate the variability due to the reconstruction or remounting of the part, as it would have to be in a laboratory test situation. It therefore underestimated the true reproducibility when applied to measurements of separating elements constructed in the field or indeed reconstructed or remounted in the laboratory (see Loverde et al. [28]). Comparison with the reference values in ISO 140-2 is erroneous. See Lang and also Hall [4,29], where this also occurs. Similar limitations, where the test specimens are identical, also occur in other research studies [30]. Their impact on the reproducibility may be acknowledged, but often it is ignored either because it is thought to be insignificant or perhaps because it is not understood. The point here is that the uncertainties calculated in these studies, because of the limitations of the DoE and analysis procedure, do not necessarily accurately reflect the specific test situation of the measurement body that is carrying out pre-completion testing on site or "in the field". This demonstrates that care is needed when attempting to draw direct comparisons between research on uncertainty using BS 5725, and the limitations of the standard and the DoE applied by the testing laboratories to the standard need to be understood. This chapter attempts to show how these anomalies can be addressed.

8.3.2 Guide to the Expression of Uncertainty in Measurement (GUM)

An alternative to the empirical method described in BS 5725 is described in the GUM [12]. A UK equivalent has been produced by the United

Kingdom Accreditation Service (UKAS) in their Document M3003 [31]. The method is based on modelling the uncertainties by constructing a combined budget that contains all input variables likely to contribute to the uncertainty in the measurement process. This method leads to the development of a comprehensive list of factors, and these factors are often referred to as "input variables." The GUM input variables likely to contribute to the combined uncertainty can be sorted under respective headings. These general uncertainty headings and the subset of associated input variables can be illustrated graphically for sound insulation in Figure 8.2 based on a simple cause and effect diagram. See also figure 8.1 of Ellison et al. [32].

Input Variables

The variables that contribute to this are many and may be difficult to quantify individually. Some are described in Table 8.2 as examples.

There are also issues with autocorrelation. Wittstock used GUM in 2005 [48] to predict measurement uncertainty and exposed autocorrelation issues between adjacent third-octave bands as a possible issue in using GUM, which assumes independence. This raised questions over the usefulness of GUM because correlation effects between third-octave bands were found to dominate the measurement uncertainty of the single-figure

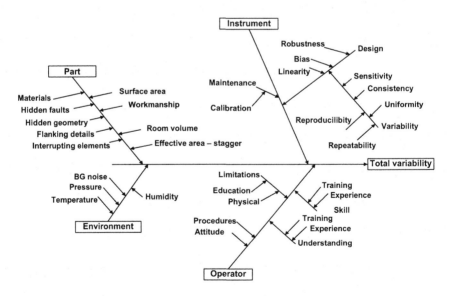

Figure 8.1 Cause and effect diagram for sound insulation testing in the field: uncertainty headings and individual input variables.

Table 8.2 Research based on individually identified variables contributing to the components of variance in measurement

Meteorological conditions: temperature, barometric pressure, humidity	The influence of temperature on measurement was highlighted by Scholes [33] and together with barometric pressure was the subject of recent research by Wittstock et al. [34]. Humidity effects are provided in manufacturers' information for the microphone (e.g. see the B&K Handbook [35]). Usually meteorological condition effects can be minimised by ensuring that the measurements are done over a short time period while the conditions are stable.
Room effects, spatial variation, discrete versus continuous sampling	The uncertainty resulting from these influences is relatively large, though predictable. Predicting the expected variability of sound pressure level is useful in assessing the consistency and reliability of the data obtained on site (see Schroeder [36], Waterhouse [37–41], Lubman [39,42–44] and Craik [45]).
Workmanship	This variability, referred to as "workmanship" by Craik et al. [46,47], was calculated for a simple concrete floor, although additional information on other, more complex constructions is not readily available.

ratings [49]. This evidence, together with its apparent tendency to significantly overestimate the measurement uncertainty shown by Lyn et al. [50], suggests that GUM does not provide a suitable framework to assess uncertainty in sound insulation testing.

Thus, to summarise, the DoE used in BS 5725 and reproduced in several large-scale studies does not necessarily provide representative measurement uncertainty information for testing sound insulation in the field, nor does it provide any detail on the uncertainty contribution from other factors, even though the experimental burden in collecting the data is high. Other recent evidence shows that the modelling method used in the GUM appears to be unsuitable for the assessment of measurement uncertainty in sound insulation.

8.4 Calculation of Measurement Uncertainty: Alternative Approach

8.4.1 Analysis of Variance (ANOVA)

An alternative analytical approach used in this chapter is careful to avoid adding to the data-collection burden but uses ANOVA and a specific DoE to identify the components of variance. Although examples are extremely rare, ANOVA has been used previously to good effect in acoustical research, two good illustrations of which are a laboratory sound insulation study by Taibo et al. [51] and a round-robin study on the measurement of absorption coefficients by Davern et al. [52,53]. The main advantages of ANOVA are listed by Deldossi et al. [54] and include the ability to determine the contribution of the operator, the part being measured and operator interaction. The special ANOVA method used is called a "gauge repeatability and reproducibility study" (GRR) and is used extensively in industry, particularly by the Automobile Industries Action Group (AIAG). The appropriate GRR, for us, is known as a "balanced two-factor crossed random model with interaction" (see Burdick et al. [55]). It is this model and additional information on experimental design provided by Montgomery [56–58], Borror [59] and Burdick [60,61] that form the analytical framework to isolate and quantify the components of variance in sound insulation measurement.

8.4.2 Gauge Repeatability and Reproducibility (GRR)

The GRR method relies on a number of gauge "operators" to measure a number of test specimens a repeated number of times. The DoE is structured to enhance the statistical robustness of the data, improve the validity and efficiency of the study and minimise the effect of systematic bias. Montgomery [62] lists three main elements in a DoE for GRR, as shown in Table 8.3.

The preferred GRR contains a "balanced, crossed" design. That is, every level of one factor is run with every level of another factor (crossed), and each measurement is repeated the same number of times (balanced); for

Table 8.3 Definition of elements in design of experiments

Replication: This is the repetition of the experiment and involves a refresh of the measurement procedure, for example, not only pressing the Record button on the sound level meter in the same position. It is expected that the setup of the meter in the room and the loudspeaker position and sound pressure level will alter as they would if you were measuring the same wall at a different time on the same day. It defines the experimental error and represents the repeatability of the measurement instrumentation.

Randomization: This is the basis of any robust statistical experiment. It means that each operator should measure the parts in a random order. Randomising reduces systematic bias in the experimental data

Blocking: This is a technique used to minimise nuisance-factor variability. For example, to reduce the variability due to room dimensions, element surface area and room volume affecting the total variability, the same room shape and volume are selected and "blocked", that is, controlled, for the test pairs. Meteorological conditions can also be blocked, if the testing can take place over a relatively short time scale where the test environmental conditions are unlikely to change significantly.

example, every part in the test sample is measured by every operator the same number of times.

8.4.3 The Model

The classical GRR study is a balanced two-factor crossed random effects model with interaction

$$Y_{ijk} = \mu + o_i + p_j + (op)_{ij} + E_{k(ij)} \tag{8.1}$$

The measurement by operator i on part j at replication k is denoted as Y_{ijk}, where $i = 1, 2, ..., p$; $j = 1, 2, ..., o$; $k = 1, 2, ..., r$; p = number of parts, o = number of operators and r = number of repetitions; μ is an overall mean common to all observations; and o_i, p_j, $(op)_{ij}$ and $+R_{k(ij)}$ are random variables representing the effects of the operator, parts, operator-by-part interaction and replications on the measurement. It is assumed that the operator, parts, operator-by-part interaction and the replication are independent random effects that are normally distributed with zero means and variances of σ_o^2, σ_p^2, σ_{op}^2 and σ_R^2, respectively, which are assumed to be constant. Assuming the mean to be zero implies the measurement system is unbiased; that is, the variability in the measurement system is due to precision only. Assuming the variances to be constant indicates that the variability of the system does not change with the magnitude of the measurement. The variance terms are defined as

$$\sigma_{r^2} = \text{repeatability (instrument variance)}$$

$$\sigma_{R^2} = \text{reproducibility (operator variance)}$$

which also can be rewritten as

$$\sigma_{\text{reproducibility}}^2 = \sigma_o^2 + \sigma_{op}^2 \tag{8.2}$$

$$\sigma_{\text{gauge}}^2 = \sigma_{\text{repeatability}}^2 + \sigma_{\text{reproducibility}}^2 \tag{8.3}$$

(see section 8.7.2 of [63])

$$\sigma_{\text{total}}^2 = \sigma_{\text{gauge}}^2 + \sigma_{\text{part}}^2 \tag{8.4}$$

$$\sigma_{\text{reproducibility}}^2 = \sigma_o^2 + \sigma_{po}^2 \tag{8.5}$$

Because reproducibility may also contain variance associated with operator by part (interaction), note that the interaction may not be present and therefore will be zero.

$$\sigma_{part^2} = \text{part-to-part variance}$$

$$\sigma_{op^2} = \text{operator-by-part variance}$$

$$\sigma_{guage^2} = \text{total gauge variance} = \sigma_r^2 + \sigma_R^2$$

$$\sigma_{total^2} = \text{total variance} = \sigma_r^2 + \sigma_R^2 + \sigma_{part}^2$$

Interaction

The presence of interaction in the measurement process is not considered in BS 5725 [2,3,13,14], UKAS Document M3003 [31] or GUM [12], although its identification and quantification are a requirement of the *European Accreditation (EA) Guidelines on the Expression of Uncertainty in Quantitative Testing* [64].

Interaction is a joint factor effect. When interactions occur, the factors involved cannot be evaluated individually. For example, an interaction occurs where a drug is given in combination with or shortly after another drug. This alters the effect of one or both drugs. In this research study, it may occur where a room type, shape or size means that an operator sets up the speaker and microphone positions in a (non-random) fixed way that is different from that of another operator. It affects the measurement and applies a particular systematic bias to some parts and not to others. One of the challenges of the DoE for any experiment is to arrive at a test sample which minimises the risk of this occurring, although the effects, if not drawn out, may in fact remain hidden and be significant.

The knowledge that there is interaction between factors can be more useful than knowledge of the main effect itself, and it is a strength of the ANOVA DoE that these effects can be included in the assessment. Presently, they are undetectable using the methods in BS 5725 and the GUM. Indeed, if there is significant interaction between factors, increasing the sample size in a traditional study will not improve the analysis results, as would normally be expected. Instead, the size of the reproducibility will be underestimated [56].

8.5 Design of Experiment: GRR

The field survey plan, method of measurement, data collection and choice of factors are described.

8.5.1 Experimental Format

The balanced two-factor crossed random model with interaction is the most appropriate GRR for our use. In essence, a number of operators measure a chosen number of parts a number of times, and everyone measures the same parts. There is a potential, with careful design, to extract the variability in the measurement process and, importantly, the variability due to the part itself. The test designs is summarised in Table 8.4.

For this experiment, each operator used his or her own test kit, which replicates a real reproducible test situation and additional variability. The five test kits were the similar and included sound level meters, calibrators, loudspeakers and wireless transmitters/receivers.

The selection of six floor elements, with three repetitions, was set by the time constraints on site, and six was the maximum number of parts that could be tested by all five operators once on the same site in a day. The number of operators, test sample and repetitions complied closely with the recommended robust GRR design suggested by Burdick et al. [55]. In all, the repetitions of six floor tests three times each took three days, with a total test time over the three days of 30 hours for each tester. The total site survey time for this experiment therefore was 150 person-hours – not a small undertaking.

8.5.2 Choice of Test Sample Construction

The example test site had a commonly used lightweight timber separating floor construction, which is described in Table 8.5.

Blocking

To minimise unwanted factors and attempt to standardise results by "part", the test samples were "blocked". The meteorological conditions were blocked by ensuring that all the testing took place over a short time period (nominally three days). There was minimal variation in internal conditions in the test rooms, that is, within ±4°C and ±20% relative humidity (RH) for the duration of the experiment.

The test room shape and size were fixed dimensionally by testing the same bedroom in identical flat-type pairs (33 m^3). By fixing the room sizes, the variability of the performance of the floor cannot be due to dimensional changes or differences in room volume. It is then due to the

Table 8.4 Testing schedule: Lightweight timber floor

Operators	Parts	Repetitions
5	6	3

Table 8.5 Test floor construction description

Floating floor	18-mm (min) tongue and groove flooring board Gypsum-based board nominal 13.5 kg/m² FFT-1 resilient composite deep battens Battens may have the resilient layer at the top or the bottom Mineral wool quilt laid between battens – 25 mm (min) 10–36 kg/m³
Structural	Floor desk: 18-mm-thick (min) wood-based board, density min 600 kg/m³ 253-mm (min) metal web joists
Ceiling	CT1 ceiling: two layers of gypsum-based board composed of 19 mm (nominal 13.5 kg/m²) fixed with 32-mm screws and 12.5 mm (nominal 10 kg/m²) fixed with 42-mm screws Resilient bar: 16-mm (min) metal resilient ceiling bars mounted at right angles to the joists at 400-mm centres (bars must achieve a minimum laboratory performance of $r_d R_w + C_{tr} = 17$ dB and $r_d Lw = 16$ dB) 100-mm (min) mineral wool quilt insulation (10–36 kg/m³) between joists

variability of the onsite construction, and the part-to-part variation will reflect that provided by the method of construction. See the influence of "workmanship" in Craik et al. [46,47].

Response Variable

The response variable in this case is the airborne sound insulation value, otherwise known in metrology as the "measurand". This experiment evaluates the measurement uncertainty of the full measured frequency range (100 Hz to 3.15 kHz) using the standardised level difference D_{nT} [65] in each of the 16 third-octave bands. An assessment of the commonly used single-figure airborne value [66] $D_{nT,w}$ and $D_{nT,w} + C_{tr}$ will also be carried out.

8.6 Results

8.6.1 Initial Data Analysis: Worked Example

The timber floor single-figure sound insulations [90 tests in total (6 × 5 × 3)] are summarised in Table 8.6 with respect to the current building regulations [10] for airborne sound insulation performance $D_{nT,w} + C_{tr}$.

Table 8.6 Mean and standard deviation: Single-figure descriptors for timber floor tests

Timber	$D_{nT,w}$	$D_{nT,w} + C_{tr}$
Mean	62.6	53.2
Standard deviation	1.0	2.3

The third-octave band frequency data for the timber GRR experiment can be compared in the same way as for the single-figure values. The mean sound insulation for the third-octave bands 100–3,150 Hz is detailed in Table 8.7. The third-octave band D_{nT} values are graphically illustrated in Figure 8.2.

8.6.2 GRR Results (Note: See Website Links for Full Data Appendix and Analysis)

The GRR data (standard deviation and variance are both in decibels) is presented in Table 8.8 for the timber separating floor (dB $D_{nT,w}$).

The ANOVA results show that the instrumentation (represented by "repeatability" σ_r) is responsible for 0.56 dB of the total standard deviation of the results; the operator (represented by σ_o) is responsible for 0.70 dB of the total standard deviation. The standard deviation of the operator in this case is equivalent to the reproducibility σ_R as the interaction term (σ_{po}) is zero. The part (timber floor, represented by σ_p) is responsible for 0.56 dB of the total standard deviation. The instrumentation and the part are contributing similar amounts of uncertainty (0.56 dB) to the total uncertainty relating to $D_{nT,w}$, but the biggest contributor is the operator in this case.

For the $D_{nTw} + C_{tr}$ single-figure value, the results are influenced by the low-frequency spectrum adaptation term:

$$D_{nT,w} + C_{tr}$$

See Table 8.9.

Table 8.7 Standardised level difference D_{nT} (90 test samples): Timber floor

D_{nT}, dB	100 Hz	125 Hz	160 Hz	200 Hz	250 Hz	315 Hz	400 Hz	500 Hz	630 Hz	800 Hz	1.0 kHz	1.25 kHz	1.6 kHz	2.0 kHz	2.5 kHz	3.15 kHz
Timber (mean)	35.3	43.6	46.0	49.6	53.3	56.5	59.4	62.1	63.8	64.9	65.8	70.3	72.7	69.5	67.2	73.4

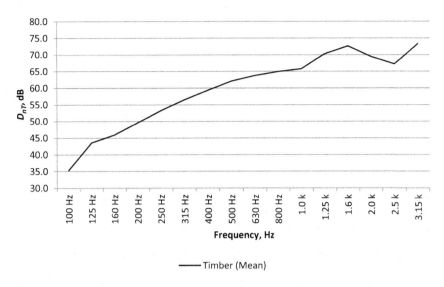

Figure 8.2 Mean D_{nT} values for timber floor samples 100–3,150 Hz.

Table 8.8 Timber lightweight floor: Major components of variance ($D_{nT,w}$)

dB=(σ^2)	σ_{GRR}^2	σ_r^2	σ_R^2	σ_o^2	σ_{po}^2	σ_p^2	σ_{total}^2
$D_{nT,w}$	0.810	0.317	0.493	0.493	0.000	0.316	1.126
dB=(σ)	σ_{GRR}	σ_r	σ_R	σ_o	σ_{po}	σ_p	σ_{total}
$D_{nT,w}$	0.900	0.560	0.700	0.700	0.000	0.560	1.060

where

$$\sigma_{r^2} = \text{repeatability (instrument variance, dB)}$$

$$\sigma_{R^2} = \text{reproducibility (operator variance, dB)}$$

$$\sigma_{p^2} = \text{part-to-part variance, dB}$$

$$\sigma_{po^2} = \text{operator-by-part variance, dB}$$

$$\sigma_{GRR^2} = \text{total gauge variance, dB} = \sigma_{r^2} + \sigma_{R^2}$$

(see section 8.7.2 of [63]; note that in the timber case, $\sigma_{po^2} = 0$ for single-figure values).

The part-to-part component (represented by σ_p) for $D_{nT,w} + C_{tr}$ has a standard deviation of 0.49 dB. This is similar to the $D_{nT,w}$ single-figure value for the part-to-part variability (0.56 dB).

Table 8.9 Timber lightweight floor: Major components of variance $D_{nT,w} + C_{tr}$.

dB=(σ^2)	σ_{GRR}^2	σ_r^2	σ_R^2	σ_o^2	σ_{po}^2	σ_p^2	σ_{total}^2
$D_{nT,w} + C_{tr}$	5.870	1.440	4.430	4.430	0.000	0.240	6.110
dB=(σ)	σ_{GRR}	σ_r	σ_R	σ_o	σ_{po}	σ_p	σ_{total}
$D_{nT,w} + C_{tr}$	2.420	1.200	2.110	2.110	0.000	0.490	2.470

The ANOVA results show that the instrumentation (representing "repeatability" σ_r) is responsible for 1.20 dB of the total standard deviation of the results; the operator (represented by σ_o) is responsible for 2.11 dB of the total standard deviation. Both these results show at least twice the variability of the $D_{nT,w}$ single-figure value for the instrument and the operator, which follows the basic analysis results for timber floors where the relatively poor low-frequency performance increases the standard deviation when the spectrum adaptation term is applied. The standard deviation of the operator in this case is equivalent to the reproducibility σ_R because the interaction term (σ_{po}) is classed as not statistically significant and set to zero. The part-to-part variability (represented by σ_p) is responsible for 0.49 dB of the total standard deviation. The operator is still the major contributor to the overall measured variability, with a standard deviation of 2.11 dB. It is the operator's choices made in the equipment setup and how they interpret the standard test procedure that affect this component-of-variance term.

It is possible to cardinally rank the importance of the components depending on size. For both single-figure values on lightweight timber floors, the operator is the factor that contributes most to the uncertainty, followed by the instrument and then the part.

Ideally, for a GRR study, the part measured would contribute the most uncertainty, and the measurement system components, that is, the operator and instrument contributions, would be low by comparison.

Where the GRR is based on a constrained sample, the part-to-part variability is likely to be small because the floor and room size have been selected to be similar, thereby minimising the potential variability. Normally, under GRR DoE, a sample with the full measurement range of the instrumentation would be chosen [56]. The relatively large variability in percentage terms of the instrumentation and the operator is therefore understandable. The variabilities, as measured in decibels, are significant because they show that the overall variability (σ_p) due to the part-to-part variability of the construction (workmanship) is 0.56 dB $D_{nT,w}$ and 0.49 dB $D_{nT,w} + C_{tr}$, which is relatively small compared with the total variability (σ_{total}) of 1.06 dB $D_{nT,w}$ and 2.47 dB $D_{nT,w} + C_{tr}$. This shows

Table 8.10 Measurement variability due to defined factors: Ordered by magnitude

(Timber floor) – measurand		$D_{nT,w}$, dB	$D_{nT,w} + C_{tr}$, dB
Order	Factor		
1	Operator σ_o	0.7	2.1
2	Instrument σ_r	0.6	1.2
3	Part σ_p	0.6	0.5

that the sound insulation performance of the floors is consistent when based on a single-figure value.

8.6.3 Frequency Data: Results

ANOVA is performed across the frequency range for the D_{nT} values. These are summarised in Tables 8.11 and 8.12, and the variances are plotted in Figure 8.3. This gives a much more detailed view of where the major regions of variability in the data lie and also indicates that there is interaction present at some frequencies, albeit relatively minor.

In Figure 8.3, the total variance for the 16 third-octave bands is characterised by a U-shaped curve showing greater variance at the low and high ends of the frequency range. There is a clear indication that the gauge, represented by the total variance minus the part-to-part contribution, is the primary component responsible for the high variance at the low-frequency bands 100 and 125 Hz. The contribution from the part is small in comparison. From 160 to 315 Hz, the contribution to total variance from the gauge is slightly higher; between 400 and 630 Hz, the variance of both is virtually equivalent and relatively small. From 800 to 1,250 Hz, the part contributes more to the total variance than the gauge; at 1,600 Hz, the part contribution drops, and the gauge contribution rises significantly to dominate the contribution, as well as for the 2,000- and 2,500-Hz bands. At 3,150 Hz, the part and the gauge variances rise significantly, and both contribute a similar amount to the total.

The important conclusions to be drawn from this graphical representation of the components of variance are at the 100- to 125-Hz and 3,150-Hz bands. The part contributes very little to the overall variance at low frequency, and in general, the part variability was indistinguishable from the variance contribution of the instrument (repeatability).

Table 8.11 Timber lightweight floor (variance): Major components of variance frequency data (D_{nT})

D_{nT} (var), Hz	σ_{GRR}^2	σ_r^2	σ_R^2	σ_o^2	σ_{po}^2	σ_p^2	σ_{total}^2
100	12.244	3.742	8.502	8.502	0.000	0.362	12.607
125	4.612	2.031	2.582	2.091	0.491	0.000	4.612
160	1.013	0.921	0.091	0.091	0.000	0.670	1.683
200	1.907	1.551	0.356	0.356	0.000	1.212	3.119
250	1.580	0.787	0.793	0.793	0.000	0.871	2.451
315	0.992	0.432	0.560	0.560	0.000	0.343	1.335
400	0.450	0.313	0.136	0.136	0.000	0.424	0.873
500	0.494	0.317	0.177	0.177	0.000	0.287	0.781
630	0.571	0.342	0.230	0.230	0.000	0.525	1.096
800	0.418	0.359	0.059	0.059	0.000	1.317	1.735
1,000	1.057	0.855	0.201	0.201	0.000	3.578	4.634
1,250	2.224	0.770	1.454	1.454	0.000	3.466	5.690
1,600	6.424	1.584	4.840	4.840	0.000	0.828	7.252
2,000	4.016	1.329	2.687	2.687	0.000	0.996	5.012
2,500	4.111	1.079	3.033	2.539	0.494	2.810	6.921
3,150	9.750	2.493	7.257	6.626	0.631	10.054	19.804

Table 8.12 Timber lightweight floor (standard deviation): Major components of variance frequency data (D_{nT})

D_{nT} (standard deviation), Hz	σ_{GRR}	σ_r	σ_R	σ_o	σ_{po}	σ_p	σ_{total}
100	3.50	1.93	2.92	2.92	0.00	0.60	3.55
125	2.15	1.43	1.61	1.45	0.70	0.00	2.15
160	1.01	0.96	0.30	0.30	0.00	0.82	1.30
200	1.38	1.25	0.60	0.60	0.00	1.10	1.77
250	1.26	0.89	0.89	0.89	0.00	0.93	1.57
315	1.00	0.66	0.75	0.75	0.00	0.59	1.16
400	0.67	0.56	0.37	0.37	0.00	0.65	0.93
500	0.70	0.56	0.42	0.42	0.00	0.54	0.88
630	0.76	0.58	0.48	0.48	0.00	0.72	1.05
800	0.65	0.60	0.24	0.24	0.00	1.15	1.32
1,000	1.03	0.92	0.45	0.45	0.00	1.89	2.15
1,250	1.49	0.88	1.21	1.21	0.00	1.86	2.39
1,600	2.53	1.26	2.20	2.20	0.00	0.91	2.69
2,000	2.00	1.15	1.64	1.64	0.00	1.00	2.24
2,500	2.03	1.04	1.74	1.59	0.70	1.68	2.63
3,150	3.12	1.58	2.69	2.57	0.79	3.17	4.45

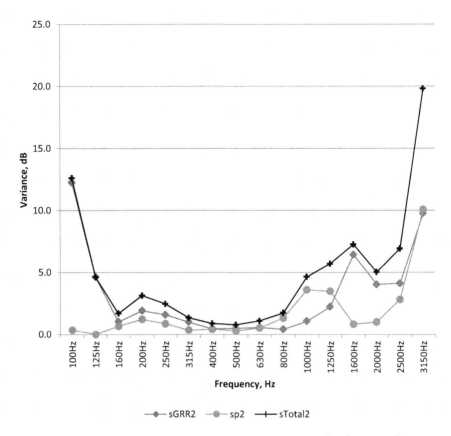

Figure 8.3 Timber lightweight floor: Components of variance σ_{GRR}^2, σ_p^2, and σ_{total}^2.

At 100–125 Hz, it is the gauge (combining both r and R, repeatability and reproducibility) that is responsible for the high level of variance, not the construction of the floor. The variance of the gauge also reflects the expected variance at the low-frequency part of the spectrum, where there is likely to be a non-diffuse sound field that can skew the sampling of measured levels in the room.

At high frequency, where there is a significant rise in variability in the measured data, both the gauge and the part contribute equal amounts. There may be a combination of factors contributing to the relatively high variance levels. Although these are only suggested reasons, they do have a good basis in logic given knowledge of typical site test conditions and the limitations of the test kit. The part variance may be due to the high performance of the timber floor at high frequency. This, combined with high levels of background noise on site, means that there are less reliable measurements of sound coming through the test element, resulting in increases in variance after 2,000 Hz. From the perspective of the gauge variability contribution, this could be interpreted to mean that the test kit could not output sufficient sound power or the background noise is increasingly influential in dominating the receive-side measured levels, resulting in increased variance in the data measured.

The gauge contribution can be broken down further by looking at the repeatability, reproducibility, operator and interaction terms individually. These are plotted in Figure 8.4.

Both repeatability and reproducibility components follow a U-shaped profile, although the repeatability contribution to the gauge overall variance

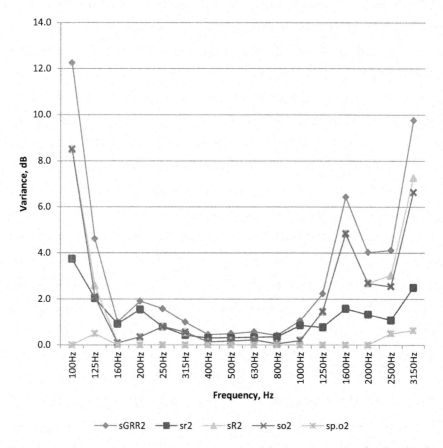

Figure 8.4 Lightweight floor: Components of variance (timber GRR): $\sigma_{GRR}{}^2$, $\sigma_r{}^2$, $\sigma_R{}^2$, $\sigma_o{}^2$ and $\sigma_{po}{}^2$.

is generally lower than the reproducibility contribution, indicating that the instrumentation used contributes least to the uncertainty in overall measurement in this experiment.

The reproducibility contribution is dominated by the operator component, and apart from a couple of minor interaction effects, the reproducibility curve duplicates the operator curve. Repeatability is dominant at 160–200 Hz; between 250 and 800 Hz, the r and R contributions are similar, and both r and R contributions rise after 1,000–1,250 Hz, but R is dominant. The increase in variance after 1,250 Hz for both repeatability and reproducibility is likely to be due to the influence of background noise.

Gauge Variability

The gauge component of variance σ_{GRR^2} is the repeatability and reproducibility components combined. It also encompasses any interaction that may be present between the part and the operator factors. The gauge variance is plotted for the timber GRR in Figure 8.5.

Visual inspection of the shape of the gauge variance curve indicates that the reproducibility variance component makes a substantial contribution and dictates the shape of the curve. The timber gauge variance is affected by the non-diffuse field in the test rooms at low frequency and by the

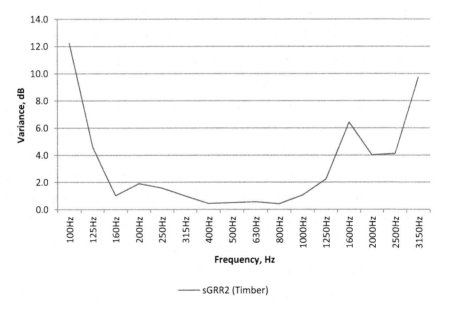

Figure 8.5 Timber gauge variance contribution.

influence of relatively high background noise and/or a combination of inadequate noise output from the test kit and high-performing floors. Between the range 160 and 1,000 Hz, the timber gauge variance is 0.4 and 1.9 dB, which is relatively low.

For the timber floor experiment σ_{GRR^2}, the reproducibility variance is most influential at 100, 125 and above 1,250 Hz; repeatability is dominant at 160, 200 and 1,000 Hz and is similar for all other frequencies. See Figure 8.6.

8.6.4 Part-to-Part Variance

Apart from quantifying the repeatability and reproducibility, the GRR data set has provided insights into other factors and their contributions to uncertainty. The part being measured has a significant contribution, even if building elements are nominally identical. This has repercussions when comparing results with the guideline reproducibility values in the international standard [25].

Apart from instrumentation variability, which is not reliant on the test sample being measured, a true comparison of reproducibility with the international standards' R [25, table A.2] must incorporate the variance of the part. This is because the variance of the part is implicit in the values for R in the standard. In inter-laboratory tests, the part-to-part

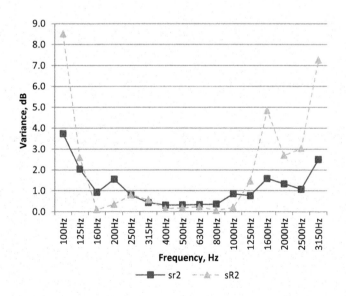

Figure 8.6 Timber GRR R and r variance.

variance results from the reconstruction of the test sample at each location (or indeed at the same location; see Loverde and Dong [28]), for example, the re-fixturing of the glazing test specimen in the laboratory wall opening or the rebuilding of the walls in each of the laboratories using standard blocks or plasterboard. This ensures that the part-to-part variability is included in the data sample.

For the timber floor, the room shape, size and volume were identical in each of the six test pairs. This allowed an assessment of the part variability resulting from the floor construction. It is shown that the floor construction contributes to the total variance measured. For timber floors, the part-to-part variance was calculated as 0.24 dB for $D_{nT,w} + C_{tr}$, with one standard deviation of 0.5 dB. The part-to-part contribution to total variance across the frequency range 100–3,150 Hz varied significantly with frequency and was between 0 and 10 dB.

This is one of the reasons the timber GRR data are more informative, because it allows the total variance to be partitioned into component parts and details the frequency region(s) where the variance caused by the replication of the construction of the part is influential. It also aligns the test scenario with the one conducted in the international standard to determine guideline values for repeatability and reproducibility. For example, the highest part-to-part variance for the timber GRR was in the 3,150-Hz band (10.1 dB), but this relatively large variance, the highest in any frequency band, was only 51% of the total variance at this frequency. The contribution of the part-to-total variance at 800 and 1,000 Hz was 76% and 77%, respectively. The part is the most influential component of variance in this region, but the variance measured in decibels was relatively low in comparison, 1.3 and 3.5 dB, respectively. If the parts or test specimens are the same (in this case, same construction and room size) and the part is seen to be the dominant element in the total variance measured, then improvements in instrumentation or measurement technique will not result in a significant reduction in the total measurement uncertainty on site.

8.7 Current Guideline Values: Standard Uncertainties

8.7.1 BS EN ISO 12999

BS EN ISO 12999, Part 1: *Sound Insulation* [67], references the GUM and follows current conventions in defining the "standard uncertainty" relating to testing sound insulation in laboratories. This is a different descriptor to that used in ISO 140-2 as it is a standard deviation, not a variance term, so a direct comparison with the ISO standard curves cannot be made without an appropriate correction, that is, taking the square root of the variance to compare it with the standard deviation (uncertainty) term.

The timber floor data are superimposed on the ISO 12999 guideline values for (*sr*) repeatability ("Max" and Situation C from ISO 12999) in Figure 8.6 and (*sR*) reproducibility given by Situation A from ISO 12999 (see Figures 8.7 and 8.8).

For repeatability, the timber floor GRR experiment data are below the maximum value curve apart from the 3,150-Hz band, where measurements again are affected by site background noise. In the timber floor GRR, the repeatability levels in some frequency bands are relatively low, and in several third-octave bands, *r* values are similar to those represented by Situation C at 160 and 315–800 Hz. Although measured standard uncertainties are higher than those described by any of the Situations A, B or C, the maximum standard should not necessarily be seen as incorrect.

The timber floor gives relatively good agreement with the ISO 12999 values for reproducibility apart from the 100-Hz band (which is possibly due to non-diffuse field and room mode effects) and above 1,600 Hz (where site background noise is more likely to influence the measured levels on site).

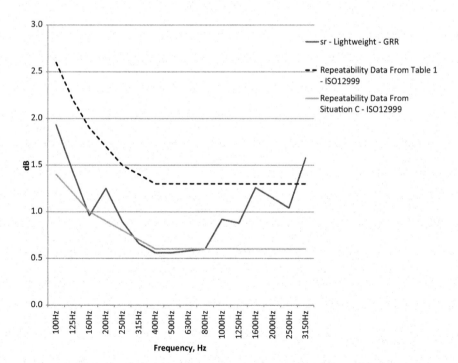

Figure 8.7 Repeatability comparison: Timber floor values for *r* and Situation C, ISO 12999.

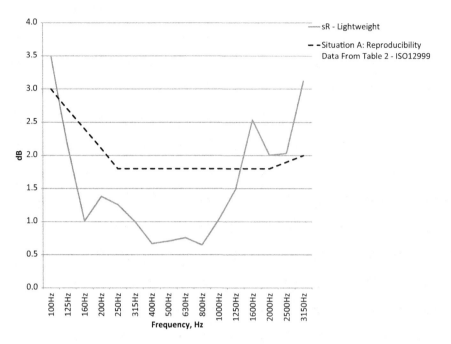

Figure 8.8 Repeatability comparison: Timber floor values for *R* and Situation A and maximum, ISO 12999.

It should be noted that ISO 12999 recommends noting levels that are above the curves. It only advises considering a higher result as invalid if an error in measurement occurred or an instrument or measurement fault is found, for example, if after the measurement the sound level meter was calibrated outside the tolerance range.

The GRR data generally give good agreement with the standard uncertainties detailed in the draft ISO 12999, Since the correct measurement and calibration procedures were followed by all operators, there is no reason to eliminate data that are above the guideline value curve, and increases in the standard deviation of results generally can be explained by the test conditions that prevailed on site. It should not be seen as a failure if there are test uncertainty values that are above the reference curves in the standard unless there is reason to suspect an error or malfunction in the test kit used

8.7.2 Single-Figure Values

The ISO 12999 draft also presents standard uncertainties for single-number values, and Situations A, B and C are used to define the experimental test scenario. The reproducibility is described by Situation A, and repeatability is described by Situation C. There is no maximum standard for repeatability detailed for the single-figure values (Tables 8.13 and 8.14).

There is no straightforward mathematical relationship between the third-octave-band sound pressure levels measured and the single-figure descriptor, and it does not necessarily follow that a data set that has lower standard uncertainties in all frequencies leads to a single-figure descriptor that has a lower standard uncertainty. Conversely, third-octave-band standard uncertainties that are higher than the ISO 12999 values do not necessarily result in a single-figure descriptor with a higher standard uncertainty. This means that two variable factors, namely $D_{nT,w}$ and C_{tr}, when added together can result in a descriptor with lower variability.

The timber floor repeatability data have individual third-octave bands that match Situation C values, but others are significantly above them. The single-number values calculated from these data are above the Situation C levels. This was expected, as BS EN ISO 12999 provides no maximum standard single-number values, as it does for the third-octave-band values.

8.8 Conclusions

8.8.1 GRR ANOVA

The available literature highlights the problems with using GUM to calculate the components of measurement uncertainty when testing airborne sound insulation in the field. The method described in BS 5725 is also inadequate, being unable to allow any partitioning of the reproducibility variability.

Table 8.13 Standard uncertainties: Reproducibility of single-number values, ISO 12999, timber floors

Single-number value: Reproducibility	Situation A, dB	Timber, dB
$D_{nT,w}$	1.2	0.7
$D_{nT,w} + C_{tr}$	1.4	2.1

Table 8.14 Standard uncertainties: Repeatability of single-number values, ISO 12999, timber floors

Single-number value: Repeatability	Situation C, dB	Timber, dB
$D_{nT,w}$	0.4	0.6
$D_{nT,w} + C_{tr}$	0.5	1.2

An alternative ANOVA design, coupled with a careful DoE, optimised the information gathered from field testing that was constrained by time. An example GRR was carried out on a lightweight timber floor construction. For the timber floor GRR experiment, the part was the least influential component of variance. The was due to the room sizes being blocked intentionally in order to scrutinise the variability resulting from construction of the floor element alone. The dominant factor in the timber GRR was the operator, and the standard uncertainty for this was three times greater for $D_{nT,w} + C_{tr}$ (2.1 dB) than it was for $D_{nT,w}$ (0.7 dB). The instrumentation also showed greater variability for the $D_{nT,w} + C_{tr}$ single-figure value (1.2 dB) as opposed to 0.6 dB for $D_{nT,w}$.

The presence of interaction can also be investigated using ANOVA, although in the case presented the interaction term was generally insignificant. The fact that it can be determined and its presence calculated gives this analysis a distinct advantage over both the GUM and BS 5725 DoEs.

The causes of interaction between the part and the operator are not always obvious but are likely driven by the operator and the choices the operator makes during the on-site survey/measurement process. Some suggested possibilities centre on operators trying to work as efficiently as possible during the survey; economising on the effort expended and coordinating their actions to take the minimum time between tests. This is likely to include the choice of test kit placement in the test room due to room geometry, power socket locations, windows, and so on that may affect operator choice and constrain the test method. As the test regime is highly regimented and repetitive, it may also be possible that after the operators "fix" on a certain arrangement for the microphones and the loudspeaker when visiting the same or similar rooms, it could even be related to the operators being left or right handed or which way a door opens into a room presenting two corners to operators as they carried heavy and ungainly equipment into the test space.

The significance of discovering interaction reinforces the conclusion that for the measurement of sound insulation in the field, GUM is not a suitable method for determining measurement uncertainty because the independence of input variables assumption is violated. It also suggests that the results of uncertainty studies that have used computer models based on GUM should be viewed with caution, unless some allowance has been made in the model for the interaction value.

In addition to interaction effects, the ANOVA also allows for an assessment of the contribution to the total uncertainty of measurement by the part, in this case the timber separating floor. This is a valuable piece of information because on-site test results are always likely to offer greater apparent variability to those acquired in the laboratory, and this provides a suitable explanation and quantification of the contribution reflected in practice on site "real-life" situations.

The results of the ANOVA were compared with the draft international standard for measurement uncertainty in building acoustics BS EN ISO 12999. The new standard employs new terminology to describe uncertainty which uses standard deviations instead of variance for the reference values. They are known as "standard uncertainties" and follow the GUM nomenclature. New research underpins their calculation, and they cannot be compared directly with the previous reference values in ISO 140 without a mathematical correction (square root of the variance terms). For these study data, Situation A from ISO 12999 describes the reproducibility test scenario that provides the most appropriate comparison, although it is more likely that the more reliable repeatability comparison is from the maximum standard deviation values for repeatability given in table 1 of ISO 12999, which better represents a pooled variance from many measurement systems. The example GRR data are comparable with the currently published repeatability and reproducibility values given in BS EN ISO 12999, which reinforces the view that the data are representative of what would be expected from a test situation on site.

8.9 Electronic Resource: Calculation of Uncertainty

The DoE for an ANOVA GRR assessment of uncertainty follows a similar on-site approach to the BS 5725 method of data collection. The difference comes in the way the data are analysed.

The experiment used in this chapter is a statistically robust method that aligns closely with the recommended sample size and part selection for a successful GRR experiment. The difference comes in the analysis of the data, and to that end, informed readers may wish to use a proprietary software program that have specific GRR tools, for example, MiniTab v15 and above, or use the "uncertainty" web resource at www.noise.co.uk/uncertainty.

The online resource also contains other useful information and explanatory notes for novice and informed users alike, together with additional information, further reading and resources to help in the determination of measurement uncertainty. I hope that you visit the site and welcome your feedback.

References

1 British Standards Institution, *General Requirements for the Competence of Testing and Calibration Laboratories*, BS EN ISO/IEC 17025. BSI, London, 2005.
2 British Standards Institution, *Accuracy (Trueness and Precision) of Measurement Methods and Results*, Part 1: *General Principles and Definitions*, BS ISO 5725-1. BSI, London, 1994.

3 British Standards Institution, *Accuracy (Trueness and Precision) of Measurement Methods and Results*, Part 2: *Basic Method for the Determination of Repeatability and Reproducibility of a Standard Measurement Method*, BS ISO 5725-2:1994. BSI, London,1994.

4 Lang J, "A round robin on sound insulation in buildings",. *Appl Acoust* 52 (314)(1997), 225–238.

5 Knudsen VO, "Measurement and calculation of sound-insulation", *J Acoust Soc Am* 2(1930), 129–140.

6 Whitfield WA, "Measurement uncertainty in airborne sound insulation: Uncertainty components in field measurement", in *ICSV 22*, Florence, 2015.

7 Whitfield WA, Gibbs, BM, "Causes of variation in field measurement of airborne sound insulation', in *Inter Noise*, Lisbon, 2010, p. 9..

8 British Standards Institution, *Acoustics: Measurement of Sound Insulation in Buildings and of Building Elements*, Part 4: *Field Measurements of Airborne Sound Insulation between Rooms*, BS EN ISO 140-4. BSI, London, 1998.

9 International Organization for Standardization, *Acoustics: Measurement of Sound Absorption in a Reverberation Room*, ISO 354: 2003.ISO, Geneva, 2003.

10 *Approved Document E: Resistance to the Passage of Sound*. HMSO, London, 2003.

11 British Standards Institution, *Acoustics: Field Measurement of Sound Insulation in Buildings and of Building Elements. Airborne Sound Insulation*, BS EN ISO 16283-1:2014+A1:2017. BSI, London, 2014.

12 Joint Committee for Guides in Metrology, "Evaluation of measurement data: Guide to the expression of uncertainty in measurement", in *JCGM 100: 2008, GUM 1995 with Minor Corrections*. JCGM,Paris, 2008.

13 British Standards Institution, *Accuracy (Trueness and Precision) of Measurement Methods and Results*, Part 3: *Intermediate Measures of the Precision of a Standard Measurement Method*, BS ISO 5725-3. BSI, London, 1994.

14 British Standards Institution, *Accuracy (Trueness and Precision) of Measurement Methods and Results*, Part 4: *Basic Methods for the Determination of the Trueness of a Standard Measurement Method*, BS ISO 5725-4. BSI, London, 1994.

15 British Standards Institution, *Accuracy (Trueness and Precision) of Measurement Methods and Results*, Part 6: *Use in Practice of Accuracy Values*, BS ISO 5725-6. BSI, London, 1994.

16 Farina A, et al., "Intercomparison of laboratory measurements of airborne sound insulation of repeatability and reproducibility values of partitions for the determination of repeatability and reproducibility values", *Building Acoustics* 6 (2)(1999), 127–140.

17 Schmitz A, Meier A, Raabe G, "Inter-laboratory test of sound insulation measurements on heavy walls, part 1", *Building Acoustics* 6(3–4)(1999), 159–169.

18 Schmitz A, Meier A., Raabe G, "Inter-laboratory test of sound insulation measurements on heavy walls, part 2",. *Building Acoustics* 6(3–4)(1999), 171–186.

19 Carvalho OPA, "Reproducibility in inter-laboratory impact sound insulation measurements", in *ICSV13*, Vienna, 2006.

20 Hoffmeyer D, Christensen J, Olesen HS, *Nordic Intercomparison Programme in the Field of Acoustics*, Part 3: *Measurement: Field Measurements of Airborne Sound Insulation*. 1995.

21 Fausti P, Pompoli R, Smith RS, "An intercomparison of laboratory measurements of airborne sound insulation of lightweight plasterboard walls', *Building Acoustics*6(2)(1999), 14.

22 Muellner H, "Inter-laboratory test: Measurement of airborne and impact sound insulation of lightweight floors, focussing on the extended frequency range below 100 Hz',, in *Forum Acusticum 2011*, Allborg, Denmark, 2011, pp. 2599–2604.

23 Kristensen SD, Rasmussen B, *Repeatability and Reprodicibility of Sound Insulation Measurements*. Danish Acoustical Institute, Kongens Lyngby, Denmark, 1984.

24 van Luxemburg LCJ, Martin H, *Repeatability and Reproducibility of Laboratory Airborne Sound Insulation Measurements: A Dutch Precision Experiment*. TPD, Ministry of Housing, 1986.

25 International Organization for Standardization, *Acoustics: Measurement of Sound Insulation in Buildings and of Building Elements*, Part 2: *Determination, Verification and Application of Precision Data*, ISO 140–2. ISO, Geneva, 1991.

26 International Organization for Standardization, *Acoustics: Determination and Application of Measurement Uncertainties in Building Acoustics*, Part 1: *Sound Insulation*, ISO/CD 12999–1. ISO, Geneva, 2011.

27 Scamoni F, Scrosati C, Mussin M, et al, "Repeatability and reproducibility of field measurements in buildings", in *EuroNoise09*, Edinburgh, 2009.

28 Loverde J, Dong W, "Categories of repeatability and reproducibility in acoustical laboratories", in *169th Meeting of the Acoustical Society of America: Session 1pAA Architectural Acoustics: Uncertainty in Laboratory Building Acoustics Measurements*,. Pittsburgh, PA, 2015.

29 Hall R, "Sound insulation measurements in buildings", in *Euronoise09*, Edinburgh, 2009.

30 Scrosati F, Scrosati C, Bassinino M, et al., "Uncertainty analysis by a round robin test of field measurements of sound insulation in buildings: Single numbers and low frequency bands evaluation – Airborne sound insulation", *Noise Control Engineering* 61(3)(2013), 291–306.

31 United Kingdom Accreditation Service, *The Expression of Uncertainty and Confidence in Measurement*,Document M3003. UKAS, Middlesex, UK, 2007.

32 Ellison SLR, Barwick VJ, "Using validation data for ISO measurement uncertainty estimation",. *The Analyst* 123(1998), 1387–1392.

33 Scholes WE, "A note on the repeatability of field measurements of airborne sound insulation?", *Journal of Sound and Vibration* 10(1)(1969), 1–6.

34 Wittstock V, Bethke C, "The role of static pressure and temperature in building acoustics",. *Building Acoustics* 10(2)(2003), 159–176.

35 Kjaer B, "Para 3.13.3: Effect of humidity", in *B&K Technical Documentation Microphone Handbook*. Berrett-Koehler Publishers, Oakland, CA.

36 Schroeder MR, "Effect of frequency and space averaging on the transmission responses of multimode media",. *J Acoust Soc Am* 46(1969), 277–283.

37 Waterhouse R, "Interference patterns in reverberant sound fields', *J Acoust Soc Am* 27(2)(1955), 247–258.

38 Waterhouse R, "Noise measurement in reverberant rooms", *J Acoust Soc Am* 54(4)(1973), 931–934.

39 Waterhouse R, Lubman D, "Discrete versus continuous space averaging in a reverberant sound field", *J Acoust Soc Am* 48(1 Pt. 1)(1970), 1–5.

40 Waterhouse RV, "Sampling statistics for an acoustic mode", *J Acoust Soc Am* 47(4 Pt. 1)(1970), 961–967.

41 Waterhouse RV, "Statistical properties of reverberant sound fields",. *J Acoust Soc Am* 43(6)(1968), 1436–1444.

42 Lubman D, "Fluctuations of sound with position in a reverberant room", *J Acoust Soc Am* 44(6)(1968), 1491–1502.

43 Lubman D, "Spatial averaging in a diffuse sound field (Letters to the Editor)", *J Acoust Soc Am* 46(3 Pt. 1)(1969), 532–534.

44 Lubman D, Waterhouse R, Chien C-S, "Effectiveness of continuous spatial averaging in a diffuse sound field",. *J Acoust Soc Am* 53(2)(1971), 650–659.

45 Craik RJM, "On the accuracy of sound pressure level measurements in rooms", *Appl Acoust* 29(1990), 25–33.

46 Craik RJM, Steel JA, "The effect of workmanship on sound transmission through buildings, Part 1: Airborne sound", *Appl Acoust* 27(1989), 57–63.

47 Craik RJM, Evans D, "The effect of workmanship on sound transmission through buildings, Part 2: Airborne sound", *Appl Acoust* 27(1989), 137–145.

48 Wittstock V, "Uncertainties in building acoustics", in *Proceedings of Forum Acusticum*, Budapest, 2005.

49 Wittstock V,, "On the uncertainty of single-number quantities for rating airborne sound insulation",. *Acta Acust United Acust* 93(2007), 375–386.

50 Lyn J, et al. "Empirical versus modelling approaches to the estimation of measurement uncertainty caused by primary sampling", *The Analyst* 132(2007), 1231–1237.

51 Taibo L, Glasserman de Dayan H, "Analysis of variability in laboratory airborne sound insulation determinations", *J Sound Vibrat* 91(3)(1983), 319–329.

52 Davern WA, Dubout P, *First Report on Australasian Comparison Measurements of Sound Absorption Coefficients*. Commonwealth Scientific and Industrial Research Organization, Division of Building Research, 1980.

53 Davern WA, Dubout P, *Second Report on Australaisian Comparison Measurements of Sound Absorption Coefficients*. Commonwealth Scientific and Industrial Research Organisation, Division of Building Research, 1985.

54 Deldossi L, Zappa D, "ISO 5725 and GUM: Comparison and comments",. *Accred Qual Assur* 14(2009), 159–166.

55 Burdick RK, Borror CM, Montgomery DC, *Design and Analysis of Gauge R&R Studies*. SIAM, 2005.

56 Montgomery DC, Runger GC, "Gauge capability and designed experiments, Part 1: Basic methods",. *Qual Eng* 6(1)(1993), 115–135.

57 Montgomery DC, Runger GC, "Gauge capability analysis and designed experiments, Part II: Experimental design models and variance component estimation", *Qual Eng* 6(2)(1993), 289–305.

58 Montgomery DC, *Design and Analysis of Experiments*, 8th ed. Wiley, Hoboken, NJ, 2013.

59 Borror CM, Montgomery DC, Runger GC, "Confidence intervals for variance components from gauge capabilitiy studies", *Qual Reliab Eng Int* 13(1997), 361–369.

60 Burdick RK, Borror CM, Montgomery DC, "A review of methods for measurement systems capability analysis", *J Qual Technol* 35(4)(2003), 342–354.

61 Burdick RK, Borror CM, Montgomery DC, "Computer programs: Excel spreadsheet for calculation of MLS confidence intervals", in *Design and Analysis of Gauge R&R Studies*. SIAM, 2005.

62 Montgomery DC, *Design and Analysis of Experiments*, 5th ed. Wiley, Hoboken, NJ, 2001.

63 Montgomery DC, *Statistical Quality Control: A Modern Introduction,*. 6th ed. Wiley, Hoboken, NJ, 2009.

64 European Co-operation for Accreditation, *EA Guidelines on the Expression of Uncertainty in Quantitative Testing*, EA-4/16, >EA, 2003.

65 British Standards Institute, *Acoustics: Rating of Sound Insulation in Buildings and of Building Elements*, Part 1: *Airborne Sound Insulation (ISO 717-1:2013)*, BS EN ISO 717-1: 2013. BSI, London,2013.

66 *Technical Handbooks 2011: Domestic Noise*, Section 5: *Noise*. Scotland, 2011.

67 British Standards Institute, *Acoustics: Determination and Application of Measurement Uncertainties in Building Acoustics*, Part 1: *Sound Insulation*, BS EN ISO 12999–1. BSI, London, 2014.

Chapter 9

Uncertainty in Measuring and Estimating Workplace Noise Exposure

Philip Dunbavin

9.1 Background

The following is a direct quotation from *A Beginner's Guide to Uncertainty of Measurement*, by Stephanie Bell of the National Physical Laboratory:

> A measurement result is only complete if it is accompanied by a statement of the uncertainty in the measurement.
>
> Measurement uncertainties can come from the measuring instrument, from the item being measured, from the environment, from the operator, and from other sources. Such uncertainties can be estimated using statistical analysis of a set of measurements, and using other kinds of information about the measurement process. There are established rules for how to calculate an overall estimate of uncertainty from these individual pieces of information.
>
> **The use of good practice** – such as traceable calibration, careful calculation, good record keeping, and checking – can reduce measurement uncertainties. When the uncertainty in a measurement is evaluated and stated, the fitness for purpose of the measurement can be properly judged.

The emboldening is mine and represents the approach I will take in discussing uncertainty in these types of measurements. Good practice is largely dictated by common sense modified with a pinch of experience. There is no substitute for having the experience gained by measuring workplace noise exposure.

9.1.1 Purpose of Measuring and Estimating Workplace Noise Exposure

In the United Kingdom, as in many countries, there is legislation to control the workplace exposure to noise. In the United Kingdom, this is the Control of Noise at Work Regulations 2005, and this requires employers who carry out work that is liable to expose any employees to noise at or above a lower-exposure action value to make suitable and sufficient assessment of the risk from that noise to the health and safety of those employees.

9.1.2 Regulatory Requirements

The main elements of the UK Control of Noise at Work Regulations 2005 are summarised as follows:
The lower exposure action values are

(a) a daily or weekly personal noise exposure of 80 dB (A-weighted) and
(b) a peak sound pressure of 135 dB (C-weighted).

The upper exposure action values are

(a) a daily or weekly personal noise exposure of 85 dB (A-weighted) and
(b) a peak sound pressure of 137 dB (C-weighted).

The exposure limit values are

(a) a daily or weekly personal noise exposure of 87 dB (A-weighted) and
(b) a peak sound pressure of 140 dB (C-weighted).

Where the exposure of an employee to noise varies markedly from day to day, an employer may use weekly personal noise exposure in place of daily personal noise exposure for the purpose of compliance with these regulations.

In applying the exposure limit values, but not in applying the lower and upper exposure action values, account is taken of the protection given to the employee by any personal hearing protectors provided by the employer. In practical terms, the most important element is the daily or weekly personal noise exposure values. The peak sound pressure levels are very rarely found in modern industry, with the obvious exception of drop forging. However, bursts of high levels of impulsive activities that do not exceed the peak sound pressure action values will still need to be measured because they can contribute significantly to the L_{Aeq} value. Consequently, this chapter focuses on good practice in reducing the uncertainty of the measurement of daily or weekly noise exposure.

Health and Safety Executive (HSE) Guidance Document L108 provides guidance on how to comply with the Control of Noise at Work Regulations 2005. Appendix 1 of that document covers measuring noise exposure in the workplace, and in paragraphs 34–37 it gives some simple guidance on reducing errors in measurements.

9.1.3 Guidance and Standards

There is a reference section at the end of this chapter that lists most relevant standards. In terms of guidance on uncertainty, the 'bible' is BS EN ISO 9612: 2009, *Acoustics: Determination of Occupational Noise Exposure – Engineering Method (ISO 9612: 2009)*. The following is a direct quote from this standard:

> This International Standard specifies an engineering method for measuring workers' exposure to noise in a working environment and calculating the noise exposure level. This International Standard deals with A-weighted levels but is applicable also to C-weighted levels.
>
> Three different strategies for measurement are specified. The methods are useful where a determination of noise exposure to engineering grade is required, e.g. for detailed noise exposure studies or epidemiological studies of hearing damage or other adverse effects.
>
> The measuring process requires observation and analysis of the noise exposure conditions so that the quality of the measurements can be controlled. This International Standard provides methods for estimating the uncertainty of the results.

Given the importance of this standard, I would recommend to all acousticians that they purchase a copy. In this chapter I will not reproduce much of that document, but rather I will concentrate on the practical aspects that will assist readers in reducing the uncertainty of their measurements to the lowest possible. A summary of the main aspects of BS EN ISO 9612:2009 are covered in Annex A to this chapter.

Author's note: Permission to reproduce extracts from Internation Organization for Standardization (ISO) publications was granted by the ISO. No other use of this material is permitted.

9.1.4 Importance of Uncertainty

Going back to fundamentals, a measurement result is only complete if it is accompanied by a statement of the uncertainty in the measurement. The value and reliability that can be placed on the results will determine what actions employers take in response to them. Thus, it is essential to have a good grasp of the sources of the uncertainty and the relative significance of each when designing your measurement procedure.

9.2 Sources of Uncertainty

The sources of uncertainty need specific consideration in order to reduce their influence as far as possible. The main sources of uncertainty are discussed in this section.

9.2.1 Operating Conditions

The practitioner should be sure that what he or she is measuring is the 'normal' operating conditions of the factory or plant. I have lost count of the times an employee has said to me, 'You should have been here yesterday(last week) when it was much noisier'. This 'observation', of course, needs to be considered, but in my experience, this is seldom a true statement. The practitioner has a duty to ensure that he or she measures 'typical/normal' noise-producing activity; failure to do so can introduce an uncertainty of an unknowable magnitude.

9.2.2 Variations in Daily Work

The starting point for planning your measurements is the work patterns of the employees. In this respect, there are three distinct types of work pattern in every workplace noise exposure survey and appropriate strategies for dealing with them.

Strategy 1: Task-based measurement. The nominal day is divided into tasks with repeatable sound pressure levels.

Strategy 2: Job-based measurement. The principle of this measurement strategy is that random samples of noise exposure are taken during the performance of jobs identified during the work analysis.

Strategy 3: Full-day measurement. This strategy employs long-duration measurements that are typically made using personal sound exposure meters or dose meters.

Management will tell you what they believe is the work pattern of each employee in the factory/plant. If you then consult each of those employees, you will discover that each believes that he or she has a somewhat different work pattern. It might be thought that the third way of assessing work patterns, which is by your direct observation, would reveal the true work pattern.

Sometimes the observed work patterns are reasonable and realistic. However, we must bear in mind Heisenberg's uncertainty principle. The uncertainty principle is one of the most famous (and probably misunderstood) ideas in physics. It tells us that there is a 'fuzziness' in nature, a fundamental limit to what we can know about the behaviour of any system. When considering work patterns, what you observe may be influenced by the simple fact that you are observing an employee. The most you can hope for is to calculate probabilities for where employees are and how they will behave.

It is important to remember that the noise exposure depends on both the noise levels and the time for which a person is exposed to those noise levels. It follows that the uncertainty in the estimate of the noise exposure levels is a combination of the uncertainty in the measurement of the sound level and the uncertainty in the estimate of the time an individual is exposed to that sound level.

It is possible to estimate the uncertainty introduced by the assumed work patterns by conducting a sensitivity analysis. Basically, this consists of playing what-if scenarios, but this needs to be tempered by the practitioner using his or her professional experience. It is not meant to be an exercise in *reductio ad absurdum* but an application of common sense and experience. In essence, this means that unreasonable worst-case conditions should not be considered, rather that you should consider 'reasonable' worst-case conditions.

9.2.3 Sampling Uncertainty

When considering the measurements you will be making, some thought needs to be given to how long your sampling period should be. Should you measure over three or four complete cycles of an operation? What we are measuring is an A-weighted L_{eq} value, and modern sound level meters have a precision of 0.1 dB. Consequently, a useful way to confirm that your sampling period is long enough is to watch the value of the sound level displayed on your sound level meter and ensure that it is stable to a precision of 0.1 dB. Logically, if this is the case, then the uncertainty in your sampling will be reduced to the lowest level practical.

9.2.4 Instrumentation and Calibration

Using a Class 2 sound level meter rather than a Class 1 sound level meter will introduce a greater uncertainty. Obviously, it is good practice to use a Class 1 sound level meter. You should demonstrate the conformity of the sound level meter and sound calibrator by means of valid certificates traceable to national standards. The certificate will state the uncertainty of the measurement equipment.

At the beginning of every measurement session, check the field calibration of the sound level meter at one or more frequencies by means of a sound calibrator, and check the calibration value at the end of the measurement. The difference between the initial calibration value, any subsequent calibration check and a final calibration check on completion of the measurement provides an estimate of the uncertainty owing to the stability of the sound level meter.

9.2.5 Microphone Location

BS EN ISO 9612: 2009 gives specific heights above the ground for your measurement microphone for standing and seated employees for fixed workstations. In practice, most employees move around, and the preferred location of the microphone is close to the ear of the employee. This means that you need to stand behind the employee with the microphone about 100 mm from the ear. You then follow the employee and try to maintain the position of the microphone relative to his or her ear. This is easier said than done and takes considerable practice.

The employee may also find this unsettling, and it is essential to explain to him or her what you will be doing and why and, more importantly, to reassure the employee that the microphone will not make contact with him or her as it could damage the delicate microphone.

9.2.6 False Contributions

Uncertainty can be introduced by false contributions from airflows, impact on the microphone or, in the case of dosimetry, rubbing on clothing. When using a sound level meter, it is good practice to use a windshield at all time. The use of a windshield will assist in preventing any impacts on the microphone and hence eliminate the potential for uncertainty owing to these sources. The possibility of noise from the microphone rubbing on cloth in dosimetry can introduce a very small measure of uncertainty that can be safely ignored in most cases. The human factors that affect the uncertainty of dosimetry are far more significant, as discussed in Section 9.2.8.

9.2.7 Contributions from Non-typical Noise Sources

Uncertainty can be introduced by non-typical noise sources such as speech, music (radio), alarm signals, and non-typical behaviour such as singing. The only way to reduce the uncertainty from such sources is to discard all measurements that are affected by them which effectively reduces the uncertainty due to the effects to the lowest level practical and in most cases this contribution can be reasonably ignored in your uncertainty budget.

9.2.8 Human Factors

Many acousticians take the view that it is simpler to use dosimetry because these measure the actually sound levels to which the ear is exposed. Regrettably this is a naive view and whilst dosimetry has a place in workplace noise exposure surveys as discussed in section 3.3 there are many human factors that can introduce significant uncertainty and these are best illustrated by actual examples. However, if you are faced with mobile work patterns in the workplace you are surveying, then a noise dose meter will probably be the best way to measure the employees noise exposure with sanity checks being performed with a sound level meter.

A noise dose meter consists of a microphone on a cable, which can be clipped to a collar. The microphone cable is then passed under the clothing to the unit itself, which is small enough to be located in a pocket or clipped to a belt. The dose meter can then be started at the beginning of the shift. If it runs until the end of the working day, the noise dose can be directly read from the instrument or downloaded without the need for calculations. Another useful feature of noise dose meters is that they will log the noise data so that when downloaded to a computer, the time history of the noise can be viewed. This provides the ability to analyze when and where high noise exposures occur. This can be even more useful when the dose meter is placed on an employee who is prepared to make a list of the times and jobs he or she was performing throughout the day. This will give the surveyor the ability to see which operations most need noise control in order to reduce exposure.

The latest 'badge' dose meters have certain advantages over traditional dose meters. Because the dose meter is small and light enough to be worn on the shoulder, it means there are no cumbersome microphone cables. If there are no cables to get in the way, not only is the device safer to wear, but also employees are less resistant to wearing it and much more likely to forget it is there. This means that the quality of the noise data collected will be improved.

It is essential to make sure that the battery is fully charged before starting the measurements and that the device is field calibrated before and after the measurement and any calibration drift is recorded. In addition, it is sensible to use a sound level meter to check that the logged sound levels correspond to the sound produced by both the job or task at the time of measurement.

As part of a workplace noise exposure assessment, I frequently have to deal with forklift truck drivers who are exposed to a wide range of sound levels in the workplace and also to the sound generated by their own forklift trucks. It is not generally possible to attached a sound level meter microphone to the truck close to the driver's ear, so dosimetry is normally adopted.

At the end of one particular shift, I downloaded the data from the dose meter and found that the forklift truck driver's exposure was apparently 105 dB L_{Aeq}. This did not make a great deal of sense until I questioned the forklift truck driver, who told me that the worst noise he was exposed to was the exhaust of his own forklift truck. He was keen to ensure that this was measured properly, so he had attached the dose meter to the frame of the forklift truck with gaffer tape and positioned the microphone on the centreline of the exhaust discharge. Thinking he was doing right, he had, of course, messed up the measurement and cooked a microphone in the process.

In that instance, the source of the 'outlier' measurement was reasonably easy to identify, and the measurement was repeated correctly using another dose meter. The human factors are sometimes quite subtle and difficult to identify. Probably the most inventive I have observed was the employee who left his dose meter behind a toilet cistern all day. Loud singing when you are not observing an employee is very difficult to detect.

Uncertainty due to human factors is possibly the most difficult area to estimate and can affect your measurements in wild and wonderful ways. The only way to safeguard against this is to constantly be on the lookout for the outlier or unusual result.

9.2.9 Repeating Measurements

It is good scientific measurement practice to repeat measurements, but this costs money that many clients will be unwilling to pay. It is sensible to build into your survey some repeat measurements and to use more than one dose meter in parallel on employees carrying out nominally similar work routines.

9.3 Procedural Approaches

The selection of an appropriate measurement strategy is influenced by several factors, such as the complexity of the work situation, the number of employees involved, the effective duration of the workday, the time available for measurement and analysis and the amount of detailed information required. BS EN ISO 9612: 2009 describes three basic approaches: task-based measurements, job-based measurements and full-day measurements. These are described in full in the standard and briefly in Section 9.2.2 of this chapter, so I will not repeat them here.

9.3.1 Noise Mapping

In many cases, the measurement of workplace noise exposure will lead to a programme of noise reduction to reduce the risk to employees. It is very helpful to put the decibel L_{Aeq} values measured on a map/drawing of the premises at their respective measurement locations. Using colour coding will turn this drawing into a visual aid making it obvious where the critical noise sources are located, which, in turn, will inform the noise control programme to reduce employee noise exposure.

A typical noise map is shown in Figure 9.1. This approach is best suited to situations where the noise patterns at various workstations, or zones, remain fairly constant from day to day.

9.3.2 Dosimetry

Dosimetry is not the ultimate measurement approach as many would expound. It can be used and is very effective, but as discussed earlier, it is subject to possible human factors, so a close watch has to be kept for outliers (see Section 9.2.8). In addition, estimation of the measurement uncertainty is more difficult.

9.3.3 Validation

The principal value of dosimetry is that it can be used to validate the results of the job-based and full-day results. This, in turn, will provide confidence in your estimation of the uncertainty of your methods.

9.3.4 Mobile Telephone Apps

There are now many mobile telephone applications that can perform very simple sound level measurements. These are not calibrated instruments but can be used by workplace personnel to identify the need for a full-exposure assessment survey; they are definitely not suitable for carrying out such a survey.

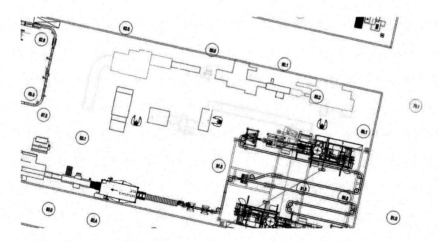

Figure 9.1 Typical workplace noise map.

9.3.5 Octave Bands

Whilst measurements of the sound levels in octave bands are not essential for an assessment of workplace noise exposure, they will certainly be required to correctly specify suitable hearing protection should the assessment results indicate that it is required. Consequently, it would be sensible and appropriate to use a sound level meter that can simultaneously measure both L_{Aeq} and L_{eq} in octave bands.

9.4 Estimation of Historical Workplace Noise Exposure

It is a logical result of exposure to high noise levels that some employees will suffer hearing loss. Claims for noise-induced hearing loss (NIHL) are very common, and in order to establish appropriate levels of compensation, if damage has been caused, courts will require expert evidence to establish what an employee's historical noise exposure would have been. This is a process that introduces many extra areas of uncertainty.

9.4.1 Sources of Data

The factory or plant where an employee worked may no longer exist, and there may also be no reliable noise exposure surveys. If an acoustician has a large database of noise levels of machines and operations, then he or she can use those measurements to estimate the probable historical noise exposure of an employee. If this is not the case, then reliance may have to be placed on historical measurements made by others with an unquantifiable level of uncertainty. This can be affected by many factors such as, is the source of the data known (e.g. HSE, published papers, textbooks or other acousticians/practitioners), and does the measurement describe the capacity of a machine under test, does it say where the measurement was made and does the measurement include many cycles of a machine/process?

9.4.2 Claimant's Evidence

In a real-life situation, it can be difficult to establish realistic work patterns. When you have to rely on a claimant's description of his or her work patterns, then this becomes much more problematical. There is obviously a natural human tendency to slightly exaggerate exposure to noise sources in claims cases. Even if such a human factor is not influencing the information you are given, it has to be accepted that human memory is fallible, and even the most honest of claimants may well misremember how it was 20 or more years ago. In establishing the most likely scenario, the documents supplied need to be examined in forensic detail. Sensitivity analysis will help you to establish the range of possible values.

9.4.3 Balance of Probabilities

I have no idea how you would realistically determine the uncertainty in a historical case, and I am not sure that it is relevant. This is because you are not really trying to say what the exact level was, but you are trying to provide an experience-based estimate of 'more likely than not it was at this exposure level or above'. The level of uncertainty when taking into account memory, work patterns, different machines and lots of other things could be huge, but in these cases, you are not looking for a certainty but for a weight of probability.

It is probably appropriate not to try to quantify uncertainty in these cases but to state that the estimation is an opinion of a most likely level based on personal experience, historical data, witness evidence, a range of assumptions (where this has occurred) and any other factors.

When you use an exposure level (with an uncertainty if you are confident enough) rather than a range, then this is more useful for the court as the figure can then be used to determine apportionment where there is more than one noisy employer.

9.5 Reporting and Communication

The structure of a report is determined by many factors, such as what the purpose of the report is and do you think that your client is likely to understand it. Basically, this comes down to knowing your client and how you were briefed/instructed.

9.5.1 Structure

In section 15, BS EN ISO 9612: 2009 provides a comprehensive list of what should be included in a report on the estimation of workplace noise exposure. Please see Annex A at the end of this chapter, where this critical list is reproduced in full.

9.5.2 What Else Should Be Included

The uncertainties associated with measuring occupational noise exposure should be determined in accordance with annex C of BS EN ISO 9612: 2009. It is doubtful whether a full explanation of the process of arriving at the uncertainty of the measurement would be understood. Personally, when I discus uncertainty budgets, most people glaze over. Thus, in many cases, anything more complicated than a simple statement of the measurement and your estimated uncertainty is a waste of paper unless, of course, your report is going to be used as expert evidence in a court case for compensation.

A full explanation of the process of arriving at the uncertainty of the measurement can be included as an annex to your report with a simple explanation in the body of the report such that it does not interrupt the flow of the report. A typical simple uncertainty budget is shown in Annex B of this chapter.

The report should state all the measures you have put in place to reduce the uncertainty of your measurement to the lowest level practical. The final results should be given as both the measured values and the value of the uncertainty. The expanded measurement uncertainty, together with the corresponding coverage factor, should be stated for a one-sided confidence interval of 95%. A full worked example is shown in Annex B of this chapter.

9.5.3 What Should Not Be Included

Your report should avoid implying levels of certainty that are not justified. Remember, your report could well be cited in a court case 20 or more years from when you write it.

9.6 Summary

This chapter has identified many of the pitfalls in measuring workplace noise exposure and reducing the uncertainty of those measurements to the lowest level practical. It is important to note that there is no such animal as an uncertainty-free measurement.

References

1 International Organization for Standardization, *Acoustics: Hearing Protectors*, Part 2: *Estimation of Effective A-Weighted Sound Pressure Levels When Hearing Protectors Are Worn*, ISO 4869-2. ISO, Geneva.

2 International Organization for Standardization, *Acoustics: Noise Emitted by Machinery and Equipment – Guidelines for the Use of Basic Standards for the Determination of Emission Sound Pressure Levels at a Work Station and at Other Specified Positions*, ISO 11200. ISO, Geneva.

3 International Organization for Standardization, *Acoustics: Noise Emitted by Machinery and Equipment – Measurement of Emission Sound Pressure Levels at a Work Station and at Other Specified Positions in an Essentially Free Field Over a Reflecting Plane with Negligible Environmental Corrections*, ISO 11201. ISO, Geneva.

4 International Organization for Standardization, *Acoustics: Noise Emitted by Machinery and Equipment – Measurement of Emission Sound Pressure Levels at a Work Station and at Other Specified Positions Applying Approximate Environmental Corrections*, ISO 11202. ISO, Geneva.

5 International Organization for Standardization, *Acoustics: Noise Emitted by Machinery and Equipment – Determination of Emission Sound Pressure Levels*

at a Work Station and at Other Specified Positions from the Sound Power Level, ISO 11203. ISO, Geneva.

6 International Organization for Standardization, *Acoustics: Noise Emitted by Machinery and Equipment – Engineering Method for the Determination of Emission Sound Pressure Levels in Situ at the Work Station and at Other Specified Positions Using Sound Intensity*, ISO 11205. ISO, Geneva.

7 International Organization for Standardization, *Acoustics: Determination of Sound Emission from Sound Sources Placed Close to the Ear*, Part 1: *Technique Using a Microphone in a Real Ear (MIRE Technique)*, ISO 11904-1. ISO, Geneva.

8 International Organization for Standardization, *Acoustics: Determination of Sound Emission from Sound Sources Placed Close to the Ear*, Part 2: *Technique Using a Manikin*, ISO 11904-2. ISO, Geneva.

9 International Organization for Standardization, *Acoustics: Definitions of Basic Quantities and Terms*, ISO/TR 25417:2007. ISO, Geneva, 2007.

10 International Electrotechnical Commission, *Sound Level Meters*, IEC 60651:2001. IEC, Geneva, 2001 (superseded by IEC 61672, all parts).

11 International Electrotechnical Commission, *Integrating-Averaging Sound Level Meters*, IEC 60804:2000 IEC, Geneva, 2000 (superseded by IEC 61672, all parts).

12 British Standards Institute, *Hearing Protectors: Recommendations for Selection, Use, Care and Maintenance – Guidance Document*, BS EN 458:2004. BSI, London 2004.

13 Grzebyk M, Thiéry L, 'Confidence intervals for the mean of sound exposure levels', *Am Indust Hyg Assoc J* 64(2003), 640–645.

14 Thiéry L, Ognedal T, 'Note about the statistical background of the methods used in ISO/DIS 9612 to estimate the uncertainty of occupational noise exposure measurements', *Acta Acust Acust* 94(2008), 331–334.

15 Health and Safety Executive, *The Control of Noise at Work Regulations 2005: Guidance on Regulation*, HSE Document L108. HSE, London, 2005.

Annex A A Summary of the Main Aspects of BS EN ISO 9612: 2009, Acoustics: Determination of Occupational Noise Exposure – Engineering Method

This international standard provides a stepwise approach to the determination of occupational noise exposure from noise level measurements. The procedure contains the following major steps:

• Work analysis,
• Selection of measurement strategy,
• Measurements,
• Error handling and uncertainty evaluations,
• Calculations and presentation of results.

The standard specifies three different measurement strategies:

- Task-based measurement,
- Job-based measurement and
- Full-day measurement.

It gives guidance on selecting an appropriate measurement strategy for a particular work situation and purpose of investigation. This international standard also provides an informative spreadsheet to allow calculation of measurement results and uncertainties.

This international standard recognises the use of hand-held sound level meters as well as personal sound exposure meters. The methods specified optimise the effort required for obtaining a given accuracy.

At section 15, it provides the following list of what information should be included in a report of noise exposure measurements carried out in accordance with this standard. Provide the following information:

a) general information:

   1) name of the client (company, department, etc.) of the investigation,
   2) identification of the worker(s) or group(s) of workers (such as name or worker number) whose exposure has been determined,
   3) name of the person(s) and company or institution who carried out measurements and calculations,
   4) purpose of the determination,
   5) reference to this International Standard and the strategy that has been applied;

b) work analysis:

   1) description of the work activities investigated,
   2) size and composition of homogeneous noise exposure groups, where relevant,
   3) description of the day(s) investigated, including the tasks comprising the nominal day when the task-based measurements have been made,
   4) measurement strategy/strategies employed, together with a reference to the statistical approach used;

c) instrumentation:

   1) identification and class of instrumentation used (manufacturer, model, serial number),
   2) configuration of the system, e.g. windscreen, extension cable, etc.,

3) calibration traceability (date and result of the most recent verification of the components of the measuring system),

4) documentation of calibration checks performed before and after each measurement;

d) measurements:

1) identification of worker(s) whose noise exposure was measured,

2) date and time of measurements,

3) instrumentation used for each measurement (if various instruments are used),

4) description of work undertaken by the worker during the course of the measurements, including duration of work activity and, if relevant, duration of cyclic events contained within the work activity,

5) report of any deviations from the normal work conditions or normal work behaviour during the course of the measurements,

6) production indicators related to the work being undertaken, where relevant,

7) description of the sources of noise contributing to the noise exposure,

8) description of any irrelevant sounds included in or deleted from the measured results,

9) description of any events observed which may have influenced the measurements (e.g. airflows, impacts on the microphone, impulsive noise),

10) relevant information on meteorological conditions (e.g. wind, rain, temperature),

11) position and orientation of microphone(s),

12) number of measurements at each position,

13) duration of each measurement,

14) duration of each task in the nominal day, and the associated uncertainty, when using the task-based approach,

15) results of each measurement, to include at least the $L_{p,A,eqT}$ and, optionally, the highest $L_{p,C\ peak}$ values;

e) results and conclusions:

1) A-weighted equivalent continuous sound pressure level $L_{p,A,eqT}$ and, optionally, C-weighted peak sound pressure level $L_{p,Cpeak}$ for each task/job,

2) when using the task-based measurement, the values of $L_{EX,8h,m}$ for each task, if relevant,

3) A-weighted noise exposure level $L_{EX,8h}$ for the nominal day(s), and the highest C-weighted peak sound pressure level $L_{p,Cpeak}$ if measured during all tasks, rounded to one decimal place,

4) uncertainty associated with $L_{EX,8h}$ and $L_{p,Cpeak}$, if available, for the nominal day(s), rounded to one decimal place (noise exposure and the measurement uncertainty shall be reported as separate values).

5) Annex C of the standard covers the evaluation of measurement uncertainties in detail and Annex D shows three examples, one for each measurement strategy, on how to calculate the daily noise exposure level and the uncertainty budget associated with each strategy.

Author's note: Permission to reproduce extracts from ISO publications was granted by the ISO. No other use of this material is permitted.

Annex B Worked Examples and a Case Study of the Calculation of the Noise Exposure and Measurement Uncertainty

BS EN ISO 9612: 2009

Part of the BS EN ISO 9612: 2009 standard is a spreadsheet that calculates the employee exposure and uncertainty associated with each of the three main measurement strategies. The spreadsheet includes options to work in the English, French and German languages.

The spreadsheet comes complete with instructions on how to use it and a sample set of data for which the employee exposure is calculated together with the uncertainty following the principles of the standard.

The following pages show how the spreadsheet looks when first opened and before entering your own data.

Author's note: Permission to reproduce extracts from ISO publications, in this case a screen shot of how the calculation spreadsheet appears, was granted by the ISO. No other use of this material is permitted.

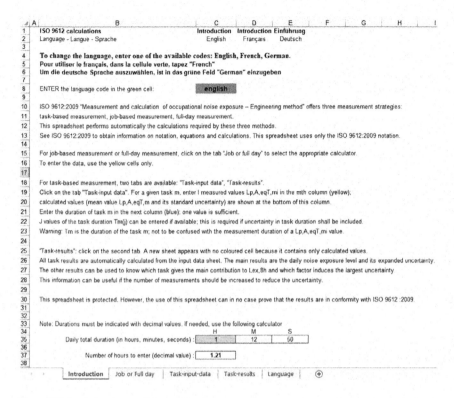

Figure 9.2 Introduction and instructions.

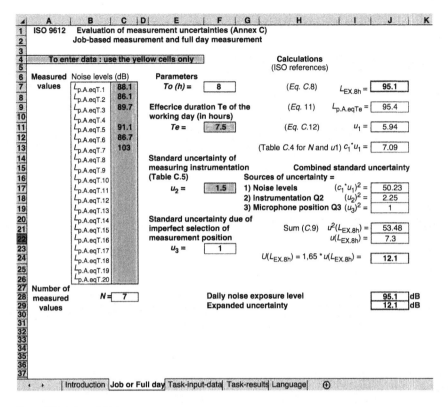

Figure 9.3 Job or full day.

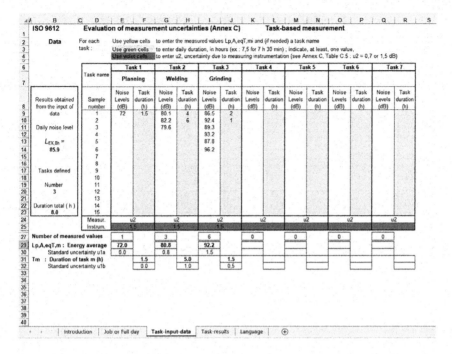

Figure 9.4 Task input data.

	B	C	D	E	F	G	H	I	J	K	L
1	ISO 9612	Evaluation of measurement uncertainties (Annex C)					Uncertainties calculations				
2		Task-based measurement					All values are calculated from the Task-input-data sheet				
3											
4	Daily noise exposure level			85.9	dB		Number of tasks		3		
5	Expanded uncertainty			3.7	dB		Total daily duration (h)		8.0		

	Uncertainty budget		(reference)	Symbols, relations	Task 1	Task 2	Task 3	Task 4	Task 5	Task 6	Task 7
7											
8	Noise level	Standard uncertainty	(C.6)	$u_{1a,m}$	0.00	0.80	1.50				
9		Sensitivity coefficient	(C.4)	$c_{1a,m}$	0.01	0.19	0.80				
10	Duration	Standard uncertainty	(C.7)	$u_{1b,m}$	0.00	1.00	0.50				
11		Sensitivity coefficient	(C.5)	$c_{1b,m}$	0.02	0.17	2.31				
12	Uncertainty contribution of noise levels			$c_{1a,m} \cdot u_{1a,m}$	0.00	0.15	1.20				
13	Uncertainty contribution of tasks durations			$c_{1b,m} \cdot u_{1b,m}$	0.00	0.17	1.16				
14	Uncert. contr. of measuring instrumentation			$c_{1a,m} \cdot u_{2,m}$	0.01	0.29	1.20				
15	Uncert. contr. of measurement position			$c_{1a,m} \cdot u_3$	0.01	0.19	0.80				
16											

					Task 1	Task 2	Task 3	Task 4	Task 5	Task 6	Task 7
17	Results										
18				Task name	Planning	Welding	Grinding				
19	Mean Noise level (dB)		(9.3 : (7))	$L_{p,A,eqT,m}$	72.0	80.8	92.2				
20	Duration (h)		(9.2 : (5))	Tm	1.5	5.0	1.5				
21	Contribution of task m to Lex,8h		(9.4 : (8))	$L_{EX,8h,m}$	64.7	78.7	84.9				
22	Uncertainty contribution	Noise level		$(c_{1a,m} \cdot u_{1a,m})^2$	0.00	0.02	1.43				
23		Duration		$(c_{1b,m} \cdot u_{1b,m})^2$	0.00	0.03	1.34				
24		Measuring instrumentation		$(c_{1a,m} \cdot u_{2,m})^2$	0.00	0.08	1.44				
25		Measurement position		$(c_{1a,m} \cdot u_3)^2$	0.00	0.04	0.64				
26		Sum per task m		$u^2 (L_{EX,8h}) m$	0.00	0.17	4.84				
28	Sum for all tasks		(C.3)	$u^2 (L_{EX,8h}) =$	5.01						
29	Combined standard uncertainty			$u(L_{EX,8h})$	2.2	dB	Expanded uncertainty				
30	Daily noise exposure level		(C.2)	$L_{EX,8h} =$	85.9	dB	$U(L_{EX,8h}) = 1.65 \cdot u(L_{EX,8h}) =$		3.7	dB	

Introduction | Job or Full day | Task-input-data | **Task-results** | Language ⊕

Figure 9.5 Task results.

Institute of Acoustics Diploma

In 2010, Mr Craig M. Lewis, an Institute of Acoustics diploma student, prepared his diploma project, 'A survey of noise exposure to employees engaged in recycling and domestic waste collections'. The following extracts are reproduced here with the permission of Mr Lewis.

Background

Residential waste is collected in bins, which are colour-coded for the following types of refuse (Figure 9.6):

- Domestic residual waste (black),
- Paper and cardboard (blue) and
- Glass, cans and plastics (brown)

Figure 9.6 The 240-litre wheelie bins (paper and cardboard, blue bin, glass, cans and platics, brown bin).

The blue and brown bins are emptied alternate weeks, and the black bin is emptied weekly. The waste containers comply with the requirements of BS 840-1:2004.

	Dimensions	
240-litre wheelie bin	Width	740 mm
	Depth	580 mm
	Height	1,100 mm
	Height (with lid open)	1,750 mm

These wheelie bins are emptied by a Dennis Eagle Elite 2 waste-collection wagon, as shown in Figure 9.7.

The collection process involves the vehicle being driven along the road and stopping every 20-30 m. The workers at the back of the vehicle bring the wheelie bins to the lifting device, where they are attached, and the worker then operates the lifting device by pressing the buttons on the rear left-hand side of the vehicle. The operator remains approximately 1 m from the rear of the vehicle. Figure 9.8 shows the rear of the vehicle and a bin ready to be attached. Figure 9.9 shows the position of the operator whilst the container is being lifted. Figure 9.10 shows the height of the bin and therefore the distance the contents will fall. At this point, the bin is 'banged' several times by the lifting device to ensure that all the contents have fallen from the waste container. Figure 9.11 shows how the bin is attached to the vehicle. It was not identified how many bin lifts each operator makes; however, it is estimated to be approximately 500 per day.

Figure 9.7 Collection vehicle.

Figure 9.8 Rear of vehicle.

The levels of noise exposure for individual employees will depend primarily on the amount (volume/weight) of glass processed, as well as on the speed of working. Workers carrying out kerbside collections of glass for recycling can be exposed to high levels of noise. It is highly likely that workers' daily personal noise exposures will exceed 85 dB (the upper exposure action value of the Control of Noise at Work Regulations, and in some cases, daily personal noise exposures may be as high as 100 dB (HSE 2007).

Figure 9.9 Operator position.

Figure 9.10 Container lift in process.

Figure 9.11 Container attachment.

Full-Day Measurements

With the use of dose meters, full-day measurements covering all work-related noise contributions and quiet periods during the workday were taken. This strategy enabled the author to ensure that the results were representative for what is defined as the relevant work situation.

Since this measurement strategy collects all contributions, it also has the highest risk of including false contributions. Observations were made of the worker during the workday whilst measurements were being made to try to identify any conditions that may give rise to adverse readings. The author discussed the workday with the worker and the supervisor at the end of the shift to identify tasks and any other areas that may have appeared adverse on the reader. The dose meter logged one-minute L_{Aeq} values, which enabled the author to scrutinise the day's events. A total of 22 full-day measurements were taken. This included 15 measurements for glass recycling, four for the domestic refuse and three for paper recycling.

The daily noise exposure level was determined for each data set. A number of corrections were made to the raw data to allow for breaks and occasions when the worker had finished the shift and the equipment was not turned off at that point.

Table 9.1 Final results for glass collection for both measured value and the value of uncertainty

N	Date	Serial no.	$L_{Aeq,T}$	$10^{0.1 \times L_{Aeq,T}}$	$L_{Aeq,ave}$	$(L_{p,Aeq,T,n} - \bar{L}_{p,Aeq,T})^2$	$L_{ep,d}$
1	24/08/10	4266	91.3	1.35E+09	89.8	2.3	92.5
2	24/08/10	4264	87.8	6.03E+08	89.8	3.9	88.6
3	24/08/10	4258	90.4	1.10E+09	89.8	0.4	91.3
4	24/08/10	4265	85.7	3.72E+08	89.8	16.7	86.8
5	24/08/10	4259	91.6	1.45E+09	89.8	3.3	92.9
6	25/08/10	4264	87.0	5.01E+08	89.8	7.7	87.9
7	25/08/10	4259	89.2	8.32E+08	89.8	0.3	90.2
8	25/08/10	4258	94.0	2.51E+09	89.8	17.8	95.1
9	26/08/10	4265	90.5	1.12E+09	89.8	0.5	91.2
10	26/08/10	4259	89.0	7.94E+08	89.8	0.6	89.6
11	26/08/10	4264	86.9	4.90E+08	89.8	8.3	87.6
12	27/08/10	4264	87.5	5.62E+08	89.8	5.2	88.1
13	27/08/10	4259	89.1	8.13E+08	89.8	0.5	89.7
14	27/08/10	4258	90.3	1.07E+09	89.8	0.3	91
15	27/08/10	4265	88.5	7.08E+08	89.8	1.6	89.2
$\sum_{n=1}^{N} 10^{0.1 \times L_{p,Aeq,T,n}}$				1.43E+10	$\sum_{n=1}^{N} 10^{0.1 \times L_{eq,d,n}}$		1.77E+10
$10\log_{10}\left(\frac{1}{N}\sum_{n=1}^{N} 10^{0.1 \times L_{p,Aeq,T,n}}\right)$ dB				89.8	$10\log_{10}\left(\frac{1}{N}\sum_{n=1}^{N} 10^{0.1 \times L_{eq,d}}\right)$ dB		90.7

(Continued)

Table 9.1 (Cont.)

N	Date	Serial no.	$L_{Aeq,T}$	$10^{0.1 \times LAeq,T}$	$L_{Aeq,ave}$	$\left(L_{p,Aeq,T,n} - \bar{L}_{p,Aeq,T}\right)^2$	$L_{ep,d}$
					$u_1^2 = \sqrt{\frac{1}{(N-1)}\left[\sum_{n=1}^{N}\left(L_{p,AeqT,n} - \bar{L}_{p,Aeq,T}\right)^2\right]}$		5.0dB
					u_1		2.2
					$c_1 u_1$ for $N = 15$ and $u_1 = 2.2$		0.8
					u_2		1.5
					u_3		1.0
					$c_2 = c_3$		1.0
					$u^2\left(L_{ep,d,8h}\right) = c_1^2 u_1^2 + c_2^2\left(u_2^2 + u_3^2\right)$		3.8
					$u\left(L_{ep,d,8h}\right)$		2.0
					Expanded uncertainty:		
					$U\left(L_{ep,d,8h}\right) = 1.65 \times u$		
					$U\left(L_{ep,d,8h}\right) = 1.65 \times 2.01$		3.2

The daily noise exposure was calculated from the following:

$$L_{ep,d} = L_{Aeq,T} + 10\log_{10}\left(\frac{T_e}{T_0}\right) dB \qquad (9.1)$$

where $L_{Aeq,T}$ is the A-weighted equivalent continuous sound pressure level, T_e is the effective duration of the workday and T_0 is the reference duration (T_0 = 8 h). The results for glass collectors are shown in full in the Table 9.1.

Therefore, the 15 glass collection operatives are subject to a daily A-weighted noise exposure level of 90.7 dB with the associated expanded uncertainty for a one-sided coverage probability of 95% (k = 1.65) of 3.2 dB.

Chapter 10

Uncertainty in the Measurement of Noise Emission from Plant and Machinery

Robert Peters

Introduction

The emission of sound from plant, machinery and other sources may be quantified, specified and measured in two ways: either as a sound power level or as a sound pressure level measured at a specified distance and direction from the source. Although the sound power level is often the preferred method, there are situations and circumstances when the sound pressure level method is the better one – for example, when it is required to find the sound pressure level close to the sound source, that is, in its near field, at workstations and machine operators' positions in workplaces.

Information about noise emission levels of sound sources is required for many reasons, including

- Comparison with manufacturers' data,
- Comparison of noise emissions from different sound sources,
- To check compliance with limits specified in a purchasing contract or a regulation,
- To diagnose and specify noise-reduction measures,
- To check the effectiveness of noise-reduction measures,
- To predict noise levels in workplaces and the environment, and
- To specify and verify compliance with European Union (EU) machinery regulations.

In any of these cases, decisions are better informed by knowledge of measurement uncertainty associated with noise emission level values.

Machinery Noise Regulations

There are two EU directives relating to machinery noise currently in operation: the machinery directive (2006/42/EC) and the outdoor directive (2000/14/EC). Both directives originate from Article 114 of the Treaty on the Functioning of the European Union relating to product requirements and the free movement of goods. In other words, these are designed to ensure that as far as machinery noise emissions are concerned, the EU requirements are the same throughout Europe. The machinery noise directive is mainly concerned with noise in the workplace and protecting the hearing of employees from damaging levels of noise exposure, and the outdoor directive is related to minimising disturbance from noise from outdoor machinery in the environment.

Both directives require that noise emissions from machinery should be declared openly and that manufacturers should reduce the noise of their products using noise-reduction techniques as far as possible. Thus, the directives are helpful in informing purchasers of equipment and machinery of noise emission levels from competing products and may thereby encourage manufacturers to seek competitive advantage by developing quieter machines.

These regulations require noise emissions from machinery to be specified in terms of a combination of sound power levels and noise emission sound pressure levels. There are three series of standards that cover these types of acoustics measurements.

Standard Methods for Measuring Noise from Machinery

The standard methods for determining sound power levels using sound level measurements are described in the series of standards BS EN ISO 3740–3747. Those for the determination of sound power level using sound intensity are described in BS EN ISO 9614, parts 1–3. Methods for measuring noise emission levels arising from machinery and equipment are described in the series of standards BS EN ISO 11200–11205. A list including full details of these standards is given at the end of this chapter which also includes standards relating to compliance with machinery noise directives.

Grades of Accuracy

The methods are divided into three classes or grades according to accuracy:

- *Precision grade (grade 1)*. These are the most accurate methods, requiring laboratory standard acoustic test facilities – an anechoic or reverberant room.
- *Engineering grade (grade 2)* and
- *Survey grade (grade 3)*.

Different levels of measurement uncertainty are associated with the different grades of measurement accuracy. Precision grade methods offer the lowest level of uncertainty, and the survey grade method has the highest.

Measurement of Sound Power Levels L$_w$

There are four standardised methods

- Involving measurements of sound levels over a prescribed test surface surrounding the source under free-field conditions (or near-free-field condition),
- Involving measurements of sound levels arising from a sound source in a reverberant space,
- Based on comparing measurements of sound levels arising from the sound source and from a sound source of known, calibrated sound power level (comparison method) and
- Involving measurements of sound intensity level.

There are seven standards in the series BS EN ISO 3740–3747. Four of them relate to measurements in reverberant spaces and three to measurement under free-field or near-free-field conditions. Two of the standards describe precision grade measurements (one for free-field and one for reverberant field-test environments), and three of them relate to engineering grade accuracy. There is one standard that is a survey grade method (BE EN ISO 3746) and one (BS EN ISO 3744) that may be either engineering or survey grade depending on the acoustic performance of the test environment. Six of the seven standards specify measurement of sound power levels in both octave bands and overall A-weighted decibel levels (dBA), but BS EN ISO 3746 is for A-weighted levels only.

Most of these standards are suitable for all types of noise (broad band, narrow band, discrete frequency, steady, non-steady, tonal, impulsive), as defined in ISO 2204. The first standard in the series, BS EN ISO 3740, gives a guide to the selection and use of the various methods for measuring sound power levels of sound sources detailed in the other standards in the series. Some of the main features of these standards are summarised in Table 10.1.

Parts 1–3 of BS EN ISO 9614 describe methods for measuring sound power levels of machinery and equipment using a sound intensity meter. No special test environments are needed for these measurements, which include precision grade, engineering grade and survey methods.

Measurement of Emission Sound Pressure Levels from Machinery

The methods for the measurement of sound emission levels described in the series of standards BS EN ISO 11200–11205 are also divided into the same three grades of accuracy. The first standard in the series, BS EN ISO 11200, gives a guide to the selection and use of the various methods for measuring sound emission levels described in the other standards of the

Table 10.1 Sound power level measurement standards

Standard	Measurement method	Accuracy grade	Test room	Character of noise from the source	Standard deviation of reproducibility σ_R, dB	ANSI/ASA equivalent
ISO 3740	This standard provides explanations and guidance about the use of the other standards in this series.					
ISO 3741	Reverberant-field method	Precision	High-quality reverberant test room	Steady, broad-band, narrow-band or discrete frequency	0.5	S12.51/ ISO 3741
ISO 3743–1	Reverberant-field method	Engineering	Room with hard, highly reflective walls, subject to require-ments in the standard	Any, but no isolated bursts	1.5	S12.53/ P1
ISO 3743–2	Reverberant-field method	Engineering	Room has to meet require-ments on reverberation time and volume	Any, but no isolated bursts	2	S12.53/ P2
ISO 3744	Free field method	Engineering	Free-field over a reflecting pane	Any	1.5	S12.54/ ISO 3744
ISO 3745	Free-field method	Precision	High-quality anechoic or semi-anechoic room	Any	0.5	S12.55/ ISO 3745
ISO 3746	Free-field method	Survey	No special room required, indoors or outdoors but with limitations on the back-ground noise level	Any	3–5	S12.56/ ISO 3746
ISO 3747	Substitution method in	Engineering or survey	In-situ method subject to		1.5 or 4	S12.57/

(Continued)

Table 10.1 (Cont.)

Standard	Measurement method	Accuracy grade	Test room	Character of noise from the source	Standard deviation of reproducibility σ_R, dB	ANSI/ASA equivalent
	a reverberant field		certain requirements	Steady, broad-band, narrow-band or discrete frequency		ISO 3747
ISO 9614–1	Any	Precision, engineering or survey	Broad-band, narrow-band or discrete frequency if stationary in time	Broad-band, narrow-band, or discrete if stationary	1, 1.5 or 4	S12.12
ISO 9614–2	Any	Engineering or survey	Broad-band, narrow-band or discrete frequency if stationary in time	Broad-band, narrow-band or discrete if stationary	1.5 or 4	S12.12
ISO 9614–3	Any	Precision	Broad-band, narrow-band or discrete frequency if stationary in time	Broad-band, narrow-band or discrete if stationary	1	S12.12

series. BS EN ISO 11200 also contains an annex giving very useful case studies (worked examples) of how to use the various standards, including how to estimate uncertainty.

Measurement Uncertainty in Machinery Noise Emission Values

The measurement of noise emissions from machinery (whether of sound power levels or of emission sound pressure levels) will be subject to uncertainties arising not only from the measurement method itself but also from variation in the noise being emitted from the source.

Variability in the Measurement Method

The first category might include variability in the performance of the measuring instrumentation and in the way it is being operated by the user; variability in the choice of microphone positions; variability in background noise level in the acoustic environment in which the measurements are occurring (i.e. proximity of sound reflecting or absorbing surfaces) and variability in the physical environment (such as changes in temperature or humidity).

Variability in Operating and Mounting Conditions

The second category includes uncertainties arising from variability in machine operating and mounting conditions. Operating conditions that could influence noise emission levels include changes in machine operating speed and fluctuations in power-line voltage, load, type of cutting tool or workpiece material being used. Many automated machines are programmed to perform over various operating cycles which may involve different speed and load conditions for various periods of times.

Also, the way in which the machine is mounted or supported (e.g. on the floor or wall or table or workbench) can influence noise emission because of the possibility of noise being transmitted to and radiated from these surfaces, as well as directly from the machine surfaces, as well as the possibility of sound being reflected from nearby surfaces. Machines mounted on a concrete floor might radiate sound differently than when mounted on other, less rigid and more sound-absorbing surfaces. Specifically, small machines may radiate more low-frequency noise when connected to large surfaces area such as floors and walls, since these can radiate structure-borne noise transmitted from the machine. The use of resilient anti-vibration mounts to support machinery may reduce sound radiation, and the use of such fixings should be reported as part of the noise emission test procedure. Ducts, conduits and cables that are rigidly connected to machinery can also radiate sound and should also be isolated. Changes in any of these conditions relating to the way a machine is mounted or supported can cause uncertainty in the noise emission levels.

Where operating and mounting conditions can cause significant changes in noise output, it may be advisable to make separate measurements of sound power level and noise emission levels for each different set of operating and mounting conditions. Careful descriptions of the operating and mounting conditions should always be included as part of any machinery noise test report.

Noise Test Codes

Noise test codes are a type of measurement standard that, for a specific class or type of machinery, describes the test methods to be used to determine noise emission levels, usually specifying methods described in one or more of the standards listed in Table 10.1. They also specify machine operating and mounting conditions to be used in such tests and may give information about measurement uncertainty. There are noise test codes for many of the different types of machinery covered by machinery noise regulations. Further details about noise test codes and the typical level of detail that they specify are provided at the end of this chapter.

Estimating Measurement Uncertainty of Machinery Noise Measurements

The *Guide to Uncertainty in Measurements* (GUM 2008) gives two alternative approaches to the determination of measurement uncertainties. The first method is to perform a series of reproducibility tests (often known as 'round-robin tests') which are sufficiently comprehensive to encompass all the variations in factors likely to cause changes in the measurement result. This may be a very time-consuming process and may be impracticable to carry out in some cases. The alternative method is to estimate the uncertainties arising from each of the factors that can affect the measurement result and then combine them to produce an estimated total uncertainty – or, in other words, to construct an 'uncertainty budget'. This second approach requires knowledge of exactly how the variability in each of the factors affects the measurement result, that is, knowledge of the functional relationship between each of the factors and the measurement result or, in the language of the GUM, an 'uncertainty model'. This could be a mathematical equation involving all the factors and the value of the measured quantity. In many cases, however, insufficient information is available to construct a full model, and the uncertainty budget must be constructed using a combination of experience, previous results, published information and noise measurements (e.g. to determine how changes in a specific factor affect the measurement result).

In the BS EN ISO 3740 series of standards, upper limits for the standard deviation of reproducibility are suggested for each of the seven standards and similarly for the BS EN ISO 9614 and BS EN ISO 11200 series of standards. These can be used as upper limits for uncertainty estimates if it is not possible to conduct reproducibility tests on specific machines. Also, the BS EN ISO 11200 series of standards suggests upper limits for reproducibility arising from variability in operation and mounting conditions and gives guidance on the development of uncertainty budgets.

Standard Deviation of Reproducibility of the Measurement Method σ_{RO}

The value of σ_{RO} can be obtained from reproducibility tests on one machine using fixed operating and mounting conditions. Alternatively, the maximum values suggested in the standards can be used. For the measured A-weighted values of sound power level or emission sound pressure level, these are detailed in Table 10.1, but in summary, they are ±0.5 dB for precision grade (grade 1) methods, between 1.5 and 2 dB for engineering (grade 2) methods and ±4 dB for survey methods.

Variability Arising from Machine Operating and Mounting Conditions σ_{OMC}

Values for the standard deviation of uncertainty arising from variability in machine operating and mounting conditions σ_{OMC} are suggested in BS EN ISO 11200 for conditions classed as *stable* ($\sigma_{OMC} = 0.5$ dB), *unstable* ($\sigma_{OMC} = 2$ dB) and *very unstable* ($2 \leq \sigma_{OMC} \leq 4$ dB).

These conditions have the following descriptions:

- *Stable.* Where the sound emission does not vary or varies only a little with time and the measuring procedure is defined properly;
- *Unstable.* Where, for example, there is a large influence of the material flow in and out of the machine or this flow may vary in an unforeseeable manner; and
- *Very unstable.* Where, for example, noise generated by the material and process varies strongly (stone-breaking machines, metal-cutting machines and presses operating under load).

Combining the Uncertainties σ_{RO} and σ_{OMC}

If the two uncertainties are uncorrelated (i.e. independent of each other), then the total uncertainty σ_{total} may be obtained by combining the two component uncertainties as follows:

$$\sigma_{total} = \pm\sqrt{\sigma_{RO}^2 + \sigma_{OMC}^2}$$

Note that the values of the standard deviation of reproducibility (σ_{RO} and σ_{OMC}) referred to in BS EN ISO 3740–3477 and in the BS EN ISO 1200 series are taken to be the same as the standard uncertainty value described in the GUM and in other chapters of this book (i.e. they can be used to determine confidence limits by multiplying by appropriate cover factor k (see, e.g., section 1.4, on measurement uncertainty, in BS EN ISO 3746).

In other words, the preceding equation can be replaced by

$$U_{total} = \pm\sqrt{U_{RO}^2 + U_{OMC}^2}$$

The total value of U_{total} for the various possible values of σ_{RO} and σ_{OMC} is shown in Table 10.2.

Expanded Uncertainties

Derived from σ_{total}, the expanded uncertainty U in decibels is calculated from

$$U = k\sigma_{total} \tag{10.1}$$

The expanded uncertainty depends on the degree of confidence that is desired. For a normal distribution of measured values, and for a confidence level of 95%, the confidence interval is $(L_p - U)$ to $(L_p + U)$. This corresponds to a coverage factor of $k = 2$.

If the purpose of determining the emission sound pressure level is to compare the result with a limit value, it may be more appropriate to apply the coverage factor for a one-sided normal distribution. In this case, the coverage factor $k = 1.6$ corresponds to a level of confidence of 95%.

Example 1

The sound power level of a machine has been measured at an outdoor site in accordance with BS EN ISO 3746, and a result of 100 dBA has

Table 10.2 Values of U_{total} for Various Values of σ_{RO} and σ_{OMC}

Accuracy grade	Standard deviation of reproducibility of the method σ_{RO}, dB	Calculated total standard deviations σ_{total}		
		Operating and mounting conditions		
		Stable, $\sigma_{OMC} = 0.5$ dB	Unstable, $\sigma_{OMC} = 2$ dB	Very unstable, $\sigma_{OMC} = 4$ dB
1	0.5	0.7	2.1	4
2	1.5	1.6	2.5	4.3
2	2	2.1	2.8	4.5
3	3	3.0	3.6	5.0
3	4	4	4.5	5.7
3	5	5	5.5	6.4

been obtained. It has not been possible to carry out reproducibility tests on the machine, so the suggested maximum value for the standard reproducibility for this grade 3 test method have been assumed. After considering the nature of the acoustic environment at the site (e.g. nearby sound reflecting and absorbing surfaces that could be included in an environmental correction factor K_2), the level of background noise and that the machine emits noise that contains prominent discrete tones, a value of 5 dB has been assumed for the standard reproducibility of the test method.

This does not allow for any variability in noise emission arising from change in operating and mounting conditions, which is found to be considerable, so an estimate of 2 dB has been included for the standard deviation due to operating and mounting conditions, as recommended for 'unstable' conditions in the BS EN ISO series of standards.

A coverage factor of 2 dB is assumed for determining the upper and lower estimates for the sound power level of the machine. Hence,

$$U = 2 \times \sqrt{\sigma_{RO}^2 + \sigma_{OMC}^2} = 2 \times \sqrt{5^2 + 2^2} = \pm\, 11 \text{dBA}$$

Therefore, the best estimate of the sound power level is 100 dBA, but it will lie between 111 and 89 dBA with a 95% confidence level.

Example 2

A noise emission level at a workstation has been measured in accordance with BS EN ISO 11201, and a result of 82 dBA has been obtained. It has not been possible to carry out reproducibility tests on the machine, so the suggested maximum value (1.5 dB) for the standard reproducibility for this grade 2 test method has been assumed. The tests results indicate some significant variation of noise emission levels with changes in machine operating and mounting conditions, so a value of 2 dB has been assumed for the standard deviation due to operating and mounting conditions. A coverage factor of 1.6 dB is assumed for determining an upper limit for the noise emission value (for a one-tailed probability distribution). Therefore, $U = 1.6 \times \sqrt{1.5^2 + 2^2} = \pm 4\, \text{dB}$, giving an upper limit to the noise emission level of 82 + 4 = 86 dBA. Therefore, it is estimated that the noise emission level will not exceed 86 dBA with a 95% confidence level.

Estimation of σ_R Using an Uncertainty Budget Based on Modelling (t

This section is taken from BS EN ISO 11201, annex C, section 11.3.3).

Modelling Approach for σ_{RO}

Generally, σ_{RO}, in decibels, depends on several partial uncertainty components $c_i u_i$ associated with the different measurement parameters such as uncertainties of instruments, environmental corrections and microphone positions. If these contributions are assumed to be uncorrelated, σ_{RO} can be described by the modelling approach presented in ISO/IEC Guide 98–3 as follows:

$$\sigma_{RO} = \sqrt{\sum_i c_i^2 u_i^2} = \sqrt{(c_1 u_1)^2 + (c_2 u_2)^2 + (c_3 u_3)^2 + \cdots + (c_n u_n)^2}$$

where u_1, u_2, ..., u_n are the standard uncertainties associated with the various components of uncertainty, and c_1, c_2, c_3, ..., c_n are the corresponding sensitivity coefficients.

Contributors to the Uncertainty σ_{RO}

The general expression for calculating the emission sound pressure level measurement L_p, including all corrections prescribed by this international standard and with all relevant uncertainties, is[1]

$$L_p = L_p\left[L'_p, \delta(B), \delta_{env}, \delta_{SLM}, \delta_{mount}, \delta_{OC}, \delta_{pos}, \delta_{met}\right] \qquad (10.2)$$

where L'_p is the measured (uncorrected) sound pressure level, $\delta(B)$ is an input quantity to allow for any uncertainty on background noise corrections, δ_{env} is an input quantity to allow for any uncertainty due to local environmental influence, δ_{SLM} is an input quantity to allow for any uncertainty in the measuring instrumentation, δ_{mount} is an input quantity to allow for any variability in the mounting conditions of the source under test, δ_{OC} is an input quantity to allow for any deviation in the operating conditions of the source under test from the nominal conditions, δ_{pos} is an input quantity to allow for any uncertainty in selection of the measuring position and δ_{met} is an input quantity to allow for any uncertainty in determining the meteorological conditions. Example values and the resulting reproducibility of measurement method uncertainty are listed in Table 10.3.

---

1 For those unfamiliar with the mathematical language and terminology (and with apologies to those who are), Equation (10.2) says that the final result of the emission sound pressure level measurement L_p is a function of (i.e. depends on) the six factors listed inside the square brackets.

Table 10.3 Uncertainty budget for determinations of emission sound pressure level (the values shown are examples related to accuracy grade 2 determinations)

	Quantity L_p	Estimate, dB	Standard uncertainty U_i, dB	Probability distribution	Sensitivity coefficient c_i	Uncertainty contribution $c_i u_i$, dB
$i = 1$	L'_p	L'_p (mean)	$S_{L'_p}$ (e.g. 0,5)	Normal	1.25	0.63
$i = 2$	$\delta(B)$	K_1	e.g. 0.7	Normal	0.25	0.18
$i = 3$	δ_{env}	0	I	Normal	I	I
$i = 4$	δ_{SLM}	0	0.5	Normal	I	0.5
$i = 5$	δ_{pos}	0	0.2	Normal	I	0.2
$i = 6$	δ_{met}	0	0.3	Normal	I	0.3
	$\sigma_{RO} = \sqrt{\sum_i c_i^2 u_i^2} = 1.3\,\text{dB}$					

Note that the estimated value from the uncertainty budget of 1.3 dB is below the maximum value of 1.5 dB for an engineering grade (class 2) accuracy, as expected. Notes on how the values of u_i and c_i were determined are given in annex C of BS EN ISO 11201. Ksub1 is the background correction factor and is mentioned again later in this Chapter.

Uncertainties Arising from Tests on a Batch of Machines

In what has been discussed so far, it has been assumed that repeatability and reproducibility tests relate to tests of one specific item of machinery. If the tests cover the use of samples from a batch of nominally identical machines, then another source of uncertainty must be considered and included. This is the variability between different machines in the batch, which may be expressed as a standard deviation of production σ_P, which can be combined with the other uncertainties.[2]

Noise Test Code Standards

In this section, some examples of noise test code standards are given, together with an extract from one of the standards to illustrate the type and level of detail that may be included.

BS EN ISO 7779:2001, *Acoustics: Measurement of Airborne Noise Emitted by Information Technology and Telecommunications Equipment*, covers equipment such as printers, teleprinters (online or keyboard operated, standalone), keyboards, copiers, card readers, card punches, magnetic tape units, disk units and storage subsystems (hard drives), visual display units (screens monitors), electronic units (processors, controllers,

---

2 For further information, please see BS 6805–4:1987.

circuits, power supplies), microform readers, facsimile machines (tele-copiers) and page scanners, personal computers and workstations, page printers (personal printers), self-service automatic teller machines (ATMs) and enclosures or rack systems.

BS EN ISO 15744:2008, *Hand-Held Nonelectric Power Tools: Noise Measurement Code – Engineering Method (Grade 2)*, is applicable to typical hand-held non-electric power tools such as drills, tappers, grinders, belt san-ders, polishers, rotary files, rotary sanders, die grinders, circular saws, orbital and random orbital sanders, jigsaws, nibblers, oscillating saws, reciprocating saws, reciprocating files and shears, reciprocating files and knives, oscillating saws and knives, hammers and riveting hammers, rammers, tampers, scaling hammers, needle scalers, drifters, plug-hole drills, rotary hammers, rock drills and stoppers, non-ratchet screwdrivers and nut-runners, screwdrivers and wrenches with ratchet clutches and pawl-type ratchet wrenches, impact wrenches and screwdrivers and air-hydraulic impulse wrenches and screwdrivers.

BS EN 60704–1:2010+A11:2012, *Household and Similar Electrical Appli-ances: Test Code for the Determination of Airborne Noise – General Requirements*, parts 2–8, cover requirements for the testing of vacuum clean-ers, fan heaters, dishwashers, washing machines and spin extractors, electric thermal storage room heaters, tumble dryers, fans and electric shavers.

As an example of the type and level of detail that may be included in these standards, a quote from EN 60704–2:2010 on requirements for fan heaters follows:

Operating Conditions

Prior to noise measurements, the appliance equipped in accordance with Subclause 6.1.1, shall have been in operation for a total period of at least 2 h for running-in at the highest speed setting with the maximum heating switched on for normal permanent use.

Oscillating function if available shall be switched on. During the running-in procedure, air filters, if any, may be removed. If filters remain in the appliances during this running-in period, they shall be clean or renewed after this period.

Mounting Conditions

For measurements on floor-standing appliances intended for placing against a wall, including those for building in, a vertical reflecting plane having an acoustic absorption coefficient of less than 0.06 must be used.

When measurements are made in a reverberation test room, a part of the wall of the room will serve for this purpose. The minimum area of this part of the wall should be determined by the projection of the appliance extended by at least 0.5 m upwards and to both sides. The minimum distance between any surface of the appliance or its cabinet and the nearest corner of the room shall be 1 m.

When measurements are made in a free-field environment, the size of the vertical reflecting plane (supported by the horizontal reflecting plane) shall be at least equal to the size of the projection of the measurement surface.

For both types of test environment, the following requirements shall be complied with:

- the appliance shall be placed in the test environment without any resilient means of support other than those incorporated in the appliance;
- care should be taken to avoid any direct contact between the appliance (including protruding parts, worktops, spacers, etc.) and the vertical reflecting wall;
- the distance between the wall and the appliance shall be established by placing the appliance in direct contact with the wall and moving it away for a distance not exceeding 10 cm ±1 cm.

Approaches to uncertainty in some other noise emission standards

This chapter concludes with a brief review of the different approaches to uncertainty estimation taken in a sample of three other standards relating to noise emission from fans, from motorcycles and aircraft in flight.

BS ISO 10302–1: 2011, Acoustics: Measurement of Airborne Noise Emitted and Structure-Borne Vibration Induced by Small Air-Moving Devices, Part 1: Airborne Noise Measurement

There are several different standards for the determination of sound power levels of various types of fans, including, for example, BS EN ISO 7779: 2010, *Acoustics: Measurement of Airborne Noise Emitted by Information Technology and Telecommunications Equipment*; BS EN ISO 5136: 2009, *Acoustics: Determination of Sound Power Radiated into a Duct by Fans and Other Air-Moving Devices – In-duct Method*; BS EN ISO 5801: 2008, *Industrial Fans: Performance Testing Using Standardized Airways*; and BS ISO 13347–1: 2004+A1: 2010, *Industrial Fans: Determination of Fan Sound Power Levels under Standardized Laboratory Conditions – General Overview*. These will all have their own approaches to 'uncertainty'.

BS ISO 10302–1: 2011 specifies, in detail, methods for determining and reporting the airborne noise emissions of small air-moving devices (AMDs) used primarily for cooling electronic equipment, such as that for information technology and telecommunications. Measurement uncertainties are dealt with in section 9 and annex E

In the absence of current knowledge sufficient to apply the methods of ISO/IEC Guide 98 to the determination of uncertainties, recommended estimates are given in section 9 of the reproducibility of measurement uncertainty in octave and one-third-octave bands and for A-weighted measurements. These vary from 1.5 dB for the mid-range octave bands from 500 to 400 Hz and for A-weighted measurements to 4 dB for the 125-Hz band and 2.5 dB for the 250- and 8,000-Hz bands.

Guidance on determination of the expanded uncertainty when more detailed knowledge becomes available is given in annex E, which lists and discusses the various uncertainties that need to be taken into account when applying the free-field method over the reflecting-plane method of ISO 3744. Apart for the uncertainties already discussed in this chapter relating to the measurement instrumentation, source operating conditions and mounting conditions, several other sources of uncertainty are discussed (with suggested uncertainty estimates given), including

- Uncertainties in measurement of the background and environmental correction factors K_1 and K_2 and of the measurement surface area. The method for determining the uncertainty in the background correction factor K_1 is very similar to that explained in Example 2 in Annex A of Chapter 2 of this book.
- A correction to account for any difference of angle between the direction in which the sound is emitted by the source and the normal to the measurement surface in decibels (because the theory underlying the method assumes that sound intensity is radiated from the machine in a direction perpendicular to the measurement surfaces);
- An input quantity to allow for any uncertainty due to the finite number of microphone positions in decibels (because the underlying theory assumes that the sound intensity is being integrated over the entire area of the measurement surface and an infinite number of microphone positions).

The standard shows how these various uncertainties may be used to draw up an uncertainty budget from which a combined standard uncertainty can be calculated and used to determine an expanded uncertainty. This could either use a k factor of 2 for a 95% confidence range or, if the purpose of determining the sound power level is to compare the result with a limit value, it might be more appropriate to apply the coverage factor for a one-sided normal distribution, in which case the coverage factor $k = 1.6$ corresponds to 95% confidence.

BS ISO 362–2: 2009, Measurement of Noise Emitted by Accelerating Road Vehicles – Engineering Method, Part 2: L Category

This standard specifies the procedure for the measurement of noise from a standard 'drive-by' test for motorcycles. The drive-by test is carried out on a single motorcycle driven under specified conditions of road speed and engine speed along an isolated stretch of road (the test track). Sound level measurements are carried out at a specified distance from the test track in order to determine the maximum sound level in A-weighted decibels as the motorcycle passes between certain marked points on the road, with their mid-point opposite the sound level measurement position. The measurement site should be free of sound-reflecting objects and have a low background noise level. Measurement uncertainties are dealt with in section 8.5 and annex B of this standard.

Many factors can lead to variability in the measured result. In the absence of complete knowledge of these factors and the way they influence the measured result, table 10.3 of section 8.5 (reproduced here as Table 10.4) gives an estimate of uncertainties based on existing statistical data, analysis of tolerances and engineering judgement.

The standard suggests that uncertainty effects may be grouped into three areas composed of the following sources:

1. Uncertainty due to changes in vehicle operation within consecutive runs, small changes in weather conditions, small changes in background noise levels and measurement system uncertainty: all these are referred to as 'run-to-run variations';
2. Uncertainty due to changes in weather conditions throughout the year, changing properties of a test surface over time, changes in measurement system performance over longer periods and changes in the vehicle operation: all these are referred to as 'day-to-day variations'; and

Table 10.4 Variability of measurement results for a coverage probability of 80%

Run- to-run. dB	Day-to-day, dB	Site-to-site, dB
0.5	0.9	1.4

Note: This is table 10.3 of BS EN ISO 362–2: 2009, Variability of Measurement Results for a Coverage Probability of 80%.

3. Uncertainty due to different test site locations, measurement systems, road surface characteristics and vehicle operation: all these are referred to as 'site-to-site variations'.

Annex B gives a framework for analysis in accordance with ISO/IEC Guide 98–3, which can be used to conduct future research on measurement uncertainty for this part of ISO 362.

Seven sources of uncertainty are identified, and it is recommended that future research should develop uncertainty budgets based on these seven factors:

*   Variations in measurement devices, such as sound level meters, calibrators and speed-measuring devices;
*   Variations in local environmental conditions that affect sound propagation at the time of measurement;
*   Variations in vehicle speed and vehicle position during the pass-by run;
*   Variations in local environmental conditions that affect the characteristics of the source;
*   Effect of variation in environmental conditions that influence the mechanical characteristics of the source, mainly engine performance (air pressure, air density, humidity, air temperature);
*   Effect of variation in environmental conditions that influence the sound production of the propulsion system (air pressure, air density, humidity, air temperature) and the roiling noise (tyre and road surface temperature, humid surfaces); and
*   Variation in test site properties (test surface texture and absorption, surface gradient).

The standard shows how these various uncertainties may be used to draw up an uncertainty budget from which a combined standard uncertainty can be calculated and used to determine an expanded uncertainty.

The expanded uncertainty U is calculated by multiplying the combined standard uncertainty u with the appropriate coverage factor for the chosen coverage probability, as described in ISO/IEC Guide 98–3.

BS ISO 20906: 2009, Acoustics: Unattended Monitoring of Aircraft Sound in the Vicinity of Airports

This standard describes requirements for monitoring the sound of aircraft operations (departures and landings) from an airport; performance specifications for sound measuring instruments and requirements for their unattended installation and operation so as to determine continuously monitored sound pressure levels of aircraft sound at selected locations.

Aircraft noise in the community close to airports is usually measured in terms of the L_{ASmax} and sound exposure level (SEL) values of individual aircraft noise events. This is usually achieved by automated monitoring of noise levels that exceed a trigger level several decibels above the typical residual noise level that occurs when there are no aircraft flying overhead. Only the captured events that correlate with aircraft radar signals by the airport's noise and track-keeping system are counted as aircraft noise events.

Selection of the trigger level is a matter of some significance. Since it is impossible to separate the residual noise and the aircraft noise occurring at the time of the event, the captured aircraft noise event signal will inevitably contain a contribution from the residual noise, and conversely, the residual noise level will contain a component of the aircraft noise level that is below the trigger level. Setting the trigger level too low will result in significant inclusion of residual noise in the measured aircraft noise level, and setting it too high will result in the omission of a significant amount of aircraft noise (below the trigger level) not being included in the measured aircraft noise level. Since both the level from individual aircraft noise events and the level of residual noise that occurs during each event will vary, there will be a contribution to the uncertainty in the measured aircraft event noise level arising from the effect of residual noise.

The average level of aircraft-related noise at the microphone may be calculated from the SEL of the individual aircraft noise level events and the number of such events that occur over any period of time (hours, days, weeks or months). Measurement uncertainties are dealt with in section 6 and annex B of the standard.

On measurement uncertainty, section 6 of the standard simply states that

> [t]he uncertainty of results obtained from measurements according to this International Standard shall be evaluated, preferably in compliance with ISO/IEC Guide 98-3. If reported, the expanded uncertainty together with the corresponding coverage factor for a stated coverage probability of 95% as defined in ISO/IEC Guide 98-3 shall be given. Guidance on the determination of the expanded uncertainty is given in Annex B.

Annex B is long and detailed, and only a brief summary is given here.

Measurement of the Noise from a Single Specific Noise Event

For the measurement of a single specific aircraft noise event, the variability in source noise emission and from sound propagation through the atmosphere that occurs from event to event is not a factor. In these circumstances, the main factors affecting uncertainty are in the measuring instrumentation and the variability in the contributions due to residual sound. A detailed analysis of the contributions to uncertainty in the sound level meter readings is given, broadly along similar lines to those given in the Chapter 3 concerned with uncertainty in measurement instrumentation. A value of ±0.74 dB is suggested.

Also, the directivity of the microphone and its variation with frequency will be an important factor. The above-mentioned figure of ±0.74 dB is for a range of angles of incidence of less than 90 degrees; the estimate increases to ±0.86 dB for a wider range of incidence sound of less than 90 degrees.

Measurement of the Noise from Many Different Aircraft Noise Events Occurring over a Period of Time

In this case, additional uncertainty will arise from the variability of the source noise emission level and of the sound transmission path due to variability in atmospheric conditions. A database referred to in the standard indicates that the effects of variability in sound propagation conditions are usually much greater than those arising from variability in noise emission levels of the aircraft.

The factors that affect the transmission of sound from an aircraft in flight to a microphone close to ground level include

- The attenuation, refraction (due to wind and temperature gradients) and scattering (due to turbulence in the air) of sound travelling through the air;
- Attenuation of sound during reflections from the ground because the sound received at the microphone will be a combination of direct airborne and ground reflected sound; and
- The frequency spectrum of the sound leaving the aircraft because the preceding factors will vary with frequency.

In addition, these factors will vary with time and so will contribute to the uncertainty arising from sound propagation from the aircraft sound source to the microphone.

To minimise measurement uncertainties, the standard gives recommendations about the selection of suitable sites for aircraft noise monitoring and about the positioning of the microphone at such sites:

- As a minimum requirement, all acoustically relevant reflecting surfaces other than the ground should be at least 10 m away from the microphone in order to provide minimum uncertainty in the sound level measurements;
- The microphone should be at least 6 m above the ground and preferably up to 9 m above the ground; and

- The residual noise level (i.e. when no aircraft noise is present) should be at least 10 dB, and preferably up to 15 dB, below the maximum level of the quietest aircraft to be measured.

However, the standard concedes, in the introduction, that

> [f]or political reasons, it is often necessary to install sound monitors in acoustically unsuitable places. For these situations, the operator of the sound-monitoring system should be aware of a potentially substantial increase of uncertainty in the results, as discussed in Annex B. In extreme situations, the uncertainty may become so large as to make an aircraft sound measurement meaningless.

The measured maximum noise level from each aircraft noise event depends on the shortest distance between the aircraft and the receiver. This shortest distance (sometimes called 'slant distance') is a valuable descriptor to characterize measurement conditions. The SEL of each event also depends on the duration of the event, that is, the time for which it exceeds the event trigger level.

For a sound measurement of aircraft elevation angles greater than 30 degrees and in the absence of obstacles affecting air-to-ground propagation, the atmospheric effects of turbulence, wind and temperature become more pronounced with increasing 'shortest distance', producing increased scatter of measured levels up to several decibels. These fast level fluctuations are the reason why aircraft sound is measured with the 'slow' time weighting S in order to reduce the influence of short-term level variations.

Similar sound events (i.e. those from the same type of aircraft) may be averaged to reduce the uncertainty of the mean value. For example, the resulting expanded uncertainty of sound monitor levels averaged over 100 similar aircraft events was reduced by a factor of 10 to between 0.3 and 0.5 dB.

Uncertainty Contributions Due to Accumulation of Sound Event Data

When the SEL data from all the individual noise events in a time periods (e.g. a day or a month) are combined to produce an overall average level of aircraft noise over the period, it is important to take into account the effect of all missing data, corrupted data, events incorrectly identified as aircraft noise and aircraft noise that is below the threshold level in order to minimise uncertainty. The standard gives advice on these matters.

Further information about uncertainty in aircraft noise measurement may be obtained from an article published by the Environmental Research and Consultancy Department (ERCD) of the Civil Aviation Authority in November 2005: S White, 'ERCD Report 0506: Precision of aircraft noise measurements at the London airports', ERCD, London, November 2005.

Standards Relating to Noise Emissions from Machinery

The BS EN ISO 3740 series for the determination of sound power levels from sound pressure level measurements

The ISO 9614 series on the measurement of sound power levels using sound intensity:

BS EN ISO 9614–1: 2009, *Acoustics: Determination of Sound Power Levels of Noise Sources Using Sound Intensity – Measurement at Discrete Points.*

BS EN ISO 9614–2: 1997, *Acoustics: Determination of Sound Power Levels of Noise Sources Using Sound Intensity – Measurement by Scanning.*

BS EN ISO 9614–3: 2009, *Acoustics: Determination of Sound Power Levels of Noise Sources Using Sound Intensity – Precision Method for Measurement by Scanning.*

BS EN ISO 11200 series for the determination of emission sound pressure levels:

BS EN ISO 11200: 2014, *Acoustics: Noise Emitted by Machinery and Equipment – Guidelines for the Use of Basic Standards for the Determination of Emission Sound Pressure Levels at a Workstation and at Other Specified Positions.*

BS EN ISO 11201: 2010, *Acoustics: Noise Emitted by Machinery and Equipment – Determination of Emission Sound Pressure Levels at a Workstation and at Other Specified Positions in an Essentially Free Field over a Reflecting Plane with Negligible Environmental Corrections.*

BS EN ISO 11202: 2010, *Acoustics: Noise Emitted by Machinery and Equipment – Determination of Emission Sound Pressure Levels at a Workstation and at Other Specified Positions Applying Approximate Environmental Corrections.*

BS EN ISO 11203: 2009, *Acoustics: Noise Emitted by Machinery and Equipment – Determination of Emission Sound Pressure Levels at a Workstation and at Other Specified Positions from the Sound Power Level.*

BS EN ISO 11204: 2010, *Acoustics: Noise Emitted by Machinery and Equipment – Determination of Emission Sound Pressure Levels at a Workstation and at Other Specified Positions Applying Accurate Environmental Corrections.*

BS EN ISO 11205: 2009. *Acoustics: Noise Emitted by Machinery and Equipment – Engineering Method for the Determination of Emission Sound Pressure Levels in Situ at the Workstation and at Other Specified Positions Using Sound Intensity.*

Standards Relating to Machinery Noise Directives

BS EN ISO 4871: 2009, *Acoustics: Declaration and Verification of Noise Emission Values of Machinery and Equipment.*

BS EN ISO 12001: 2009, *Acoustics: Noise Emitted by Machinery and Equipment – Rules for the Drafting and Presentation of a Noise Test Code.*

ISO 5725–2: 1994, *Accuracy (Trueness and Precision) of Measurement Methods and Results*, Part 2: *Basic Method for the Determination of Repeatability and Reproducibility of a Standard Measurement Method.*

ISO 7574 series on statistical methods for determining and verifying stated noise emission values of machinery and equipment.

BS 6805–1: 1987, ISO 7574–1: 1985, EN 27574–1: 1988, *Statistical Methods for Determining and Verifying Stated Noise Emission Values of Machinery and Equipment*, Part 1: *Glossary of Terms.*

BS 6805–2: 1987, EN 27574–2: 1988, ISO 7574–2: 1985, *Statistical Methods for Determining and Verifying Stated Noise Emission Values of Machinery and Equipment – Method for Determining and Verifying Stated Values for Individual Machines.*

BS 6805–3: 1987, EN 27574–3: 1988, ISO 7574–3: 1985, *Statistical Methods for Determining and Verifying Stated Noise Emission Values of Machinery and Equipment – Method for Determining and Verifying Stated Values for Batches of Machines Using a Simple (Transition) Method.*

BS 6805–4: 1987, EN 27574–4: 1988, ISO 7574–4: 1985, *Statistical Methods for Determining and Verifying Stated Noise Emission Values of Machinery and Equipment – Methods for Determining and Verifying Stated Values for Batches of Machines.*

All these standards, which are interrelated, provide information about measurement uncertainty.

Further Reading

A guide to measuring sound power: A review of international standards', white paper issued by Siemens PLM Software, www.siemens.com/plm.

Ambrosini M, 'The uncertainty in standardised sound power measurements: Complying with ISO 1702', Research Doctorate in Agricultural Engineering, Cycle XX, Disciplinary Scientific Sector AGR/09, University of Bologna, http://amsdot torato.unibo.it/1243/1/marco_ambrosini_tesi.pdf (accessed 20/03/20)

Keith SE, Bly SHP, 'Changes to ISO standards and their effects on machinery noise declarations', *Can Acoust* 39(3 (2011), 198.

Keith SE, 'ISO/IEC: The GUM applied to estimation of sound power measurement uncertainties', *Can Acoust* 36(3) (2008), 90.

Kurtz P, 'Is the airborne sound power level of a source unambiguous?', in *Internoise 2014.*

Melbourne: Australian Acoustics Society, 2014, www.acoustics.org.au (accessed 20/03/20).

Mehrgou M, 'A method to apply ISO 3745 for the sound power measurement of I.C. engines in a limited space', Master's degree project, Royal Institute of Technology, Stockholm, Sweden, 2012.

Payne RC, 'Uncertainties associated with the use of a sound level meter', NPL Report DQL-AC 002, National Physical Laboratory, Teddington, UK, April 2004.

PayneR C, Simmons DJ, 'Measurement uncertainties in the determination of the sound power level and emission sound pressure level of machines, National Physical Laboratory Report CIRA (EXT) 007. ISSN 1361–405, Teddington, UK, February 1996.

Peppin RJ, Putnam RA, 'Uncertainty in sound power determination and implications for power plant acoustics', *Inter-noise 2000*. Nice, France: Societe Francaise d'Acoustique, 2000, pp. 1–6.

Rodrigo PB, Costa, F, 'Measurement precision under repeatability conditions of a batch of sound power assessment for blenders in reverberation room', *Arch Acoust* 41(3)(2016),591–597.

Taraldsen G, Berge T, Haukland F, Jonasson H, 'Uncertainty of decibel levels', J Acoust Soc Am 138(3)(2015), 264–269.

Chapter 11

Uncertainty in the Measurement of Vibration Levels

Jorge D'Avillez and Robert Peters

In the next two chapters, you may find the term "variation" being used when discussing uncertainty. Although variation may lead to uncertainty, these two concepts are not interchangeable and should not be treated in the same manner. "Uncertainty" is associated with lack of control, whilst "variation" simply reflects the fact that the results will vary according to some known factor, which may or may not be within the control of the analyst. For instance, room modes cause low-frequency noise levels to vary substantially across the room being analyzed. If we have knowledge of the room shape and so forth, then this "variation" should not necessarily translate into "uncertainty". If, by contrast, for some unknown reason there is no way of gathering the necessary information, such as room shape and size, then this "variation" will be translated into "uncertainty." Uncertainty is highly dependent on the chosen modelling technique. At a scoping stage, where you want to assess whether a detailed model is justified, the assessment trades in accuracy for uncertainty, and we accept the variation in order to save resources spent on collecting all the necessary data. Or even better, we allow for some variation so that our assessment covers a vast set of possibilities.

11.1 Introduction

Levels of vibration in an environment (e.g. dwelling or workplace) emitted from mechanical sources, such as fixed plant or vehicles, may be measured by the acoustician or engineer for many reasons, including

- To determine compliance with maximum acceptable levels of vibration at vibration-sensitive locations – for instance, when carrying out pre-installation tests to establish whether the conditions (typically influenced by the building design and occupation proposed layout) in which the equipment is to operate are adequate.
- To establish compliance with occupational exposure limits – for example, for whole-body or hand-arm vibration limits of the 2005 control of vibration at work regulations.

- To compare with predicted levels of vibration – such as those from construction sites or from road or rail traffic.
- To determine and rank vibration emissions from machinery or operations – for example, to establish the source term for prediction purposes (i.e. characterising the source in terms of vibration emissions).
- To determine transfer functions relating vibration levels at different locations – for instance, between the outside and inside of a building.
- To carry out empirical modelling – this is sometimes done to establish the contribution of specific structural elements such as the ground.
- To calibrate a theoretical model of a situation (e.g. analytical or numerical model of a given scenario).
- To check the effectiveness of measures that have been introduced to achieve a reduction in vibration levels.
- As part of a machinery condition-monitoring programme.

In all these cases, decision making can only be improved by an assessment of the uncertainty of the vibration measurements.

Among the most common vibration indices to be measured either directly or post-processed from measurements are root-mean-square (RMS) levels of vibration displacement, velocity or acceleration [either overall broadband values, frequency-weighted values, octave band or one-third octave band values (maximum, minimum or averaged values)], peak levels, vibration dose values (VDVs) and A(8) values.[1] Sometimes vibration levels are expressed in decibels. This chapter does not claim to give an even or comprehensive coverage of all these situations but an introduction to some of those which are most commonly used.

The first part of this chapter addresses sources of uncertainty associated with vibration measurements. The second part aims to illustrate how to estimate uncertainties in the measurement. For such, it takes up practical examples on rail-induced vibration assessment and occupational vibrational levels assessment.

11.2 Part I: Sources of Uncertainty Associated with Ground-Borne Noise and Vibration Measurements

Vibration assessment is an extensive theme requiring a wide range of assessment strategies to fit the scenario being evaluated. The characteristics of the vibration that the source generates, the propagation medium, the incident wave field, the receiving structure and even the selected criteria are aspects

---

1 A(8) represents the total value (vector sum) of the frequency-weighted acceleration values measured in each orthogonal axis and normalized to eight hours. This metric is commonly used for assessing the daily hand-arm vibration exposure.

that may influence the assessment procedure. Given the extensiveness of the theme, this first part of the chapter discusses sources of uncertainty related to rail-induced ground-borne noise and vibration assessment procedures. Not only is rail-induced ground-borne vibration taken to be the most common and widespread source of perceptible environmental vibration, its assessment, when establishing ground-borne noise, involves structure-borne vibration and radiated noise (see Figure 11.1), therefore touching on a wide range of different assessment methods.

Airborne sound propagates, as compressional waves, through a somewhat homogeneous medium, and therefore, as it travels, its relationship with space is reasonably predictable once the atmospheric fluctuations throughout its propagation path (such as the humidity, temperatures and wind) are accounted for. Ground-borne vibration, by contrast, which propagates through a layered heterogeneous structure promoting a change in wave type as it travels, is highly affected by the propagation medium not only in magnitude but also in characteristics (i.e. frequency content). Changes along its propagation path therefore become difficult to predict, unless there is a precise and comprehensive geotechnical knowledge of the ground structure through which it propagates. When studying the response of a building, for

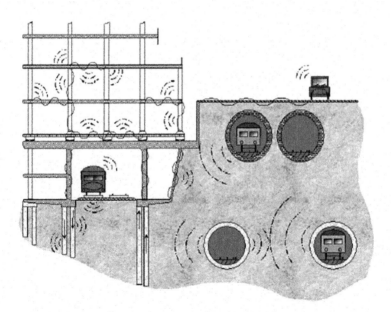

Figure 11.1 Schematic of the propagation and structural response of rail-induced ground-borne vibration. (Reproduced from [1] by kind permission of J. P. Talbot.)

example, structure-borne vibration can be simplified by breaking down its propagation path into various homogeneous domains or continuous systems (e.g. structural components), in which all the movement of mass being considered is directly linked together, such as concrete columns. Furthermore, given that in most of the cases of structure-borne vibration the structure is manmade, knowledge of its properties should be readily available from engineering drawings. As such, uncertainty tends to be greater when dealing with ground-borne vibration. Given the difficulty in attaining distinct ground parameters for the site being evaluated, the best alternative is to experimentally characterise the dynamic behaviour of the ground via vibration measurements. To excite the ground, the assessor commonly makes use of trains running nearby and, if they are unavailable, then impact tests are used (e.g. using an instrumental sledgehammer or a falling weight deflectometer). Vibration measurements, therefore, constitute a critical part of the assessment process.

When measuring ground-borne vibration, the main sources of uncertainty may include the following:

- Equipment and its calibration,
- Measurement settings,
- Ambient vibration,
- Sensor mounting and
- Ground conditions.

As for equipment and its calibration, refer to Chapter 3. The next subsection addresses these remaining items in turn.

11.3 Measurement Settings

For the purpose of uncertainty, the term "measurement settings" refers to aspects that may affect the reading of vibration levels. If the user is to record the raw waveform, then only the sampling frequency and the dynamic range of the measuring equipment would be of relevance. However, if the user is to record statistical data (e.g. RMS one-third-octave bands), then the equipment setup should consider the frequency resolution, which is intimately linked with the duration time block from which the spectrum is derived. Note that some points addressed in this sub-section may also be relevant for post-processing.

RMS is a statistical descriptor based on an average of cycles; as frequency decreases, fewer cycles per unit of time will be present. When characterising a steady-state random signal, the duration of the sample being analysed may affect how accurate the signal is represented at low frequencies. According to ISO/TS 14,837-32: 2015 [2], low frequencies around 1 Hz require a long integration time of around 3–5 s. However, this rule of thumb should be used with care, as the low-frequency content of the signal must be sufficiently stationary over the measurement period.

Rail movements tend to give rise to random stationary vibration signals. Depending on the size and speed of the train, its resulting vibration time period can typically vary from 2 to 40 s, and therefore, only a few cycles at low frequencies (e.g. <6 Hz) are contained within the time block being analysed, reducing its statistical significance.

Impact tests carried out to characterise the ground structure tend to have all the energy concentrated within a very short period. As demonstrated in De Avillez [3], at the load cell of the instrumented hammer, depending on the soil and tip hardness, the energy from the falling load is concentrated within a time block of approximately 0.04 second. At the receiving point, depending on the ground properties, distance and ambient vibration, the effective energy is typically spread within a 0.3- to 0.8-s time block. Beyond that time frame, the ambient vibration is expected to mask the tail of the response. The shorter the time block, the higher are the levels of uncertainty, especially at low frequencies.

11.3.1 Frequency Resolution

Fast Fourier transform (FFT) analysis (which can be presented in terms of RMS, power spectral density, etc.) is formed of discrete frequency components. The magnitude of the frequency components may be misrepresented when carrying out a "low-frequency resolution" analysis (i.e. an analysis having a low number of FFT lines per Hz). The frequency resolution is directly proportional to the time block (i.e. signal segment being analysed) and inversely proportional to the sampling frequency. Therefore, the greater the time block, the higher becomes the frequency resolution, increasing the precision on how the energy is distributed throughout the frequency spectrum. Depending on the selected algorithm (i.e. time-to-frequency-domain routine), this may have a direct impact on the broadband RMS spectrum analyses, such as the one-third-octave band, which is commonly used when assessing rail-induced vibration. When deriving the one-third-octave-band spectrum by applying the FFT routine directly to the signal (i.e. calculating each one-third-octave band by combining all frequency components within each band), this lack of resolution may cause errors due to having fewer data points within a given frequency range. This becomes relevant at low frequencies because the octave scale is logarithmic and the FFT lines are evenly spread over the frequency axis. Thus, high-frequency bands are the synthesis of a large number of FFT components as opposed to low-frequency bands, which will be computed with fewer frequency components. To reduce the spectral uncertainty associated with low-frequency resolution, the third-octave bands can be predefined and split using digital bandpass filters.

11.4 Ambient Vibration

When evaluating the impact from a specific vibration source, the ambient vibration needs to be established and accounted for so that the vibration resulting from the actual event in question can be characterised. There will be situations in which the vibration levels produced by the source will not significantly exceed the ambient levels. Establishing the spectral content of the vibration being generated by the source in question enables the analyst to evaluate whether or not the ambient vibration is capable of impacting on the source characterisation. Failing to remove the ambient vibration from the measurements may lead to an erroneous verdict. A special case where ambient vibration may lead to unreasonable levels of uncertainty occurs when evaluating vibration dosage values (VDVs) from passing trains.

Figure 11.2 contrasts the vertical and horizontal spectra for a set of London Underground train movements with the W_b and W_d weighting curves, respectively. The figure suggests that for conventional passenger trains, very little energy is generated at the frequencies that carry the most weight when evaluating VDV. Next, Figure 11.3 contrasts the measured rail-induced vibration (black solid line) with the ambient vibration (grey solid line) measured between train events. Note that a significant part of the vibration energy below 12.5 Hz is not produced by rail-related events. In this situation, which is representative of many areas within London, the VDV would be highly affected by the ambient vibration, so any attempt to mitigate at the track would be ineffective. This suggests that where levels of ambient vibration are high, VDV will most likely be driven by the ambient vibration itself.

To address uncertainty arising from ambient vibration, a set of background measurements taken between the events being characterised should be recorded and contrasted with the measured event in question, as illustrated in Figure 11.3.

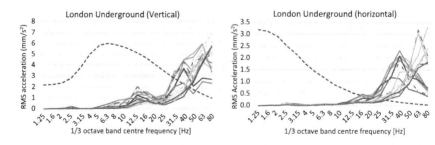

Figure 11.2 *(Left)* Representative vertical vibration acceleration levels generated by London Underground train movements in contrast with the W_b weighting curve (dashed line). *(Right)* From the same data set, representative horizontal vibration acceleration levels in contrast with the W_d weighting curve (dashed line). (Reproduced with permission from D'Avillez and Saife [4].)

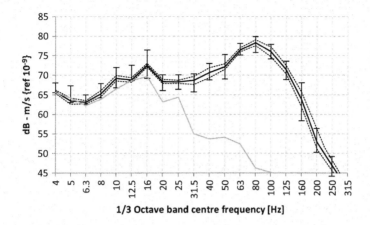

Figure 11.3 One-third-octave-band RMS values of the vertical velocity – statistical spectral analysis contrasting the London Underground vibration velocity levels with the ambient levels. The solid black curve shows the average velocity level, dashed curves show the standard deviation below and above the average, the error bars refer to the minima and maxima and the grey curve shows the ambient vibration velocity levels averaged throughout the measuring period. (Reproduced with permission from D'Avillez and Saife [4].)

11.5 Sensor Mounting

11.5.1 Structure Loading

When mounting a transducer to a structure, such as a concrete slab or a wooden floor, its coupling along with its orthogonal orientation and loading impact should be considered. The act of mounting a sensor on a structure will, to some extent, load the structure being assessed, which may influence the vibration readings, especially when the measured vibration is of a very low magnitude, because the micro-displacements of the structure may be influenced by an exogenous load. To minimise such effects, the sensor and its mounting apparatus should be as light as reasonably possible.

11.5.2 Orthogonal Orientation Misalignment

Alignment of the transducer's orthogonal orientation ensures that each axis of motion is adequately represented, avoiding energy spillage from one axis to another. For example, assessing the vertical motion requires, in principle, for the transducer to be mounted at 90 degrees to the ground surface.

Cross-axis sensitivity (or transverse sensitivity), which is a measure of error on the signal produced by a transducer if it is vibrated at the right angle to its working axis, may also contribute to orthogonal reading misrepresentation. Modern sensors (Class A), however, tend to have low cross-axis sensitivity (<5%).

To evaluate the impact of sensor orthogonal misalignment, a practical test was carried out and presented in D'Avillez et al. [5]. Figure 11.4 shows the likely misreading, in terms of vibration magnitude, for 10- and 20-degree misalignments. There is very little variation up to around 250 Hz, which is typically taken to be the highest frequency of interest when dealing with rail-induced ground-borne vibration. It is also worth noting that a 10-degree misalignment is easily detected without the use of a spirit level.

11.5.3 Couplant

Poor coupling, which results from the inability of the transducer to maintain proper contact with the structure, may significantly distort the incoming signal, contributing to the measurement uncertainty budget. Beeswax is one of the most commonly used and reliable couplant. ISO 5348: 1998 [6] considers it to be an effective option whenever coupling a small sensor onto flat, rigid surfaces. Dental plaster, which is a solid and long-lasting equivalent to cement, is commonly used for coupling sensors to irregular surfaces. Blu-tack,

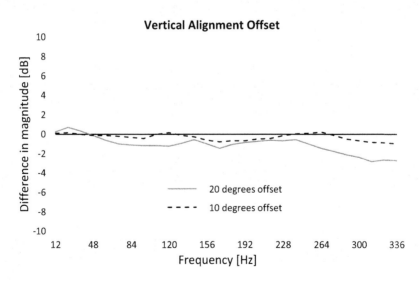

Figure 11.4 Insertion gain for 10- and 20-degree misalignments (curves are normalised to 0 degree).

which is a reusable putty-like pressure-sensitive adhesive, has shown to be highly practical when coupling small transducers to a dry, firm surface. The study presented in D'Avillez et al. [5] evaluated the impact of each of the mentioned couplants. As shown in Figure 11.5, the study reveals that all couplants perform similarly up to 250 Hz. In this region, divergence among them is kept to less than 1 dB. A 2.5-dB divergence around 375 Hz reveals that plaster couplant overestimates compared with the others. It is important to note that Blu-tack, which is not referred to in standards, performs within the frequencies of interest for rail-induced ground-borne vibration, very similarly to beeswax, which is taken to be a reliable couplant.

A special case of transducer mounting is the transducer-to-soil mounting. Its coupling effectiveness, among other things, is dictated by the composition of the soil to which the transducer is being mounted.

11.5.4 Transducer-to-Ground Mounting Techniques: Mechanically Fixing the Sensor onto the Soil

Good transducer-to-ground coupling, which is essential for measuring the free-field response, requires the transducer to maintain proper contact with the ground. Poor coupling can cause friction and slippage of the transducer, resulting in distortion, altering the amplitude and phase of the signal, often yielding higher measured vibration levels.

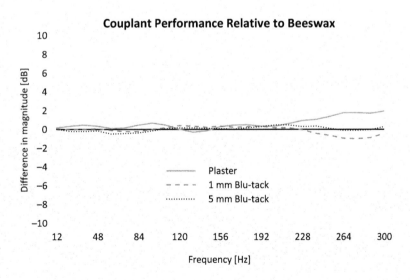

Figure 11.5 Comparison of couplants (spectral amplitude difference curve normalised to beeswax coupling).

An effective coupling between the transducer and the ground is often difficult to achieve, especially when the transducer mounting is restricted to the limited options that the ground structure offers. There is no unique transducer mounting method that can satisfy all types of ground structures.

Literature presented in D'Avillez et al. [5] reveals that different coupling methods suit different ground conditions better than others. It falls on the surveyor to evaluate field conditions and to obtain a good transducer-to-ground coupling. National and international standards (e.g. see references [7] and [8]), along with professional body guidance (e.g. see reference [9]), state some accepted mounting techniques and what should be avoided and provide alternatives where the requirements cannot be met. Figure 11.6 shows representative mounting techniques often adopted for either academic research or commercial technical reports.

Figure 11.7 compares the measured free-field response readings of five commonly used transducer planting mechanisms:

- The "buried transducer method", which is considered by a vast number of experts as the method that minimises ground coupling distortion, consists of boring a cavity with suitable dimensions, allowing the sensor to fit and leaving the top of the vertical axis at ground level.
- The "spike method", which is the most commonly used method for ground vibration surveys, consists of a small transducer mounting disc welded to a steel spike.
    - Round spike (O-spike), following the recommendations given in references [9] and [10], is a 250-mm-long stainless-steel spike with a 30-mm diameter.

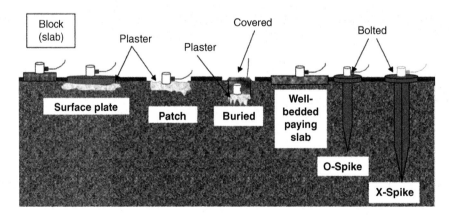

Figure 11.6 Most common transducer-to-ground mounting systems. (Reproduced with permission from De Avillez [3].)

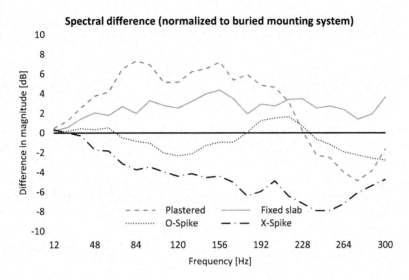

Figure 11.7 Normalised free-field response as a function of sensor mounting technique. Zero magnitude represents the response captured via a buried transducer. Each curve derives from an average of six pass-by measurements.

o   Cross-spike (X-spike) is a 500-mm cross-spike that follows the recommendations in reference [8].

• Fixed slabs and portable small slabs were also tested. Although not recommended by some, it is common practice to use small slabs as a base for small, light transducers, increasing their mass to make use of gravity for coupling purpose. In some situations, such as where the site is floored with fixed slabs, these become unavoidable.

• Plaster directly on the soil was also tested. For this, the loose soil and vegetation were removed prior to pouring the dental plaster.

Synchronised measurements of train vibration were carried out 25 m alongside the train line. The measurements in Figure 11.7 (based on the findings in reference [5]) have been normalised to the buried transducer, which is taken by many to be the most reliable mounting setup. (Note that this may be true for a transducer with a large base, but when dealing with small transducers, there is the likelihood of their base losing contact with the ground as the transducer is being covered up.)

Because there were a limited number of cycles per unit of time at low frequencies (due to the length of the excitation signal, which in this case is a function of train length and its traveling speed), results below 10 Hz could be misleading and therefore not presented herein. However, within the frequency range of interest (~10– 300 Hz), the figure suggests that there may be a 10-dB inconsistency arising from the transducer coupling technique.

Depending on the ground conditions (i.e. soil composition), the coupling of the supporting structure that holds the sensor, like spikes, may vary accordingly (e.g. dry soil might facilitate the coupling).

As the study by D'Avillez et al. [5] shows, measurements made with an X-spike in a granular ground composition, which approximates a brittle ground surface, demonstrated a variation of 20 dB when compared to the buried transducer technique. This shows the impact which the soil composition has on the effectiveness of the transducer-to-ground mounting technique. The study by D'Avillez et al. [5], which assesses the mechanical mountings scheme's consistencies at different sites with different soil compositions, also concluded that the spikes are the most consistent method, yielding less uncertainty when measuring rail-induced ground-borne vibration.

11.6 Ground Conditions Affecting Measurement Uncertainty

11.6.1 Uncertainty Associated with the Ground's Heterogeneity

When capturing the dynamic behaviour of a structural element, the location of the transducer becomes critical, as the levels of vibration vary significantly as a function of the geometry, supports and mechanical properties of the structure being assessed. When considering a manmade structure, the surveyor, by carrying out a theoretical analysis of the dynamic behaviour of the structure being assessed, may be able to identify in advance the measurement locations of interest. When assessing natural structures such as the soil, however, the likelihood of variation is more difficult to anticipate. Natural soils are inherently heterogeneous, and vibration levels may vary even over small distances.

As the complexity of the soil composition (including strata profile) rises, the representativeness of each measurement location diminishes. If the aim is to quantify vibration levels at a single point on the ground, then strata arrangement (including the degree of heterogeneity) is not a significant issue. However, if the aim is to establish the vibration climate within a defined area – to evaluate the incident wave field affecting the response of a proposed development or to characterise the dynamic behaviour of the intervenient propagation path – then the composition of the soil at the measurement point plays a critical role.

The implications of having a heterogeneous medium of a random nature may lead the analyst to consider the following: is the measurement location representative of the site being assessed? And if not, how many measurement locations do I need to consider to obtain a representative ground response, and where should they be located?

A way of characterising the influence of the ground when modelling rail-induced vibration is to establish the line source transfer mobility (LSTM), which relates the source with the receiver at that specific site. Typically, the LSTM is evaluated by performing a set of impact tests on the ground at multiple points along the proposed railway route. Figure 11.8 presents the LSTM evaluated using a calibrated impact source measured at 9 m at seven urban sites located within a 2-km radius to illustrate how the ground, as a propagation path, varies depending on the soil type. The figure suggests that a 20-dB variation within the frequency range of interest can be expected.

11.6.2 Uncertainty Associated with Local Soil Variations

For a detailed ground-borne vibration assessment, FTA-VA-90-1003-06 [11] provides a well-established assessment procedure and proposes carrying out the impacts that define the LSTM every few metres along the proposed railway route. If the ground, like air, is assumed to be homogeneous, one impact point would suffice to compute the LSTM of the ground, as the transfer function would be unaffected by arbitrary ground stretch disrupting source receiver continuity. However, if the composition of the ground is highly uneven, in terms of its mechanical properties, the LSTM would be affected accordingly.

To assess the uncertainty that arises from reducing the number of impact points, D'Avillez et al. [12] carried out a study at 10 urban sites within an approximately 2-km radius. When characterising the intervening ground using a LSTM derived from only one impact location (i.e. only the impact location at the central position of the line segment being assessed), the estimation will

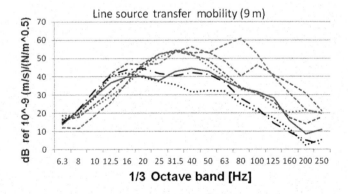

Figure 11.8 Ground surface LSTM spectra (measured 9 m from the nearest impact point) derived from 11 impact points at seven nearby sites (within a 2-km radius).

yield some deviation from the true LSTM due to the ground's heterogeneity. As the number of impact locations increases, the deviation between the speculative and true values should diminish, reducing the level of uncertainty.

Figure 11.9 shows the deviation between the speculative LSTM (extrapolated from a single-impact-point location) and true LSTM (using multiple impact locations every 3 m along the line) for each one-third-octave band. If we concentrate on the frequency region between 8 and 125 Hz, as frequencies outside this range have been deemed "noise contaminated" (as not enough energy above ambient noise was present at the receiving location), we see a deviation, on average, for all 10 sites of around 4 dB. On one of the 10 sites, a 15-dB discrepancy was found in the 63-Hz one-third-octave band, which is a highly relevant as rail-induced vibration tends to be concentrated around this band. Figure 11.10 shows the average error, along with the 95% confidence interval (shown by the error bars), as a function of the number of impact locations.

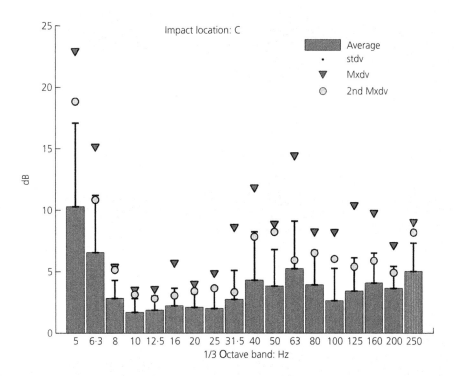

Figure 11.9 Variation between the extrapolated LSTM (using only the centre impact location) and the true LSTM (established through 11 impact points spaced every 3 m along the line of impacts). (Reproduced with permission from D'Avillez et al. [12].)

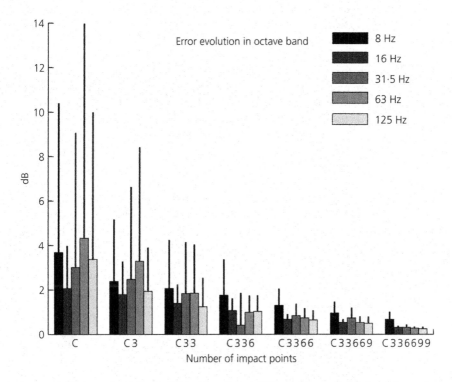

Figure 11.10 Average error (solid bars) and the 95% confidence level (line above the bars) as a function of the number of impact points for each octave band of interest. (Reproduced with permission from D'Avillez et al. [12].)

It is worth noting that a higher variation in ground response is expected within urban areas in comparison with rural areas. In urban areas, the ground tends to be extremely irregular, particularly the top layer, where utility services (e.g. drains, cables, etc.) and manmade structures interrupt the ground's composition.

11.7 Spatial Variability in Manmade Structures

When carrying out sound measurements, the sound pressure levels at the microphone are affected by structures such as reflective surfaces that can be visually identified and accounted for. When measuring the vibration response of simple structures, such as floor slabs, knowledge of the geometry and supports should suffice to make an informed decision as to where to locate the transducer. For example, when assessing rail-induced ground-borne noise, professional guidance tends to recommend mounting the sensor slightly off-centre but towards the middle of the floor plate [e.g. the Association of Noise Consultants (ANC) [9] suggests that the "slightly off-centre" location approximates the spatial mean). Assuming floor plates to be somewhat rectangular, which for conventional dwelling they tend to be, the slightly off-centre position will represent the worst affected location in terms of the environmental vibration climate. If the slightly off-centre location is consistently adopted, experimental data can be used for comparison, which is critical when assessing the radiated noise via empirical means.

The uncertainty that derives from sensor location can be evaluated through modal analysis. Knowledge of the floor slab materials, shape and supports is sufficient to map out the expected variation in measured vibration levels as a function of measurement location. Figure 11.11, which represents a floor plate of a room being assessed for rail-induced ground-borne noise, shows the variation in vibration levels for the 200-Hz one-

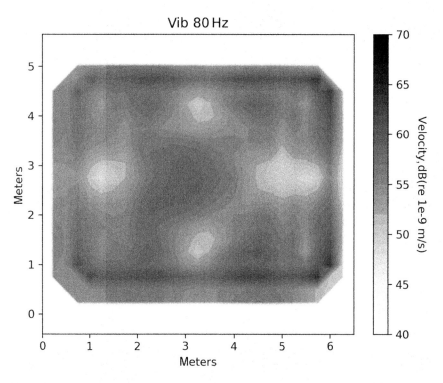

Figure 11.11 Rail-induced ground-borne vibration surface plot predicted at the floor slab of a room measuring 4 × 5 m.

third-octave band. It shows a potentially large variation, exceeding 5 dB, within a small area span (<0.5 m^2 area) around the 80-Hz frequency.

Similarly, when presenting the predicted radiated noise levels, consideration must be given to where the reported levels will be felt. Figures 11.12 to 11.14 show the distribution of the A-weighted sound pressure levels throughout a room at three different heights. When evaluating rail-induced ground-borne noise within a conventional room (16–50 m^2), research (e.g. see Villot et al. [13]) acknowledges a variation of approximately 16 dB(A) as a function of assessment location.

11.8 Part 2: Practical Examples of Estimating Uncertainty in Vibration Level Measurements

The approach to estimation of uncertainties in vibration levels has much in common with the methods explained in Chapters 1 and 2 (and other chapters as well) for sound and noise levels without, in many cases, the complications arising from the use of decibels. As explained in Chapters 1 and 2, uncertainties can be estimated using the methods of the GUM, either by (1)

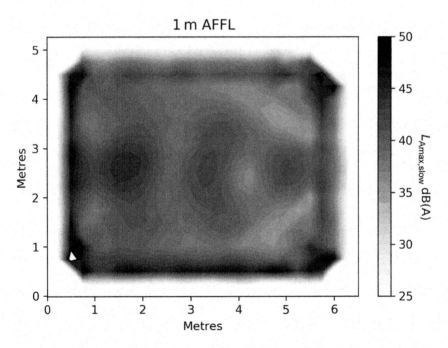

Figure 11.12 Rail-induced ground-borne noise surface plot predicted 1 m above the finish floor level of a room measuring 4 × 5 m.

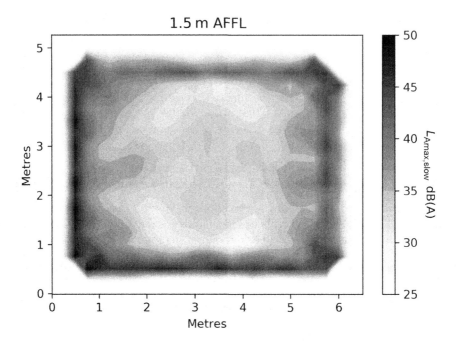

Figure 11.13 Rail-induced ground-borne noise surface plot predicted 1.5 m above the finish floor level of a room measuring 4 × 5 m.

constructing an uncertainty budget from all the factors affecting the vibration measurements that contribute to uncertainty (if these are known) or, more commonly, (2) using repeatability and reproducibility measurements. Most often, a combination of both methods would be used.

As with sound levels, sources of uncertainty in vibration measurements can be conveniently grouped as relating to those arising from the instrumentation and measurement procedures and those arising from the vibration source, transmission path and measurement location. These various issues have been discussed in much greater detail in the first part of this chapter. This second part will concentrate on estimating the uncertainty associated with them.

11.8.1 Estimating Uncertainties in Measuring Rail-Induced Vibration Levels

As with sound level measurements, the uncertainty arising from the vibration measurement instrumentation and measurement procedures is often small compared with the other components. One of the simple example calculations included at the end of Chapter 2 demonstrates how uncertainties from

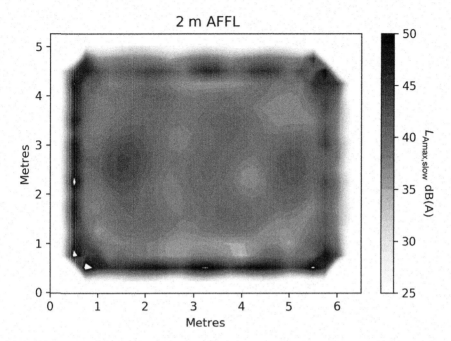

Figure 11.14 Rail-induced ground-borne noise surface plot predicted 2 m above the finish floor level of a room measuring 4 × 5 m.

various factors can be combined for noise levels for construction site noise, and a very similar approach can be taken to estimate vibration uncertainty. The following example illustrates how repeatability of vibration from trains can be used to provide an estimate of uncertainty in vibration levels.

Example: Estimates of Uncertainty Based on Simple Repeatability Measurements of Vibration from Trains

We have assessed the repeatability of vibration arising from train-to-train variability at the same position and on the same day. Free-field measurements were taken from 10 nominally identical trains with an accelerometer fixed to the ground 10 m from the train line. The accelerometer signal was fed into a vibration meter that measured VDV and met the requirements of ISO 8041. The site was a rural one on the English-Welsh border. The purpose of the vibration survey was to inform a planning application to build residential properties on this site, close to the railway line.

The site had been used as railway sidings, but all railway lines had been cleared from the site. The ground was still covered with compacted ballast and unsuitable for directly fixing an accelerometer to measure ground vibration. A metal spike was driven into the ground to its full length (~0.5 m), and the accelerometer was attached with a magnetic fixing stud on the top of the spike. The VDVs varied between 0.017 and 0.034 m/s$^{1.75}$, with an arithmetic mean of 0.0252 m/s$^{1.75}$. The standard deviation was 0.0044, which is about 17.5% of the mean, or ±1.5 dB. The background vibration level (i.e. when no trains are passing) corresponds to a VDV of 0.002 m/s$^{1.75}$ (i.e. well below the vibration levels from passing trains).

The corresponding standard uncertainty is obtained as shown in Chapters 1 and 2 by dividing the standard deviation by the square root of the sample size, which, in the present case, gives a value of 0.0014 m/s$^{1.75}$ (i.e. a standard uncertainty, which is about 5.5% of the mean, or about ±0.5 dB).

Uncertainties arising from the measurement instrumentation system are discussed in Chapter 3. A reasonable assumption for the expanded uncertainty is ±5% with a 95% confidence limit. Therefore, assuming a coverage factor of two, the corresponding standard uncertainty is ±2.5%, or about ±0.2%.

These two standard uncertainties arising from repeatability and instrumentation may now be combined to give a combined or total standard uncertainty of $U_t = \sqrt{5.5^2 + 2.5^2} = \pm 6\%$, or about ± 0.5%. Hence, the resulting expanded uncertainty for a coverage factor of two is ±12%, or about ±1 dB. As such, the expanded uncertainty, which takes into account repeatability measurements and uncertainty arising from the measurement equipment, is 0.0252 ± 0.0003 m/s$^{1.75}$.

Further variability could be expected if measurements had been carried out on different days and at different distances from the track. The variability that will arise between different measurement positions has been addressed earlier in this chapter.

Measurements carried out for a similar purpose and under similar conditions for a sample of 14 trains at another site in south-west London a few years earlier using RMS-based signal processing produces a slightly larger train-to-train variation, giving an eVDV value[2] of 0.0121 ± 0.0030 m/s$^{1.75}$. This gives rise to a combined extended value of 0.0121 ± 0.0017 m/s$^{1.75}$, that is, ±17%.

The next section illustrates how to estimate uncertainties in the measurement of occupational vibration levels (whole-body and hand-arm vibration – which is very important in assessing the risk to health for people exposed to these forms of vibration at work).

---

2 Evaluated VDV (eVDV) is a method based on spectral analysis proposed in BS 6472–1 to approximate the actual VDV.

11.8.2 Estimating Uncertainty in the Assessment of Human Exposure to Vibration

According to EC Directive 2002/44/EC and the UK Control of Vibration at Work Regulations, the assessment of vibration exposure at work is based on calculation of the daily exposure expressed as equivalent continuous acceleration determined over an eight-hour work period, either from a continuous all-day measurement or calculated from activity samples, and is denoted by the symbol A(8). In the case of hand-arm vibration (HAV), A(8) is defined in ISO 5349–1 as the total value (vector sum) of the frequency-weighted acceleration values measured in the orthogonal axes aw_x, aw_y and aw_z and normalized to eight hours. The frequency weighting is performed by the W_h filter on all three axes, since the risk of damage is considered equal in all directions.

For whole-body vibration (WBV) exposure, A(8) is obtained from the highest RMS value of the frequency-weighted acceleration determined in three orthogonal axes. It is defined in ISO 2631–1 for a seated or standing worker. Since the risk of damage is not equal in all axes, a multiplying factor and a specific weighting filter must be applied to balance the contributions of the different directions. Therefore, the RMS acceleration values for the x- and y-axes are multiplied by 1.4 and weighted by the W_d filter, whereas the z-axis acceleration is multiplied by 1.0 and weighted by the W_k filter.

Extracts from ISO 5439–2, Mechanical Vibration: Measurement and Evaluation of Human Exposure to Hand Transmitted Vibration, Part 2: 2001 Practical Guidance for Measurement at the Workplace

Section 9 of ISO 5439 requires that reports of evaluation of daily vibration exposure should contain details of vibration measurement equipment and procedures and evaluation of results and uncertainties. Section 7 of the standard gives guidance on sources of uncertainty and their evaluation.

Determination of the value of A(8) involves measurement of both average values of frequency-weighted acceleration and the estimation of exposure times, and the total uncertainty arises from the combination of uncertainty in both level and duration. Uncertainty in the measurement of vibration acceleration levels is mainly that which arises from the measurement procedure and that which arises from variability in machine operating conditions. Uncertainty related to the measurement procedure includes the instrumentation, calibration procedures, cable movement and vibration, electrical interference and other environmental influences, as well as the location of the accelerometer and its mounting method.

Uncertainties arising from these sources will usually be much smaller than those arising from variabilities in the hand-operated machine and the way that it used by the operator. These uncertainties include variability in the way an operator uses the tool, including changes in hand posture and applied force; variability between different workers performing the same task; variability between different but nominally identical machines, including variation in machine wear and condition of maintenance; and variability in work procedures and practices, including in the materials being processed.

When the purpose of the measurement is to evaluate the vibration exposure of a particular employee, then measurements should be performed with the same employee performing the same task with at least three different machines. When the purpose of the measurement is to evaluate the vibration exposure of operators performing a particular task, then measurements should be carried out with at least three different operators performing the same task.

Uncertainty related to the estimation of exposure duration can arise because of variability in the time taken to perform a particular task or work cycle and in the estimate of the number of such tasks or work cycles performed in any one day. Calculation of the daily vibration exposure is carried out by combining the measurements of average acceleration levels and estimations of duration to determine values of A(8), with the evaluation of uncertainty based on the variabilities (e.g. standard deviations) of these measured or estimated values.

It is noted in the standard that the uncertainty associated with the evaluation of A(8) is often high (e.g. 20%–40%). Therefore, values of A(8) should not normally be presented with more than two significant figures. Annex E of the standard gives useful practical examples of the estimation of daily vibration exposure A(8) but does not include estimates of uncertainty. The methods for the assessment of both WBV and HAV exposure, including the estimation of uncertainties, are well illustrated in an experimental study by Adamo et al. [14].

An Experimental Study Assessing the Uncertainty in Human Exposure to Vibration

A paper by Adamo et al [14] reports the results of measurements of HAV and WBV exposures and an estimation of uncertainties in the A(8) values determined from the results. In both cases, RMS acceleration levels were measured using the appropriate frequency weighting r in three orthogonal directions using PSB shear-type tri-axial accelerometers containing integrated signal conditioning circuitry. Rather than measuring vibration levels continuously over an eight-hour period, it was acceptable to estimate values of A(8) from representative sample measurements, provided that minimum measurement times were at least one minute for HAV and three minutes for WBV.

For HAV, the accelerometers were mounted on a specially designed adaptor held between the fingers of the operator and the grip of the vibrating tool. In the case of the WBV, the accelerometer was attached to a semi-rigid adaptor mounted on the seat of the vibrating machine. For the HAV tests, measurements were carried out for the following activities: grass cutting with a brush cutter, cutting large tree trunks with a chain saw, cutting tree branches with a chainsaw and cutting a hedge with a hedge trimmer. Measurements were carried out for both hands in accordance with the ISO standard. For the WBV tests, the following activities were measured: handling a heavy load with a lorry crane, handling wood with a backhoe loader and mowing brush with a sickle bar mower.

In all cases, measurements were performed with operators in seated positions according to the practical procedures described in ISO 2631 and ISO 5349. For each HAV and WBV test, a minimum of 30 measurements of the RMS accelerations value were acquired in order to enable uncertainty to be evaluated.

Preliminary statistical tests (box-plot tests) were carried out to eliminate outliers not considered to be representative, followed by a Shapiro-Wilk test to evaluate how well the remaining data fit a normal (Gaussian) distribution. High outliers can arise where very high acceleration values are acquired because of occasional shock, load variation, wrong instrument use or malfunction of machinery. Low outliers may occur when the measurement starts before the effective exposure to vibration or in the presence of long waiting times or load variation. The results show that if the outliers were removed, the experimental data exhibit a bell-shaped distribution that approximates the Gaussian distribution.

The final stage of the analysis leading to the evaluation of uncertainty follows the procedures given in the *Guide to the Expression of Uncertainty in Measurement* (GUM). For type A uncertainties, the standard uncertainty was evaluated from the mean and standard deviation values of the sample of measured acceleration results for each axis. Uncertainties due to the measuring equipment (type B uncertainties) were estimated from instrument calibration certificates. The combined total standard uncertainty was converted to an expanded uncertainty at a coverage probability of 95% confidence level, giving the following final results: for HAV:

$$\text{Right HAV: } 6.22 \pm 0.23\text{m/s}^2 \text{ (i.e. } \pm 4\%)$$

$$\text{Left HAV: } 6.40 \pm 0.23\text{m/s}^2 \text{ (i.e. } \pm 4\%)$$

$$\text{For WBV: } 0.87 \pm 0.13\text{m/s}^2 \text{ (i.e. } \pm 15\%)$$

Further analysis of the results showed that these uncertainty estimates remain more or less the same, provided that the number of measurements in the sample was greater than 20.

References

1 Talbot JP, "On the performance of base-isolated buildings: A generic model", PhD dissertation, University of Cambridge, Cambridge, UK, 2002.

2 International Organization for Standardization, *Mechanical Vibration: Ground-Borne Noise and Vibration Arising from Rail Systems*, Part 32: *Measurement of Dynamic Properties of the Ground*, ISO/TS 14837-32. Geneva, 2015.

3 D'Avillez J, "Routine procedure for the assessment of rail-induced vibration" PhD dissertation, University of Loughborough, Loughborough, UK, 2013.

4 D'Avillez J, Saife B, "Assessing rail-induced vibration in terms of VDV", *Proc Inst Acoust* 40(Pt. 1)(2018), 317–324.

5 D'Avillez J, Frost MW, Cawser S, et al., "The influence of vibration transducer mounting on the practical measurement of railway vibration", in *Proceedings of the 39th Inter-Noise International Congress*, Lisbon, Portugal, 15–16 June 2010.

6 International Organization for Standardization, *Mechanical Vibration and Shock: Mechanical Mounting of Accelerometers*, ISO 5348:1998. ISO, Geneva, 1998.

7 British Standards Institute, *Evaluation and Measurement for Vibration in Buildings*, Part 1: *Guide for Measurement of Vibrations and Evaluation of Their Effects on Buildings*, BS 7385-1:1990. BSI, London, 1990.

8 German Institute of Standardization *Measurement of Vibration Emission*, Part 2: *Measuring Method*, DIN 45669-2:2005. DIN, Berlin, 2005.

9 Association of Noise Consultants, *Measurement and Assessment of Ground-Borne Noise and Vibration*. Fresco, Uckfield, UK, 2001.

10 British Standards Institute, *Evaluation and Measurement for Vibration in Buildings*, Part 1: *Guide for Measurement of Vibrations and Evaluation of Their Effects on Buildings*, BS 7385-1:1990 BSI, London, 1990.

11 Hanson CE, Towers DA, Meister LD, "Transit noise and vibration impact assessment", FTA-VA-90-1003-06. Federal Transit Administration, Washington, DC, 2006.

12 D'Avillez J, Frost M, Cawser S, et al., "Ground response data capture for railway vibration prediction", *Proc Inst Civil Eng Transp* 168(1)(2015),83–92.

13 Villot M, Guigou C, Jean P, Picard N, "Railway-induced vibration abatement solutions collaborative project", SCP0-GA-2010-265754. RIVAS, August 15, 2012.

14 Adamo F, Attivissimo F, Saponaro F, Cervellera V, "Assessment of the uncertainty in human exposure to vibration: An experimental study", *EEE Sensors J* 14(2)(2014),474–481.

Chapter 12

Uncertainty in Prediction of Rail-Induced Ground-Borne Vibration

Jorge D'Avillez

This chapter starts by making a distinction between prediction methods based on empirical (or semi-empirical) prediction modelling, which is partly based on experimental measured data, and more completely theoretical or analytical prediction modelling methods based on calculations derived from fundamental equations of physics and engineering. One of the most commonly used empirical models – the Federal Transit Administration of the U.S. Department of Transportation (FTA) detailed vibration analysis [1] – is discussed in some detail. This rail-induced vibration model procedure breaks the prediction process down into three sub-systems: source-track interaction, propagation path via the ground and the building's response to the ground vibration.

The modelling assumptions adopted for each of these three sub-systems, along with their associated uncertainties, will be discussed in detail. For the source-track interaction sub-system, the effects of various loading assumptions (interaction between the moving train and the soil) are examined via theoretical studies. Next, the uncertainty introduced by the coupling losses between the rail track and ground are considered. Focusing on the third sub-system (i.e. modelling the response of the building), uncertainty associated with the effects of soil conditions in which the building rests, which are found to affect the coupling losses between the ground and building, are also examined in terms of uncertainty. Whilst drawing from the same set of theoretical studies, the impact, in terms of uncertainty, on the chosen assessment location is also addressed. Uncertainties drawn from representative fieldwork studies are then presented. The section on FTA method ends with a detailed summary.

Consideration then moves on to uncertainties associated with analytical or theoretical prediction methods, first with factors associated with the vibration source such as rail discontinuities, rail support, sleeper support stiffness, discontinuous slab tracks, voids at the structure-soil interface, wheel-rail roughness, type of rolling stock and train speed. Next, factors affecting the prediction of ground propagation are considered such as soil homogeneity, the profile of soil layers and soil water content.

This chapter ends with a discussion of how the theoretical model should be calibrated by comparison with vibration measurements. By addressing the uncertainty arising from theoretical modelling, a practical way of quantifying the uncertainty range is presented. Finally, since the assessment criteria tend to be frequency dependent, consideration is given to the weighting that should be applied to different parts of the vibration spectrum.

12.1 Introduction to Rail-Induced Ground-Borne Vibration Predictions

Rail-induced vibration may adversely affect the serviceability of buildings, in the form of both mechanical vibration and radiated noise. "Serviceability" of a building refers to its suitability for occupations of different types. An effective rail-induced ground-borne vibration prediction is essential for proposing cost-effective mitigation measures. Designers and engineers use semi-empirical and theoretical models (i.e. analytical and numerical models) to predict vibration levels when assessing isolation requirements and mitigation strategies. Knowledge of input, such as track support stiffness and soil density, is necessary to obtain reliable predictions. In general, for commercial projects, empirical predictions are often used for their simplicity and swiftness. Academic research, by contrast, tends to rely on theoretical models when analysing the behaviour and impact of specific railway, ground and building components.

Rail-induced ground-borne vibration levels in buildings are, to a large extent, determined by the nature of the vibration characteristics that the source generates (excitation at the wheel-rail interface), along with the mechanical properties and design of the intervening structures (track bed, ground and building), the coupling between the track and its supporting ground, the distance between the source and receiver and the coupling between the ground and the building. Figure 12.1 is a schematic of the intervening elements when considering rail-induced vibration.

12.1.1 Assessment Stages

For estimating the likely impact from rail-induced ground-borne noise and vibration on buildings, the following three levels of assessment, differing in complexity and expected accuracy, are commonly used at different stages of the development:

- *Scoping*, to be carried out at an early stage of the development (e.g. at the planning stage), sometimes forming part of feasibility studies [Royal Institute of British Architects (RIBA) stages 0–1] – typically consists of a high-level assessment considering source type, distance

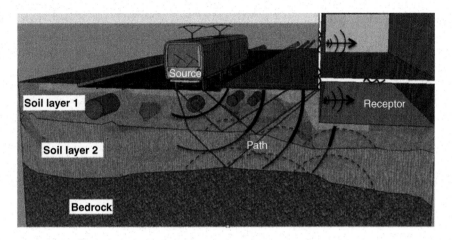

Figure 12.1 Schematic of the overall rail-induced vibration system, including: source, path and receiver. (Reproduced with permission from D'Avillez et al. [2].)

and "receiver category". Historical data are used to evaluate whether a more detailed assessment is required.

- *General assessment*, intended to be carried out at the concept design and development design stages (RIBA stages 2–3) – usually relies on a set of empirically derived generic functions, which, in turn, can be selected to meet the type of rail track, ground and development expected to form part of the scenario being assessed. The assessment is sometimes supplemented with on-site measurements.
- *Detailed design assessment (or detailed vibration assessment)*, to be carried out at the detailed design stage, with the aim of quantifying, with a higher degree of precision, the impact and, if found necessary, to assess and inform the isolation strategy (usually carried out at RIBA stage 4 in tandem with the structural team or at stage 3 to provide cost information).

The first two assessment levels are used for general screening purposes. They are prone to high uncertainty, so worst-case assumptions are usually adopted, producing highly conservative outcomes. This ensures that the project goes, within reason, through the necessary assessment stages and that only the scenarios that are clearly demonstrated to be unaffected by rail-induced vibration will be exempted from the successive, more elaborate prediction procedure.

For the first two assessment stages, the procedure is typically based on semi-empirical modelling. The advantage of semi-empirical assessment, compared with parametric models, is that it inherently takes into account all relevant parameters. To enhance precision, however, the second-stage procedure can be complemented with analytical modelling blocks, for example, evaluating insertion gains arising from a change in configuration through analytical models such as pipe-in-pipe (PiP) [3] – to assess the impact from specific changes to the base case. For the third stage, a comprehensive empirical procedure or numerical modelling is expected. Numerical models, such as Findwave [4], allow for a great flexibility in dealing with different track and/or building geometries, whilst, for the cases where vibration propagation measurements are carried out at the site in question, empirical models allow for an accurate assessment of the vibration propagation through the soil. In hybrid predictions, the advantages of both approaches are combined.

12.1.2 Sources of Uncertainty

As empirical modelling relies on experimental data, which inherently take into account all parameters, it tends to be adopted, in a generalised form, for the first two assessment stages given its simplicity and speed of usage. The accuracy can be improved, and therefore used for the detailed design stage, by characterising the actual scenario experimentally, through specific measurements.

The Federal Transit Administration (FTA) of the US Department of Transportation and the Federal Railroad Administration (FRA [5]) have developed a set of empirical procedures to predict vibration levels due to railway traffic. The guidance is based on empirical research and does not refer to uncertainty. For the first two stages, the prediction procedures are structured around a nominal vibration level characterising a typical source-path-receiver configuration (taken as the base case). For flexibility, the procedure makes use of a table that contains insertion gain values reflecting the change in vibration levels caused by introducing a circumstantial modification to the base case; such modifications can then be combined to simulate the proposed scenario.

For the detailed design stage, the vibration velocity level in a building is predicted based on a separate characterisation of the source, the propagation path – determined directly based on wave propagation tests undertaken at the site being evaluated – and the receiver, which are characterised experimentally by a force density, a line transfer mobility and a coupling loss factor, respectively. The coupling loss factor characterises the modification of the vibration velocity level as a result of the dynamic soil-structure interaction at the receiver side. This empirical modelling procedure can be extended through insertion gain function curves, a frequency-dependent coupling loss factor, obtained from pre-established curves or determined indirectly through analytical modelling, where the relative impact from each intervenient component is evaluated individually. Limitations to this modelling technique will be addressed in Section 12.2.3. Apart from uncertainties arising from measurement procedures (discussed in Chapter 10), the main sources of uncertainties associated with the empirical model being considered are as follows:

- *Vibration source.* This includes rail roughness (associated with the rail maintenance regime), track form, rolling stock, operation profile, and the consistency, in terms of parametric values, of the track form throughout a section of track.
- *Propagation path.* This includes soil composition (including layer profile) and manmade underground structures (such as utility services) and voids between structures and soil.
- *The coupling between the structure (track bed and/or building foundation) and the ground.* This is characterised through the "coupling loss factor", which is taken as a measure of the modification of the vibration spectrum resulting from the dynamic interaction between the structure of the track bed or building and the ground.
- *Receiver.* This is the response of the building being assessed, which is a function of its materials, design and, as mentioned earlier, coupling loss.

12.2 Uncertainty Associated with Empirical Modelling (FTA)

The FTA detailed assessment procedure breaks the system into three sub-systems (see Figure 12.1): source (train-track interaction) propagation path, intervening ground, and receiver building response. It then assesses each sub-system independently and combines them to predict the vibration levels expected at the receptor.

12.2.1 FTA-Proposed Assessment Procedure

By carrying out free-field vibration measurements (L_V, vibration velocity in decibels) alongside a similar train-track configuration, the method characterises the source term in a normalised form by removing the ground's contribution from L_V. To relate the source to the receiver at that specific site, the line source transfer mobility (LSTM), which establishes vibration levels that is transmitted through the soil relative to the power per unit length radiated by the source, is evaluated. In practice, to evaluate the LSTM, a set of point-source transfer mobilities needs to be estimated through impact tests carried out at multiple points, evenly distanced, along the proposed train-line route. The LSTM is then inferred by integrating the multiple-point-source transfer mobility along the line of impacts based on the assumption that the train can be modelled as an incoherent line source. The required number of impact points is left to the analyst's discretion (usually every 5 m covering the length of the train being modelled). By subtracting the LSTM from the pass-by-induced L_V, the method then gives a quantitative description of the source as a normalised force density level (FDL), which represents the power per unit length radiated by the source. This is assumed, in the method, to be independent of the ground characteristics. As such,

$$FDL = L_V - LSTM$$

where the FDL is expressed in decibels (ref. N/m$^{0.5}$), L_V is the train-emitted vibration velocity level expressed in decibels (ref. 10^{-9} m/s) measured during a train passage and LSTM is the line source source transfer mobility expressed in decibels [ref. 10^{-9} (m/s)/(N/m$^{0.5}$)]. The test procedure requires these quantities to be expressed as the root-mean-square (RMS) value in one-third-octave bands.

In the FTA method, depending on where the line of impacts is performed when carrying out the LSTM, the source term (FDL) can either represent the train-track interaction or the train-rail interaction. By undertaking the impact tests adjacent to the track, FDL will inherently represent the vibration levels emitted by the pass-by not only as a function of train speed, vehicle dynamics and rail roughness but also as a function of track design (i.e. all track components such as rail pads, sleepers and ballast). If, however, the impacts are conducted at the rail, then the FDL will represent the train-rail interaction, as the track contribution to L_V will be subtracted in the LSTM evaluation process.

The source-term FDL can then be employed accordingly at the proposed location, having first acquired its LSTM, to evaluate the free-field vibration levels L_V emitted by the pass-by by employing:

$$L_V = FDL + LSTM$$

To estimate the vibration entering the building, the ground-to-building coupling loss at the foundation is applied to the free-field vibration levels L_V. To be used as insertion gain functions when carrying out a detailed analysis, FTA offers a set of building coupling loss curves corresponding to a building category and foundation type. Floor response curves representing typical room responses within a dwelling, which FTA provides, can then be used to determine the vibration level at the floor plate of the room being assessed. Associated radiated noise levels can then be estimated using the Kurtzweil method [6]. Likely uncertainties arising from the assumptions adopted by the FTA empirical model, which is representative of most empirical models, will be discussed next.

12.2.2 Theoretical Assessment of Potential Uncertainties Associated with FTA Detailed Prediction Procedure

For any train formation, a set of moving axles unevenly spaced with different load values running over the same random unevenness is expected to produce a unique vibration pattern shaped by constructive and destructive interference at the free field. Therefore, to what extent can a set of fixed point loads with equal load values evenly spaced represent a moving train, as is assumed by the FTA procedure?

Uncertainty When Assuming Incoherent and Equal Point Loads

The FTA procedure assumes incoherent fixed-point loads with equal load values and therefore may miss out on the interference between the contributions from different axles under different load values (e.g. from motor and trailer cars) when running over a random unevenness, as a real train does. Yet, a theoretical study [7] that contrasts the resulting free-field response from a series of coherent fixed-point loads to a series of incoherent fixed-point loads suggests limited variation between the two (see Figure 12.2). Similar agreement was also found when comparing the impact axles with

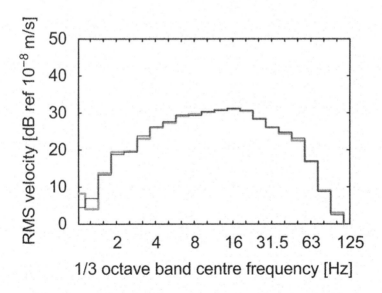

Figure 12.2 One-third octave band RMS values of the vertical velocity due to a series of coherent (grey line) and uncoherent (black line) fixed-point loads evaluated at the surface 60 m away from the tunnel. (Reproduced with permission from Verbraken, Lombaert and Degrande [7].)

individually different load values with trains where all axles had the same load values.

As also established by Wu and Thompson [8], it can therefore be assumed that the incoherent and equal point loads assumption will not generate any meaningful uncertainty when evaluated in sufficiently broad frequency bands, such as one-third octaves.

Uncertainty When Assuming Fixed-Point Loads

As with the previous study, the extent to which the fixed-point load assumption (used in the FTA model) corresponds to the actual moving load has been assessed through numerical modelling. Figure 12.3 illustrates the discrepancy between modelling methods using moving and fixed-point loads. In general, as seen in the figure, there is a relatively good correlation at frequencies above about 6 Hz.

Drop-outs in Figure 12.3 are observed at low frequencies in the frequency spectrum for the moving train. These drop-outs are attributed to the interference between delayed contributions of different axles to the free-field response. Note that in the narrowband frequency analysis, this excitation characteristic, which is due to the spatial variation of axle loads, manifests

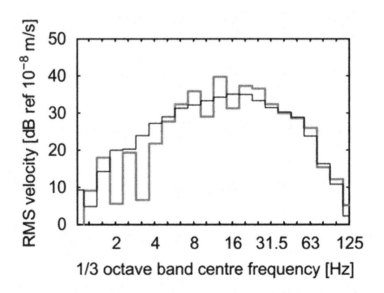

Figure 12.3 One-third-octave band RMS value of the vertical velocity due to a train running at 48 km/h (grey line) and a train at a fixed position (black line) evaluated at the surface 60 m away from the tunnel. (Reproduced with permission from Verbraken, Lombaert and Degrande [7].)

as narrow lobes. The frequencies in which these drop-outs occur are expected to vary with axle spacing and train speed. The reason for drop-outs being more pronounced in the low end of the spectrum is the way the frequency bands are defined in the octave-band base spectrum – an octave-band spectrum will imply a narrowing bandwidth with decreasing frequency. Accordingly, as the bandwidth of the frequency band broadens, the greater is the potential for these lobes to be averaged out.

Assumption of Equidistant Impact Points

LSTM, which forms part of the FTA procedure, is inferred from point-source transfer mobilities obtained from equidistant impact points, yet, in reality axle loads are spaced unevenly. By examining the degree of uncertainty arising from adopting the equidistant impact points assumption, it has been demonstrated through numerical modelling [7] that the effect of neglecting the axle spatial configuration is limited, especially if the analysis is carried out in one-third-octave bands.

12.2.3 Uncertainty Arising from decoupling the system into Sub-systems Irrespective of the soil conditions

"Coupling loss" refers to the changes in vibration levels arising from the dynamic interaction between two coupled elements, such as soil and building foundations. As with most empirical models, when breaking up the assessment into modules (or sub-systems) and combining them irrespective of how they interact when coupled, the FTA framework assumes that structures, such as track form and building, can be analysed independently with no effect on the overall synthesis.

It has been demonstrated [9], however, that the dynamic response of a building is significantly affected by the ground conditions on which it rests. If the ground has a certain springiness, then the superimposed structure, the mass, will respond to the incoming vibration accordingly. As such, it would not be appropriate to assume, as most empirical models (including the FTA model) do, to relate building coupling losses only with the building design (i.e. geometry, materials and load). It can therefore be argued that the methodology requires similar soil conditions at the site where the coupling loss factor has been determined and at the site where it will be applied. Additionally, when carrying out free-field measurements at the location where the building is being proposed, the ground is unloaded, and therefore, it will likely respond differently from the loaded ground from which the coupling loss was derived. The same applies to railway track structures because the vibration that the train track generates is intrinsically related to the ground on which the track rests; thus, the force density is also influenced by the soil properties.

Impact of Track-Soil Coupling on Uncertainty

When assessing sound, power is conveniently used to describe the source term because it is independent of its mounting scheme and site conditions. Similarly, the advantage of using the force density, when considering empirical models of rail-induced vibration, is to have a source term that is independent of the site conditions. As such, a force-density spectrum should be identical for the same train-track formation regardless of soil composition.

The influence of the soil characteristics on the force-density estimation has been theoretically investigated [10] using three sites with different soil characteristics: stiff, medium and soft. These are defined by the shear wave speeds of 300 m/s (to represent stiff soil), 150 m/s (to represent medium soil) and 100 m/s (to represent soft soil).

The potential variation arising from the different soil conditions from which the source term has been experimentally derived and those being modelled is shown for both FTA assessment methods: (1) impacts carried out along side of the track (in the case where the proposed track has not been built) and (2) impacts carried out at the track (e.g. to assess the proposed rolling stock).

Figure 12.4(a) shows, for the same train, a variation in force density in respect to the soil conditions of more than 20 dB when derived from impacts adjacent to the track. Figure 12.4(b), however, shows force densities converging at frequencies above 4 Hz when the impacts are carried out on the track.

From this study, it can be concluded that whenever ground conditions are not accounted for, uncertainty arising from force-density evaluations can vary from 7 dB, if derived from impacts on the track, to 20 dB, if derived from impacts adjacent to the track. These levels of uncertainties are significant and should be addressed when following the FTA detailed design procedure.

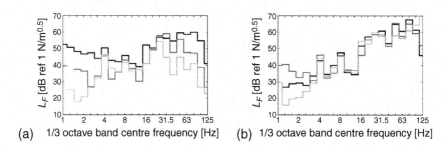

Figure 12.4 Force density (a) derived from impacts alongside the track and (b) derived from impacts on the rails on soft (light grey), medium (dark grey) and stiff (black) soil. (Reproduced with permission from Verbraken et al. [10].)

*Combining the Source Term with the Ground Response when
Independently Evaluated*

The theoretical study just presented goes on to assess the degree of uncertainty that the FTA detailed prediction scheme may yield when evaluating the free-field vibration levels by combining the source term with the ground response when independently determined. The free-field vibration is evaluated on medium soil and compares the results when using force density obtained on soft, medium and stiff soil derived from impacts alongside the track and on the rails.

When applying the FTA detailed assessment procedure to predict free-field vibration levels, with the force density derived using impact on the rails above a medium soil composition, it was shown (see Figure 12.5) for the three types of soil that, above 8 Hz, results derived from the FTA method differ less than 6 dB when compared with exact solution (i.e. derived via train movement simulation). However, when the impacts are carried out alongside the track, at frequencies above 8 Hz, a 15-dB variation is observed.

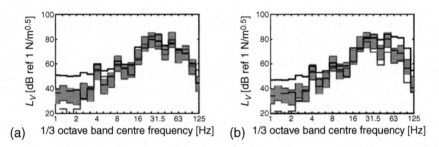

Figure 12.5 Free-field vibration levels 32 m from the track using force density (*left*) derived from impacts on the rails and (*right*) derived from impacts alongside the track on soft (light grey), medium (dark grey) and stiff (black) soil types compared with a 12-dB (±6 dB) interval around the exact solution (grey region). (Reproduced with permission from Verbraken et al. [10].)

In the case where impacts are carried out on the rail, the predictions of the vibration velocity level in the free field are relatively good, with differences generally below 6 dB, even with force densities determined at sites with significantly different soil characteristics. In the second case, where impacts are applied on the soil adjacent to the track, the influence of the soil characteristics on the force density is very large, resulting in differences of up to 15 dB.

Building Coupling Loss

As a way of assessing possible inaccuracies in the FTA detailed prediction due to a mismatch in the soil conditions, this section makes use of a theoretical study [9] to show how the coupling loss associated with the change in vibration level between the free field and the foundation of a structure is affected by the relative stiffness of the building and the soil. The study examines the variation in coupling loss due to soil composition and assessment location on a two-storey masonry building founded on a concrete strip foundation over a 12- × 6-m area (see Figure 12.6) located 30 m from the track. The study varies the soil composition from stiff, medium to soft, defined by compressional wave speeds of 600, 300 and 200 m/s, respectively.

Impact of Soil Conditions on Building Response

Figure 12.7 shows the predictions made for the vibration velocity level at two opposite extreme points of the foundation of a building (points A and B in Figure 12.6). Each curve, corresponding to the resulting vibration for the building on a medium soil, has been estimated by employing a specific coupling loss factor evaluated on the same building design but on soft, medium and hard soils. A 12-dB region around the exact solution for the medium soil is shown for comparison.

In the case where the soil conditions from which the coupling loss was derived match the soil conditions in which the building is being proposed, the prediction is generally within 6 dB of the exact solution. In the case where the soil conditions diverge, however, the accuracy is significantly reduced at higher frequencies. An overestimation of up to approximately 20 dB is found when applying the stiff soil–derived coupling loss factor to a building resting on a medium-stiff soil, whereas an underestimation of up to approximately 15 dB is obtained with the coupling loss factor is derived for soft soil.

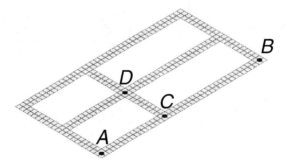

Figure 12.6 Foundation of the building and location of the measurement points. (Reproduced with permission from Verbraken et al. [9].)

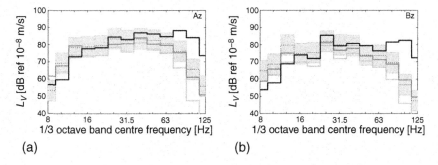

Figure 12.7 Vibration velocity level evaluated at two opposite sides of the building (a) assessed at a point of the building (point *A*) fronting the rail track and (b) assessed at the back of the building (point *B*) using the average coupling loss factor determined on a soft (light grey line), medium (dark grey line) and stiff (black line) soil. A 12-dB region around the exact solution (dotted black line) is shown by the grey region. (Reproduced with permission from Verbraken et al. [9].)

The study [9, section 8, p. 13] concludes with the following remark:

The dependency of the coupling loss factor on the soil conditions leads to an overestimation of the response in case of a coupling loss factor determined on a stiffer soil and an underestimation in case of a coupling loss factor determined on a softer soil.

Building Response Spatial Variation: Impact on the Assessment Location

Depending on where on the foundation the coupling loss factor is being assessed, a variation in insertion loss should be expected due to the kinematics of the building and local variations in the geometry. Considering the strip foundation given in Figure 12.6, the variation in coupling loss factor with assessment location taken from a theoretical study [9] is presented to describe the likely levels of uncertainty when carrying out an empirical model based on generalised insertion gain curves.

Figure 12.8 shows, for different soil compositions, the average value of the coupling loss factor, which is strongly dependent on the location of the measurement point, over all points in the foundation and the 90% interval of the coupling loss factor values over the foundation.

In the three cases presented, the uncertainty interval ranged from roughly 10 dB at low frequencies (i.e. 8–25 Hz), 6 dB around the mid-frequency range (i.e. 25–50 Hz) and 20 dB across the high-frequency range (i.e. 63–100 Hz). The study therefore suggests that the coupling loss

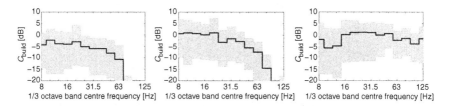

Figure 12.8 Average value (black line) and 90% interval (grey region) of the building coupling loss factor on a soft (left), medium (centre) and stiff (right) soil. (Reproduced with permission from Verbraken et al. [9].)

factor is strongly dependent on the location of the measurement point, as a large spread, up to 20 dB, can be observed.

12.2.4 Assessing Potential Uncertainties Associated with the FTA Detailed Prediction Procedure through Field Work

So far, the theoretical study presented earlier confirms that the coupling loss factor for a building is highly dependent on the assessment location, on the characteristics of the soil on which the building rests and on the foundation type and building design. Therefore, a generalised insertion loss, which is based solely on a building type, will lead to high levels of uncertainty. In practice, however, the uncertainty associated with the FTA detailed assessment methodology can be higher than theoretically established, especially because it is difficult to carry out the assessment in a controlled environment, as with the theoretical study presented earlier.

A fieldwork study [2] shows a diminishing confidence level when predicting absolute levels of rail-induced vibration. The highest degree of uncertainty came from determining the source terms, force density and building response. A difference was observed of almost 10 dB in force-density levels derived from impacts adjacent to the track at two nearby test locations (same track approximately 500 m apart) having the same soil conditions. This suggests a weakness in the FTA detailed prediction procedure. The study attributes the discrepancy to a possible mismatch of track conditions and to the subsoil below the track, which may have local variations.

For relatively small buildings, a 15-dB variation at the room floor plate was found when comparing similar building configurations. As established in the preceding section, a large variation should be expected whenever different measurement locations are being used. In practical terms, it becomes difficult to guarantee equivalent building measurement locations when characterising the building response. As such, empirical assessments based on generalised insertion gain curves become highly vulnerable to uncertainty. Only when proposing a new track alongside an existing building can such a high level of uncertainty be ameliorated. In this situation, the building can be assessed by extending the LSTM to capture the response at the room in question. However, the impact force required to generate an effective response at such a great distance may turn out to be considerable – almost impracticable – when considering the apparatus that would be needed to produce an impact capable of generating sufficiently high vibration levels to be measurable at the receiving location. In essence, when modelling the response of a building, the FTA method may be deemed impractical mainly when considering buildings at a significant distance from the proposed track (e.g. >20 m).

12.2.5 Concluding Remarks on the Likely Degree of Uncertainty Associated with the FTA Methodology

Compared with theoretical models, the advantage of empirical models is that they inherently consider all relevant parameters. Given the difficulty in obtaining physical parameters for the ground throughout the area of interest (which would be needed for a theoretical model), the LSTM is shown to be an ideal approach when capturing the dynamic behaviour of intervening ground. The source term, although it could rely on an empirical determination for the first two assessment stages once uncertainty is accounted for, would significantly benefit from a parametric model at the detailed design stage (e.g. analytical modelling such as the one provided by PiP [3]). The receiver, which is the building response, seems to pose the highest levels of uncertainty. Numerical modelling (e.g. Findwave [4]) would be highly advisable if the ground in which the building rests is adequately accounted for.

The advantage of numerical prediction of the coupling loss factor is greater flexibility in dealing with different building and foundation configurations and soil properties, whilst the advantage of using experimental data for the source and transfer of vibration is that the local conditions are correctly accounted for. A hybrid experimental-numerical prediction combines both.

12.3 Uncertainty When Employing Theoretical Models

12.3.1 Source Term: Railway Transportation Infrastructure and Vehicle

When addressing the degree of uncertainty that the source offers, you may ask the following questions: Is the track design predictable throughout the entire stretch that will affect the receiver in question? Will there be any discontinuities (switches and crossings, disjointed rail, discontinuous slab) throughout the track segment in question? What is the track and vehicle maintenance regime?

Rail Discontinuities

Transient excitation due to rail joints, switches and crossings may vary significantly, and therefore, their contribution to the overall rail-induced vibration becomes difficult to predict. However, as these relate to point-source rather than line-source excitation mechanisms, the effects of these discontinuities tend to diminish with increasing distance at a higher rate than the effects associated with rail-wheel roughness. Therefore, their contribution to the overall vibration level decreases with increasing track-receiver distance. The uncertainty impact from rail discontinuities can be addressed, in part, by contrasting the rates of decay for the point and line sources A_l/A_p via the following expression:

$$\frac{A_l}{A_p} = \left(\frac{r_0}{r_1}\right)^{(l-p)}$$

where r_0 and r_1 represent both receptor point distances from the track, and l and p are the Lamb geometric coefficient for the line and point sources, respectively (from Table 12.1, it can be established that the $l - p > 0.5$ for both surface rail and underground rail systems). It follows that the contribution from the point source to the overall vibration is reduced by at least 3 dB with the doubling of distance.

For example, if the receiver is 10 m from the rail track, then the A_l/A_p ratio works out, when contrasting to observations at 1 m away from the track, to be approximately 0.3, meaning that transient vibration arising from the rail discontinuity has, due to geometrical spreading, lost 70% of its magnitude in comparison with the excitation process produced by the rail-wheel roughness, which translates to a 10-dB reduction.

Table 12.1 Lamb's geometric attenuation coefficients

	Case a (point source)		Case b (line source)	
	R waves	P&S waves	R waves	P&S waves
At surface	$n = -\frac{1}{2}$	$n = -2$	$n = 0$	$n = -1$
Interior	—	$n = -1$	—	$n = -\frac{1}{2}$

Rail Supports

In addressing rail-induced ground-borne vibration at the source, the track design can be tuned to alter the coupled wheel/track resonance frequency. A range of design options, from ballasted track to floating slab, can be employed to reduce the emissions of railway vibration. Each system's performance is mainly dependent on its stiffness. In this section, only uncertainty associated with the modelling of ballast and rail pads will be addressed. All other systems, which are specifically designed to limit the emission of vibration (anti-vibration track), are too unique to be considered in a general overview such as this.

A variable that is sometimes assumed, often on the basis of limited data, is the rail pad stiffness. When modelling existing tracks to establish the effects on proposed developments, the assessment needs to represent the existing conditions as accurately as possible. There may be situations where the properties of track elements, which were installed many years ago, are unavailable. For instance, the stiffness of the installed rail pad is not always available. In such cases, a prediction model will assume a typical rail pad stiffness. Conventional rail pad stiffness may vary from 50 to 1,000 MN/m per meter of track. Conventional track designs, when not specifically designed for ground-borne vibration attenuation, tend to employ a soft rail pad of around 200 MN/m per metre of track; if only rail pads are used to support the rail on a concrete slab, softer rail pads, of around 50–150 MN/m per metre of track, could be employed. If rail pad parameters are not given, then, assuming a conventional track, the range of likely variation in vibration transfer, which makes up the uncertainty interval, may be more than 3 dB around the frequency of interest.

Figure 12.9 illustrates a dynamic model of a commuter train running on a conventional track that ensures that all relevant parameters, except for the rail pad stiffness, are known. Assuming the rail pad stiffness to be 200 MN/m per metre of track and accepting that the rail pad stiffness may range from 50 to 500 MN/m per metre of track, a 17-dB uncertainty interval (–10 to 7 dB, as shown in Figure 12.9) should then be expected. The figure maps out the potential variation in terms of free-field vibration levels for a range of rail pad stiffness. It is important to note that these insertion gain curves are dependent on the base case and may therefore be less representative if the base case is to vary substantially.

Sleeper Support Stiffness

Not all track elements perform consistently throughout a section of track. The track support stiffness – which may vary significantly within a stretch of track due to ground heterogeneity and varying ballast composition (including variation in its depth, density and the irregular nature of

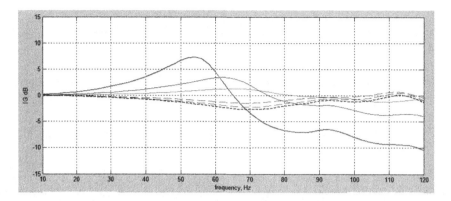

Figure 12.9 Potential response variation due to a change in rail pad stiffness: solid thick dark grey line: 50 MN/m/m; solid grey line: 100 MN/m/m; light grey line: 150 MN/m/m; dashed light grey line: 300 MN/m/m; dashed dark grey line: 400 MN/m/m; short dashed grey line: 500MN/m/m.

the stones) – is an important factor in the variation of ballast settlement under load. France [11] reported a standard deviation of 25% (within the 50- to 100-Hz range) when measuring the sleeper support stiffness (i.e. referring to everything supporting the sleeper, including the ground, the thickness of ballast and the degree of contact between the ballast and the sleeper) along a relatively short track stretch. Oscarsson [12], when measuring sleeper support stiffness at two sites, reported a standard deviations of 6% to 12%. Temple and Block [13] found track support stiffness to vary from a standard deviation of 15% to 60% of mean stiffness at different sites.

These studies show the extent to which the stiffness of the track supports may vary. To limit the uncertainty arising from the potential range of stiffness variation on existing tracks, it would be recommended to carry out vibration level measurements of pass-bys and calibrate the model accordingly.

Discontinuous Slab Tracks

One of the most effective ways of reducing rail-induced vibration at the source is to employ a floating slab track. The slab can be cast in situ, resulting in a track with a continuous slab, or it can be assembled as discrete precast sections, leading to a track with a discontinuous slab. A track with a continuous slab can be adequately modelled empirically by using experimental data taken from a continuous-slab track. Analytical modelling tools, such as PiP, can also be employed to model the effects, in terms of vibration levels, of a continuous-slab track. If the impact of a discontinuous-slab track requires evaluating, then the modelling complexity rises considerably; numerical modelling, such as Findwave [4], where the track structure is discretised, can be used to adequately represent a discontinuous-slab track. Given the resources required to carry out a numerical model, it becomes tempting to relax assumptions and trade in a continuous- for a discontinuous-slab track. However, as demonstrated in Jones et al. [14], the levels of uncertainty increase considerably. When comparing the underground free-field rail-induced vibration arising from a continuous-slab track with a discontinuous-slab track, Jones et al. showed (see Figure 12.10) a potential variation in vibration levels of up to about 10 dB around the 150-Hz frequency region.

Voids: Bond Discontinuity at the Structure-Soil Interface

Another assumption commonly adopted is that the soil is in continuous contact with manmade structures (e.g. no voids or gaps at the tunnel-soil interface are assumed). However, research shows that subsidence and frost cause significant tunnel movement, potentially allowing voids to form over a section of the tunnel, affecting the assumed tunnel-soil interface. Since there is no force transmission at the void, motion of the tunnel at this location will not directly result in wave propagation.

The extent of the effect, in terms of vibration spread, has been assessed by computing a 4-m-long void spanning 45 degrees on either side of the

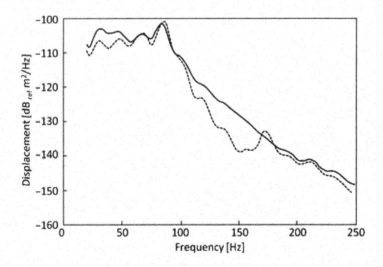

Figure 12.10 Power spectral density for continuous (solid line) and discontinuous (dashed line) floating-slab tracks. (Reproduced with permission from Jones et al. [14].)

crown of the tunnel (see Jones et al. [14]). The results are shown in insertion gain (IG), where positive values signify an increase in RMS velocity when the void is present, compared with a continuous tunnel-soil bond. The study conducted by Jones et al. [14] demonstrates how the void effect gains relevance with increasing frequency. For frequencies around 50 to 100 Hz, an 8-dB variation in vibration levels may occur for vibration travelling away from the tunnel in the x-direction (i.e. perpendicular to the tunnel). Less variation is noted when considering the vertical axis (i.e. vibration travelling up to the surface). Figure 12.11 shows a tendency for the void effect to diminish with increasing height, suggesting that vibrations arising from deeper tunnels tend to be less affected by the void. Across the horizontal plane located 10 m above the origin, the study shows a potential for a 3-dB variation, and at 15 m above the origin, only a 1-dB variation is observed.

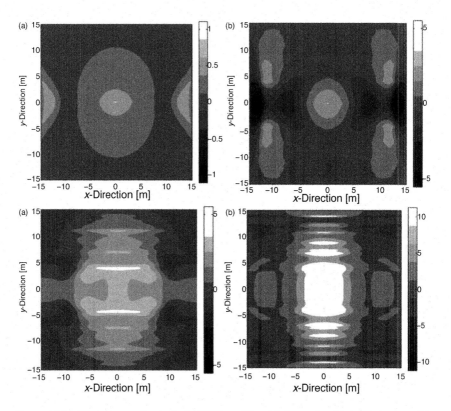

Figure 12.11 IG response in decibels (RMS, ref 1 m/s) for a 4-m-long 90-degree-wide void; load at x = 0 in tunnel: (a) 25 Hz, (b) 50 Hz, (c) 100 Hz and (d) 160 Hz passband. (Reproduced with permission from Jones et al. [14].)

For the given scenario, 160 Hz corresponds to a wavelength of about 5 m, approximating the void dimensions. As such, a significant amount of energy is not transmitted to the soil directly above the tunnel void, causing [as shown in Figure 12.11(d)] the highest ground response spatial irregularity. In this example, a variation of approximately 12 dB is observed.

It is worth noting that, in general, vibrations from underground trains tend to concentrate their energy around the 30- to 80-Hz frequency range when measured within a building. As such, voids would need to be substantially larger than the 4-m-long void noted earlier to cause a meaningful disruption to the expected vibration transfer.

Additionally, study shows that the relationship between the foundations (or piles) and tunnel depth is an important factor and indicates that if the foundations are within 15 m of the tunnel, then a 5-dB variation could be expected at the secant pile wall. However, in practical terms, when determining the building response, a smaller variation in vibration levels due to voids should be expected as the building response at low frequencies tends to be an average response to the incident wave field. The best way to limit the uncertainty arising from possible voids is to calibrate the model representing the existing rail tunnel using a calibration curve derived from an average of a set of site measurements (calibration for reducing uncertainty is addressed at the end of this chapter in Section 4).

Wheel/Rail Roughness

The principal source of excitation that gives rise to vibration and ground-borne noise is the existence of variations in the height of the rail running surfaces and the wheel tread (see Figure 12.12), generally referred to as "roughness". For well-maintained disc-braked stock, the contribution of wheel roughness is much less than that of rail roughness, and the exciting

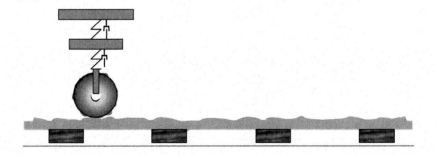

Figure 12.12 Schematic representing wheel and rail roughness.

force is effectively provided by the rail. In the case of wheels with flat spots caused by wheel slide and in the case of tread-braked wheels, the roughness contribution of the wheels may be more significant.

For rolling stock for which braking is entirely provided by tread brakes with cast-iron blocks, roughening and uneven heating of the steel tread cause roughness variations in steel hardness which, together, may increase excitation by as much as 10 dB more than that which is due to rail roughness alone. Modern tread-braked rolling stock does not exhibit wheel roughness effects as strongly as older steel-built coaching stock. As the tracks get used, rail roughness tends to worsen, causing the entire track bed system along with the rolling stock travelling on it to increase its deterioration process. To limit the deterioration, the roughness is controlled through a maintenance regime that involves the grinding of the rail every so often.

For modelling purposes, a representative wheel/rail roughness can be derived from rail roughness measurements from a wide range of contemporaneous systems (i.e. train and track). Whenever predicting the effects of an existing railway system on a proposed building, uncertainty between the modelled and actual wheel/rail roughness can be addressed through back-calibration based on site measurements (see Section 4).

Rolling Stock

The generation of rail-induced vibration is also determined by the rolling stock. The main component is the unsprung mass of the vehicle (mainly the wheels of the train). Sprung components of the vehicle (i.e. any vehicle element above its main wheel suspensions, such as the vehicle body) do not influence the excitation process as much as the unsprung mass of the train. Factors such as stiff primary suspensions on the vehicle and flat or worn wheels will increase the vibration excitation at the wheel-rail interface. Electric multiple units (EMUs) such as the ones used in urban underground systems tend to generate less vibration than diesel units. Given that not much effort goes into the suspension design of freight trains, as there are no passengers to complain about occupant comfort, they tend to generate higher levels of vibration at lower frequencies, especially if the train is loaded.

Trains belonging to the same "class" share the same vehicle design, and therefore, similar induced vibration levels should be expected from any set of commuter trains belonging to the same class when running on the same track stretch at similar speeds. When considering passenger trains with soft primary suspension, there should not be significant inconsistencies in terms of induced vibration between loaded and unloaded trains. Assessing for peak-hour and for off-peak-hour traffic should produce similar results; and therefore, any time period could be used to represent both on- and off-peak scenarios with limited impact in terms of uncertainty. The same may not be true for freight trains because their stiffer primary suspensions are less forgiving to load changes in terms of the force exerted at the rail.

One way to reduce uncertainty when characterising a train class that comprises the source term of the prediction process is to record a set of pass-bys and visually inspect both the raw data and spectrum to exclude potential outliers, which are typically due to worn wheels and/or defects in the primary suspension. To represent an existing scenario where trains are in operation, models can be calibrated by using measured pass-by vibration levels.

When calibrating a model, the "train class" most likely to adversely affect the proposed development should be taken to represent the source. If only one measurement is used for the calibration process, the levels of uncertainty would be very high, as there would be no guarantee that it would be representative of the typical scenario, and all adjustments arising from the calibration process would be biased to fit that particular train operation. By contrast, variation in rail-induced vibration associated with vehicle maintenance regime (the degree to which wheel flats and worn suspensions are tolerated before action is taken) should be expected and accounted for. The degree of uncertainty, or in this case of variation, can be accounted for by producing a statistical description of the set of measurements that forms the characterisation of the class being considered. Figure 12.13 shows the statistical characterisation of a train class potentially affecting the proposed development being assessed. The 10 measured pass-bys are represented by average spectrum, standard deviation, minimum and maximum levels affecting each third-octave band and ambient vibration (grey curve).

Operation Profile: Speed

For the case where all vehicles running on a track are of the same class, the only operational profile that may have a meaningful impact on the resulting vibration levels is the train's speed. Assuming the train to be at a steady speed as it passes by the point of interest, the variation in vibration levels can be evaluated from the following expression:

$$\text{Variation in vibration level} = 20 \log\left(\frac{\text{assumed speed}}{\text{actual speed}}\right)$$

As such, a 6-dB increase is expected for a doubling of speed. Therefore, a slight change from the assumed speed will not significantly alter the resulting vibration levels. There is, however, also a slight change in spectral

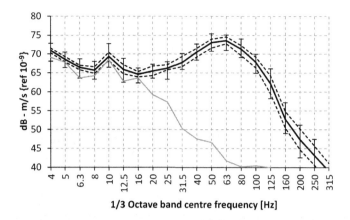

Figure 12.13 One-third-octave band RMS values of the vertical velocity: statistical spectral analysis contrasting the London Underground vibration velocity levels with ambient levels. Solid black line shows the average velocity level; dashed lines show one standard deviation below and above the average; the error bars refer to the minimum and maximum; the grey line shows the ambient vibration velocity levels averaged throughout the measuring period.

shape. As the speed increases, the vibration energy will transfer into higher frequencies in line with

$$f_{\text{dyanmic}} = \frac{V_{\text{pass-by}}}{\lambda_{\text{roughness}}}$$

where $V_{\text{pass-by}}$ is the speed of the train, and $\lambda_{\text{roughness}}$ is the roughness profile. For trains travelling at low speeds (e.g. <20 km/h), experience shows that this relationship will not hold, suggesting that factors other than speed have a greater influence at these speeds.

12.3.2 Ground as the Propagation Path

Soil Homogeneity Assumption

Modelling can be facilitated by assuming a homogeneous half-space or, when considering layered ground, assuming homogeneous layers. However, as shown empirically in Chapter 10, the ground can vary significantly over the area of interest, and the dynamic properties of soils are difficult to measure. Although it is difficult to establish uncertainty arising from soil heterogeneity, a theoretical study [14] proposes, under the caveat of limited research, an uncertainty range of at least ±5 dB when simulating ground vibration using simplified soil models.

Layer Profile

Real ground structures are usually stratified, possessing discontinuities forming layers. In layered ground, some of the vibration energy is refracted through to adjacent layer(s), and some is reflected. A change in speed of the reflected and refracted waves is expected in line with the density ratio between adjacent layers and the angle of incidence at their boundary.

When reducing the complexity of the model, horizontal layers of homogeneous soils are typically assumed. However, in real soil, a slight inclination of the layers with respect to the surface could reasonably be expected. To investigate the potential variation in RMS vibration vertical velocity at the surface, two scenarios based on a layer on a half-space ground are considered in one study [14]: scenario (a) top layer having a shear wave speed twice that of the underlying half-space and scenario (b) top layer having a shear wave speed half that of the half-space (more in line with typical ground compositions).

Figure 12.14, for each of the scenarios, combines the modelled schematic (lighter shapes) with the evaluated RMS vibration levels at the surface along a line perpendicular to the tunnel (darker curves). The model schematic shows the tunnel (circle) in relation to the resulting vibration curves along with the inclination profiles of each of the layer boundaries (lighter lines) being assessed: 1 degree (solid line), 3 degrees (dashed line), 5 degrees (dash-dotted line). The evaluated vibration curves represent the surface RMS vibration levels within the 15- to 200-Hz frequency range in respect to the ground layer profile and tunnel location. For the 5-degree layer inclination, a variation of up to 7 dB is observed when comparing

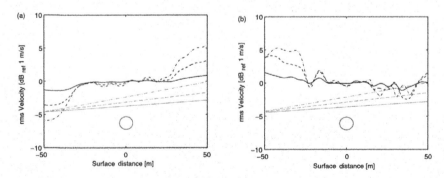

Figure 12.14 RMS velocity along a surface with a 5-m layer inclination at various angles (solid line: 1 degree; dashed line: 3 degrees; dash-dotted line: 5 degrees). (a) $C_{layer} = 2C_{half-space}$ and (b) $C_{layer} = 0.5C_{half-space}$. The circle represents the tunnel location, and diagonal lines represent the layer inclinations. (Adapted with permission from Jones et al. [14].)

the surface vibration levels around the 50-m limits. Variations of 2–5 dB are observed for the 1- and 3-degree inclinations, respectively.

Uncertainty Arising from Variations in Soil Water Content: Water Saturation

Water saturation affects the soil's density and has a large effect on its wave-propagation properties, especially when dealing with soft soils. For a fully saturated loose soil, the compressional wave speed is usually about 1500 m/s, approximating the speed of sound in water; the Poisson's ratio tends to approximately 0.5, and subsequently, the shear-wave speed generally turns out to be less than 200 m/s (values suggested in ISO/TS 14837-32:2015 [15]). It is important to note that this situation requires the soil to be completely saturated and that a slight reduction in the degree of saturation drastically reduces such pronounced changes in soil parameters.

For dry and partly saturated soils, the value of the low-strain Poisson's ratio is usually in the range of 0.2–0.3. In this range, the V_p/V_s ratio is not particularly sensitive to the Poisson's ratio and therefore should not introduce more than about ±7% uncertainty in the V_p estimate. This holds true for degrees of saturation up to about 0.99.

12.4 Addressing Uncertainty via Back-Calibration

Uncertainty is associated with possible errors (i.e. disparity between actual and predicted values) that comes from the set of theoretical assumptions that makes up the modelling tool, and the authenticity of the adopted parametric values describes the actual conditions being modelled. A considerable number of rail-induced ground-borne noise and vibration assessments, especially in urban areas, are carried out to evaluate the impact of an existing railway line on a proposed development. For such cases, a three-dimensional numerical modelling assessment can make use of the prevailing vibration environment arising from existing train movements to calibrate, to some extent, the source and local ground conditions being modelled.

As established earlier, within a certain surface area, vibration may vary according to the actual mechanical properties of the medium (e.g. ground properties). Therefore, in line with the ground's heterogeneity, vibration response may vary locally across the entire area of interest. When calibrating the model, inconsistencies between calibration curves become a function of generalisation that is associated with the degree of modelling detail. For modelling purposes, the entire ground domain is commonly generalised in line with the geotechnical report (which only reports on the ground conditions at the borehole locations) and/or on-site vibration measurements. As such, localised discrepancies between modelled and actual ground are to be expected. The impact that such variation has on the prediction certainty depends on the calibration technique. If enough measurements are carried out, a high variation in calibration curves can be addressed through averaging.

Inconsistencies between actual site properties (including rail infrastructures and operations) and model input parameters can be assessed and accounted for by calibrating a model of the existing conditions. For the case where there is an existing rail line affecting the vibration environment across the site being assessed, calibration can be done effectively by contrasting the output of a model depicting the existing scenario with the actual site measurements.

Experienced analysts will perform the calibration by adjusting the modelled parameters of the soil or train/track to match the measured spectrum and by interpreting spectral features of the modelled and measured vibration levels. The extent to which the discrepancy should be attributed to the track, vehicle or ground parameters may depend on the characteristics/features of the resulting spectra. If the aim is to assess the vibration response of the proposed building, then only the exact ground parameters become relevant; whether a spectral feature is attributed to the track or vehicles will not affect the incident wave field as long as the source spectral features are accounted for at either the track or the vehicle. For example, when accounting for excessive levels around the 40- to 100-Hz frequency region, once the ground parameters are fixed, the calibration process can be completed in the model by either softening the rail pads or by raising the mass of the vehicle wheelset, as shown in Figure 12.15. It is important to emphasise that the calibration process should be based on a set of measurement locations, as each measurement location may be influenced by local ground or structural features, resulting in a spectrum that cannot be extrapolated to represent the entire intervenient ground structure.

In the example given in this figure, the uncertainty associated with the source term has fallen from 3 to 1 dB by tuning the source assumptions to meet the measured spectrum, either by adjusting the wheelset mass or rail pad stiffness. Once the source term has been approximated through source calibration, all other site parameters (e.g. ground parameters) can be calibrated by contrasting the model of the existing conditions with actual site vibration measurements. This will reduce uncertainties arising from the site characterisation.

Figure 12.16 presents, in the form of insertion gain, the inconsistencies between the model output and the vibration measurement readings at nine representative locations across the site. The dashed curve represents the average calibration curve. As seen in this figure, there is no single standard calibration curve; each measurement location produces its individual curve, raising the level of uncertainty. Whenever the calibration process relies on

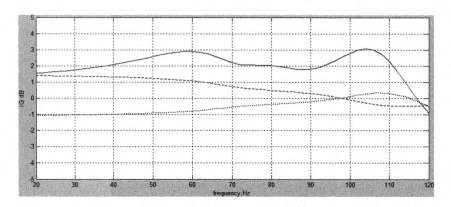

Figure 12.15 Insertion gain representing the difference in vibration levels between the modelled and actual free-field spectrum produced by a moving train (solid curve); the dashed curve represents the calibrated process with a rail pad stiffness increase of 220 MN/m/m (from 180 to 400 MN/m/m); the dotted curve represents the calibrated process with a wheelset mass correction of −250 kg (from 1,000 to 750 kg).

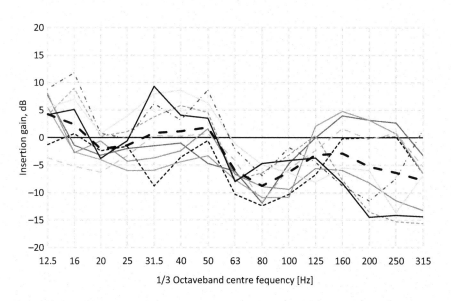

Figure 12.16 Insertion gain curves contrasting measured with predicted values of the existing conditions.

a set of measurement locations, different calibration curves may be expected (i.e. each measurement location will produce its own calibration curve). If, however, we accept the system to be linear and time invariant[1] and assume (1) a homogeneous ground structure and (2) that the line source is generating vibration consistently throughout its effective length, then the calibration curves should converge. Therefore, in principal, any discrepancy between calibration curves should be attributed to inconsistencies in parameters between the actual and modelled ground or to the heterogeneous nature of the ground structure being modelled. As the building response at low frequencies tends to be an average response to the incident wave field, at least within a certain ground surface area, the average calibration curve could be applied to the output of the model in the form of a correction curve as a way of reducing uncertainty.

But what if not all measuring points are valid? In other words, suppose that the transducer mounting is affecting the vibration measurements used to calibrate the model, or suppose that the vibration levels incident on the building will change once the site works have finished. This can happen, for example, as some temporarily buried ground structures may change the measured vibration levels. In this scenario, you have a range of uncertainty, given by the spread around the average calibration curve, defined by the lowest and highest values at each one-third-octave frequency band.

Not all frequencies carry the same degree of relevance in terms of the overall $L_{Amax,slow}$ noise rating. Typically, rail-induced vibration energy is most significant around the 31.5- to 125-Hz frequency range.

To assess the degree of significance that each one-third-octave band has on overall human perception, the resulting average A-weighted spectrum (taken from the model's output) should be investigated to establish the most influential bands and quantify their impact on the rated value (i.e. $L_{Amax,slow}$).

12.4.1 Weighting the Degree of Uncertainty When Assessing Ground-Borne Noise and Vibration

Up to this point, impacts arising from uncertainty have been assessed within a wide frequency range of approximately 12–250 Hz, which is where the most vibration energy lies when referring to commuter trains. Yet, depending on the criteria, not all frequencies carry the same weight when establishing the impact; furthermore, some descriptors may be limited within narrower a frequency interval. For example, VDV is limited within the 1- to 80-Hz frequency range, with frequencies around the 4 Hz carrying more weight than

---

1 Note that rail-induced vibration phenomenon is conventionally treated as a linear time-invariant system.

frequencies around 60 Hz. Therefore, any uncertainty around the 60- to 120-Hz range will not directly translate into the VDV uncertainty budget. Similarly, when assessing radiated noise, as measured by the $L_{Amax,slow}$ descriptor, the noise impact is directly related to human hearing sensitivity, which is unevenly distributed throughout the frequency spectrum. To obtain equivalent levels of perception across the audible spectrum at moderate sound pressure levels, the A-weighting coefficients, which form part of the $L_{Amax,slow}$ descriptor, are employed to form the A-weighted decibel rating level. Therefore, uncertainty in the 2- to 30-Hz range does not influence the predicted $L_{Amax,slow}$ as much as it does in the 50- to 120-Hz region.

Not only the weighting coefficients are of relevance to establish the most meaningful frequencies, but the manner in which the building responds to the incoming vibration will also dictate the frequencies of interest. If the building or its radiating elements lose responsiveness at frequencies beyond 80 Hz, then the large uncertainty interval beyond 120 Hz shown in Figure 12.16 will not have a significant impact on the overall uncertainty budget.

After establishing the influence that each one-third-octave band frequency has on the overall rating level, the uncertainty levels arising from calibration discrepancies (see Figure 12.16) can be used to assess uncertainty arising from differences between actual conditions and modelled site parameters. Once this is done, a limit on the likely uncertainty range may be suggested.

12.5 Conclusion

This chapter has presented the main sources of uncertainty when predicting ground-borne noise and vibration levels through both empirical and theoretical prediction modelling. By analysing a set of theoretical and practical studies, this chapter illustrates the impact these main sources of uncertainty can have on prediction accuracy.

After identifying inherent weaknesses associated with standard prediction routines, this chapter concluded with a practical suggestion for addressing such a complex uncertainty budget. It proposes using the back-calibration process as a way of quantifying the likely range of uncertainty. The discussion herein also emphasises how critical it is to establish the weighting that should be afforded to different regions of the predicted vibration spectrum before claiming an uncertainty interval.

References

1. Hanson CE, Towers DA, Meister LD, "Transit noise and vibration impact assessment", FTA-VA-90-1003-06. Federal Transport Agency, Washington, DC, 2006.

2. D'Avillez J, Frost M, Cawser S, et al., "Ground response data capture for railway vibration prediction", *Proc Instit Civil Eng: Transp* 168(1)(2015), 83–92.
3. Hussein MFM, Hunt HEM, "The PiP model: A software for calculating vibration from underground railways", in *Proceedings of the 14th International Congress on Sound and Vibration (ICSV14)*, Australian Acoustical Society, 2007, pp. 2627–2632.
4. Thornely-Taylor RM, "The use of numerical methods in the prediction of vibration", *Proc Inst Acoust* 29(Pt. 1)(2007), 14–21.
5. Hanson CE, Ross JC, Towers DA, Harris M, *High-Speed Ground Transportation Noise and Vibration Impact Assessment*, DOT/FRA/ORD-12/15. Federal Railroad Administration, Office of Railroad Policy and Development, Washington, DC, 2012.
6. Kurzweil LG, "Ground-borne noise and vibration from underground rail systems", *J Sound Vib* 66(3)(1979), 363–370.
7. Verbraken H, Lombaert G, Degrande G, "Verification of an empirical prediction method for railway induced vibrations by means of numerical simulations", *J Sound Vib* 330(8)(2011), 1692–1703.
8. Wu TX, Thompson DJ, "Vibration analysis of railway track with multiple wheels on the rail", *J Sound Vib* 239(1)(2001), 69–97.
9. Verbraken H, François S, Degrande G, Lombaert G, "Verification of an empirical prediction method for ground borne vibration in buildings due to high speed railway traffic", in *COMPDYN 2011, ECCOMAS, Thematic Conference on Computational Methods in Structural Dynamics and Earthquake Engineering*, Corfu, Greece, 25–28 May 2011.
10. Verbraken H, Eysermans H, Dechief E, et al., "Verification of an empirical prediction method for railway-induced vibration", in *Noise and Vibration Mitigation for Rail Transportation Systems*. Springer, Tokyo, 2012, pp. 239–247.
11. France G., "Railway track effects on rail support stiffness on vibration and noise", MSc dissertation, ISVR University of Southampton, Southampton, UK, 1998.
12. Oscarsson J, "Dynamic train-track interaction: Variability attributable to scatter in the track properties", *Vehicle Syst Dyn* 37(1)(2002), 59–79.
13. Temple BP, Block JR, "RENVIB II, phase 2, task 4: Vibration mitigation for surface lines", AEA Technology Rail Report No. RR-TCE-98-196 for the European Rail Research, March 1999.
14. Jones S, Kuo K, Hussein M, Hunt H, "Prediction uncertainties and inaccuracies resulting from common assumptions in modelling vibration from underground railways", *Proc Inst Mech Eng*, Part F: *J Rail Rapid Transit* 226(5)(2012), 501–512.
15. International Organization for Standardization, *Mechanical Vibration: Ground-Borne Noise and Vibration Arising from Rail Systems*, Part 32: *Measurement of Dynamic Properties of the Ground*, ISO/TS 14837-32:2015. ISO, Geneva, 2015.

Chapter 13

Managing Uncertainty in Noise Assessment Processes

Colin Cobbing and Charlotte Clark

13.1 Introduction

Major schemes can have an impact on people, buildings or equipment as a result of noise and vibration generated during their construction or operation. However, our main focus is on the assessment of the effects of noise on people because this is the main impact of concern for most schemes. That said, a lot of the principles covered can be applied to the assessment of vibration and other receptors or resources as well as to the assessment of noise on people.

This chapter describes:

- the different components of uncertainty associated with the noise assessment process that can occur at different stages during the promotion, design, construction and operation of major schemes; and
- the different principles and approaches that can be used to manage uncertainty.

In order to understand why and when noise assessments are carried out, it is important to have some understanding of the planning process and the legal and policy context in which assessments are undertaken. A description of the regulatory requirements and policy context is therefore provided in Section 13.2. However, experienced practitioners may not need to read this section.

The human response to noise is one of the main sources of uncertainty in the assessment process. Consequently, Section 13.3 provides an explanation of the different types of evidence that may be used in the assessment process and the strengths and weaknesses of different studies or types of evidence.

We explain why human response is often highly uncertain and show that it is not possible to predict the effects of noise, especially a change in effect resulting from a proposed change in the noise environment, on specific communities with a high degree of confidence. Accordingly, Section 13.4

describes the general approaches and principles that can be used to manage uncertainty and improve decision making.

Section 13.5 covers the reporting of effects. Section 13.6 considers how post-consent assurance mechanisms and processes can help to manage and reduce the uncertainty by comparison with the level of uncertainty associated with the assessment undertaken before consent is granted.

Section 13.7 addresses monitoring and ongoing validation and verification measures and how ongoing monitoring can:

- help to reduce the risk that the adverse effects of noise resulting from a scheme could be under-estimated;
- be used to identify additional noise protection measures if the adverse effects are greater than those predicted before consent was granted; and
- help to improve the evidence base that can be used to assess the effects of noise from future schemes.

Practical steps should be taken to manage uncertainty in the noise assessment process. Aspects of uncertainty that may affect the findings and conclusions of the assessment should be systematically evaluated and reported so as to provide decision makers with a proper appreciation of the risks and limitations associated with the assessment of the likely significant effects. Aspects of uncertainty include:

- uncertainty as to human responses to noise;
- uncertainty in the prediction of noise levels; and
- uncertainty in measurement of noise and vibration levels.

13.2 Background

Most assessments are undertaken as part of a development or other consenting process. For example, noise and vibration assessments are typically required to support planning applications for noise-generating development or sensitive buildings. Applications for noise-generating or noise-sensitive development are often determined by the relevant local authority in their capacity as the local planning authority. However, other types of applications may be determined by different consenting bodies. For example, applications for a minerals development will be made to the minerals planning authority, and applications for an airspace change will be made to the Civil Aviation Authority. Different consenting mechanisms, such as the Transport and Works Act, Development Control Order and Hybrid Bill processes, apply to major projects and nationally significant infrastructure schemes [1]. However, one thing that is common across the various consenting mechanisms for major schemes is that an environmental impact assessment (EIA) must be carried out, and the findings of the assessment must be reported in an environmental statement (ES), including a description of the envisaged mitigation.

This chapter focuses on noise and vibration assessments that are needed to support the planning process. That said, many of the principles discussed in this chapter may well apply to other processes.

Most assessments are undertaken to identify the impacts on people in their use and enjoyment of land or other areas. However, impacts on animals and other resources may be relevant. For example, vibration generated during construction or operation of a scheme could potentially interfere with sensitive equipment. In extreme cases, building damage may be a relevant issue.

13.2.1 Regulatory Requirements

The EIA regulations [2] (the 'assessment regulations') require a 'description of the likely significant effects of the development on the environment' and a 'description of the measures envisaged to prevent, reduce and where possible offset any significant adverse effects on the environment.' Updated assessment regulations were introduced in 2017 to transpose EU Directive 2014/52/EU [3].

EIA is described as a process that

> must identify, describe and assess in an appropriate manner, in light of each individual case, the direct and indirect significant effects of the proposed development on the following factors:
>
> (a) population and human health;
> (b) biodiversity, with particular attention to species and habitats protected under Directive 92/43/EEC and Directive 2009/147/EC;
> (c) land, soil, water, air and climate;
> (d) material assets, cultural heritage and the landscape;
> (e) the interaction between the factors referred to in sub-paragraphs (a) to (d).

The explicit requirement to assess the effects on human health was first introduced by Directive 2014/52/EU, which amended EIA Directive 2011/92/EU [4]. This directive was transposed into UK regulations in 2017 (the assessment regulations) [2]. Prior to that, the focus was on the effects on people without any specific reference to human health.

The assessment regulations also specify what must be included in an ES. The assessment regulations state:

1. A description of the development, including in particular:

    (a) a description of the location of the development;
    (b) a description of the physical characteristics of the whole development, including, where relevant, requisite demolition works, and the land-use requirements during the construction and operational phases;
    (c) a description of the main characteristics of the operational phase of the development (in particular any production process), for instance, energy demand and energy used, nature and quantity of the materials and natural resources (including water, land, soil and biodiversity) used;
    (d) an estimate, by type and quantity, of expected residues and emissions (*such as water, air, soil and subsoil pollution, noise, vibration, light, heat, radiation and quantities and types of waste) produced during the construction and operation phases.

2. A description of the reasonable alternatives (for example in terms of development design, technology, location, size and scale) studied by the developer, which are relevant to the proposed project and its specific characteristics, and an indication of the main reasons for selecting the chosen option, including a comparison of the environmental effects.

3. A description of the relevant aspects of the current state of the environment (baseline scenario) and an outline of the likely evolution thereof without implementation of the development as far as natural changes from the baseline scenario can be assessed with reasonable effort on the basis of the availability of environmental information and scientific knowledge.

4. A description of the factors specified in regulation 4(2) likely to be significantly affected by the development: population, human health, biodiversity (for example fauna and flora), land (for example land take), soil (for example organic matter, erosion, compaction, sealing), water (for example hydromorphological changes, quantity and quality), air, climate (for example greenhouse gas emissions, impacts relevant to adaptation), material assets, cultural heritage, including architectural and archaeological aspects, and landscape.

5. A description of the likely significant effects of the development on the environment resulting from, inter alia:

    (a) the construction and existence of the development, including, where relevant, demolition works;
    (b) the use of natural resources, in particular land, soil, water and biodiversity, considering as far as possible the sustainable availability of these resources;

(c) the emission of pollutants, noise, vibration, light, heat and radiation, the creation of nuisances, and the disposal and recovery of waste;

(d) the risks to human health, cultural heritage or the environment (for example due to accidents or disasters);

(e) the cumulation of effects with other existing and/or approved projects, taking into account any existing environmental problems relating to areas of particular environmental importance likely to be affected or the use of natural resources;

(f) the impact of the project on climate (for example the nature and magnitude of greenhouse gas emissions) and the vulnerability of the project to climate change;

(g) the technologies and the substances used.

The description of the likely significant effects on the factors specified in regulation 4(2) should cover the direct effects and any indirect, secondary, cumulative, transboundary, short-term, medium-term and long-term, permanent and temporary, positive and negative effects of the development. This description should take into account the environmental protection objectives established at Union or Member State level which are relevant to the project, including in particular those established under Council Directive 92/43/EEC(a) and Directive 2009/147/EC(b).

6. A description of the forecasting methods or evidence, used to identify and assess the significant effects on the environment, including details of difficulties (for example technical deficiencies or lack of knowledge) encountered compiling the required information and the main uncertainties involved.

7. A description of the measures envisaged to avoid, prevent, reduce or, if possible, offset any identified significant adverse effects on the environment and, where appropriate, of any proposed monitoring arrangements (for example the preparation of a post-project analysis). That description should explain the extent, to which significant adverse effects on the environment are avoided, prevented, reduced or offset, and should cover both the construction and operational phases.

8. A description of the expected significant adverse effects of the development on the environment deriving from the vulnerability of the development to risks of major accidents and/or disasters which are relevant to the project concerned. Relevant information available and obtained through risk assessments pursuant to EU legislation such as Directive 2012/18/EU(c) of the European Parliament and of the Council or Council Directive 2009/71/Euratom(d) or UK environmental assessments may be used for this purpose provided

that the requirements of this Directive are met. Where appropriate, this description should include measures envisaged to prevent or mitigate the significant adverse effects of such events on the environment and details of the preparedness for and proposed response to such emergencies.

9. A non-technical summary of the information provided under paragraphs 1 to 8.

10. A reference list detailing the sources used for the descriptions and assessments included in the environmental statement.

It can be seen from this that engagement with uncertainty is central to the assessment process. The fundamental requirement of the assessment regulations is to identify the likely significant effects and so the probability and likelihood of the effects must be considered. It is also plain that any limitations encountered when compiling the required information and the main uncertainties involved must be considered and reported.

It is important to recognise that EIA can by necessity attempt to look considerably into the future. For major infrastructure schemes with long consenting processes and long construction programmes, which can be more than ten years, it is not unusual for the assessment to be carried out several years before the start of construction and the start of operation. In addition, assessments are also undertaken for some development, especially transportation schemes, to cover a future design year to reflect the fact that some schemes may not reach full capacity or have the greatest environmental impact until sometime after when the scheme becomes operational. A future design year set at 15 years after opening is quite common. Thus, it can be seen that forecasts are made well into the future and well before any detailed design is undertaken. In general, the greater the timescales involved the greater the uncertainty there will be at the beginning of the EIA process.

The assessment regulations require any proposed monitoring arrangements (for example the preparation of a post-project analysis) to be described and reported in the ES. A 'monitoring measure' is defined as a provision requiring the monitoring of any significant adverse effects on the environment of proposed development. Later in this chapter we will explain how the EIA process can continue beyond the time when the development is consented, and that ongoing assessment can provide an important means of managing uncertainty.

13.2.2 Policy

When development is proposed, the applicant or sponsor must demonstrate how the proposals conform with the relevant policies. The UK government and the devolved governments have all developed planning policies and guidance relating to noise. A number of them cover noise assessment principles and relate to health and quality of life.

The Noise Policy Statement for England (NPSE) provides the overarching noise policy in England. The aims of national noise policy, set out in the NPSE [5], are defined to promote good health and quality of life:

> Through the effective management and control of environmental, neighbour and neighbourhood noise within the context of Government policy on sustainable development:
>
> - avoid significant adverse impacts on health and quality of life;
> - mitigate and minimise adverse impacts on health and quality of life; and
> - where possible, contribute to the improvement of health and quality of life.

In explaining these aims, the NPSE describes the meaning of 'adverse' and 'significant adverse' using the following terms:

NOEL: *No observed effect level*, the level below which no effect can be detected. In simple terms, below this level, there is no detectable effect on health and quality of life due to the noise.

LOAEL: *Lowest observed adverse effect level*, the level above which adverse effects on health and quality of life can be detected.

SOAEL: *Significant observed adverse effect level*, the level above which significant adverse health effects on health and quality of life occur.

A fundamental principle of the NPSE and National Noise Policy is that noise should not be considered in isolation. There is a need to integrate consideration of the economic and social benefit of the proposed activity with proper consideration of the adverse environmental effects, including the impact of noise on health and quality of life. Any determination of an application therefore should consider not just the effect on the acoustic environment but also the context, including the benefit to society, of the noise-making activity, integrating economic and social benefits with consideration of adverse environmental effects.

Most planning policies in England refer to and embrace the objectives and principles set out in the NPSE. For example, the National Planning Policy Framework (NPPF)[6] sets out the government's planning policies for England and how these are expected to be applied. Paragraph 180 of NPPF states:

> Planning policies and decisions should also ensure that new development is appropriate for its location taking into account the likely effects

(including cumulative effects) of pollution on health, living conditions and the natural environment, as well as the potential sensitivity of the site or the wider area to impacts that could arise from the development. In doing so they should:

a) mitigate and reduce to a minimum potential adverse impacts resulting from noise from new development – and avoid noise giving rise to significant adverse impacts on health and the quality of life;

b) identify and protect tranquil areas which have remained relatively undisturbed by noise and are prized for their recreational and amenity value for this reason.

Consideration of noise policy is relevant to our discussion because noise assessment reports are typically used to demonstrate how relevant policies are satisfied. Indeed, it could be argued that the aims of noise policy are now incorporated as a requirement of the EIA because the assessment regulations require environmental protection objectives established at the EU or Member State level to be taken into account.

13.2.3 Standards and Guidance

There are a number of standards and guidance that deal with the assessment of noise and vibration. These include:

1. The *Design Manual for Roads and Bridges* (DMRB) [7,8];
2. The *IEMA Guidelines for Environmental Noise Impact Assessment* [9];
3. The ANC guidelines (second edition): *Measurement and Assessment of Groundborne Noise and Vibration* [10];
4. British Standard (BS) 4142: 2014[11];
5. *Planning Practice Guidance: Minerals* (PPG-M) [12];
6. *Planning Practice Guidance: Noise* (PPG-N) [13];
7. BS 6472: 2008 [14];
8. ISO 14837–1: 2005 [15]; and
9. BS 5228–1: 2009+A1: 2014: *Code of Practice for Noise and Vibration Control on Construction and Open Sites* [16].

These guidance documents and standards are often used in noise assessments. We will refer to them in this chapter where they assist with particular aspects of managing uncertainty. However, for now, it is worth noting that some of the guidance and standards specifically require uncertainty to be addressed. For example, the BS 4142: 2014 contains a section on the steps to be taken to reduce uncertainty in the use of measurements and calculations.

13.3 Effects of Noise and Vibration

Before we consider how to assess the effects of airborne noise on people, it is worth considering the types of effects that may be relevant and how the strength of the evidence for each of the effects should be properly considered. Our focus is on the effects of airborne noise as opposed to vibration or ground-borne noise because:

- this represents the main effect of concern for the majority of noise assessments undertaken; and
- the ANC *Guidelines on the Assessment of Groundborne Noise and Vibration* [10] already provides advice on the assessment of ground-borne noise and vibration.

Evidence for the effects of environmental noise exposure on health and quality of life continues to grow [17]. It is widely accepted that there is sufficient evidence that noise exposure leads to annoyance, disturbs sleep, increases cardiovascular disease and impairs children's learning [17,18], with evidence emerging for other health outcomes (e.g. diabetes [19,20], low birth weight [21,22] and cancer [23]). There is some evidence for other outcomes, such as diminished mental health and quality of life [17,24], but these are areas where there is generally less agreement between research studies and where further research is required.

Exposure-response functions (ERFs) between noise exposure and health outcomes estimate how the impact of noise exposure increases the effect on the health outcome, for example, how each 5- or 1-dB increase in noise increases the risk for the health outcome. Many studies are available that provide ERFs for average noise metrics and key health outcomes such as annoyance, sleep disturbance and cardiovascular disease [25–27].

Most studies of noise and health use annual average noise metrics such as $L_{Aeq,16h}$, $L_{Aeq,8h}$ or L_{den}. Studies of objective sleep disturbance (where physiological assessments are made of sleep) use L_{pmax} metrics to link noise events to impacts on sleep. Advances in modelling have enabled a broader range of metrics to be calculated for individual studies, and some studies now also use 'number of event' or 'time above' metrics. As with any noise modelling, the relationship demonstrated between noise and health may be influenced by the quality of the noise exposure assessment for the study. When the evidence is interrogated, it is often the case that some metrics are better correlated with certain health outcomes or that the evidence for some health outcomes is stronger for certain noise sources than others. For example, there is much more evidence for the effects of aviation noise on children's learning than for road traffic or railway noise [28]. Often, evidence from one source is applied to an assessment of another, which again may introduce further uncertainty into assessments. For example, in the United

Kingdom, the aviation noise ERF from the RANCH study [29] is often applied to assessments where environmental noise from roads or railways may have an impact on schools. Whilst it is preferable to use source-specific evidence, often we are forced to apply evidence from other sources in an assessment, which may introduce uncertainty into the assessment.

However, it is important to appreciate that not all evidence is equal: different research methodologies are used to examine noise effects on health that may influence the quality of the study and the nature of the findings. The main types of study methodologies used to assess the effects of environmental noise exposure on health are as follows:

- *Cross-sectional studies.* These measure noise exposure and the health outcome at the same point in time. Most studies of annoyance use this approach, with studies assessing annual average noise exposure related to noise annoyance, which is often assessed for the past 12 months using the standard ISO question [30]. The major disadvantage of cross-sectional study designs is that noise and the outcome are assessed at the same point in time, which means that we cannot be clear that noise is leading to the health outcome observed. This leads to debate about whether these types of studies can demonstrate cause and effect between noise and the health outcome.
- *Longitudinal studies (sometimes called cohort or prospective studies.* These link noise exposure to health outcomes at a later point in time. Most types of health outcomes, for example, cardiovascular disease, diabetes and disturbed mental health, can be studied using this approach. These studies exclude subjects with the health condition at the start of the study in order to assess those who develop the condition over time. At present, these types of studies tend to evaluate steady-state noise exposure rather than changing noise exposure. This study design is seen as stronger than cross-sectional study designs, as a link can be made between previous exposure to noise and the development of a health outcome.
- *Ecological studies.* In these studies, the relationship between noise exposure and health outcomes is assessed for a specific population geographically (e.g. the population of greater London) or temporally. Here there may or may not be temporal separation between the exposure and the outcome. The outcomes are usually from medical records/hospital admissions and therefore are often used to evaluate outcomes such as heart attacks and stroke.
- *Meta-analyses.* A meta-analysis is a statistical analysis that combines the results of multiple individual scientific studies to provide an estimate of the effect observed across the studies. Combining estimates across studies increases the statistical power (a greater amount of data improves provision of the estimate) and gives a more robust estimate

of the effect or the ERF. Several studies of this type over the years have combined study estimates of annoyance to produce annoyance ERFs [26,31,32].

Each study methodology has its own strengths and weaknesses, but it is important to appreciate that studies that are able to take account of temporal separation between the noise exposure and the health outcome and meta-analyses are viewed as providing stronger evidence for causality than cross-sectional studies. Most evidence to date for noise effects on health and quality of life come from cross-sectional studies, with evidence from cohort studies, ecological studies and meta-analyses increasing over the past few years.

Studies that examine the association between an exposure and an effect are not able to examine causality between the exposure and the effect. However, there are several criteria that can be applied to research findings to try to establish whether there is likely to be a causal relationship between the exposure and an effect. These criteria are known as the 'Bradford Hill criteria' [33] and include, among others:

- consistency of findings across different research studies;
- temporality, where the effect is demonstrated to occur after the exposure;
- biological gradient, where the greater the exposure, the larger the effect;
- plausibility, where there is a plausible mechanism to explain how the exposure leads to the effect; and
- strength of the association, where the stronger the association, the more likely it is to be causal.

It is important to bear these criteria in mind when evaluating evidence of noise effects on most health and quality-of-life outcomes that will develop over time. Uncertainties may be introduced into the assessment by use of a specific ERF for the assessment or by the methodologies of the research evidence used.

Caution should also be exercised against the risk of over-interpreting research evidence or extrapolating research evidence beyond that presented in the paper. Research papers undergo a rigorous peer-review process and are extremely careful in what they do and do not state, so take care to understand what is and what is not being said in the paper.

Ideally, evidence should be relevant to the population involved in the assessment, but many research studies are not representative of national populations but instead are selected based on certain criteria, which may make them less than ideal for application to EIA assessments. There is also huge individual variation in how individuals within a population respond to noise exposure. Figures 13.1 and 13.2 both illustrate ERFs

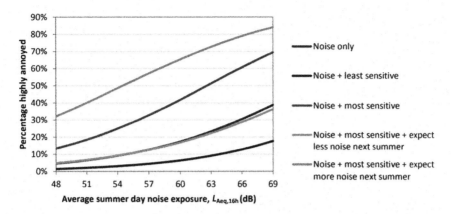

Figure 13.1 Variation in dose-response relationships for different logistic regression models [34].

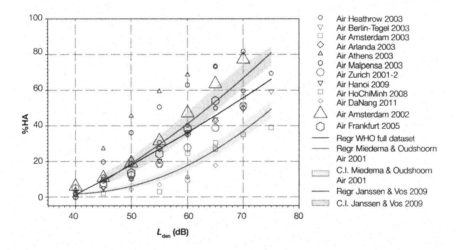

Figure 13.2 WHO exposure-response relationship for annoyance [26].

between aviation noise and noise annoyance. In Figure 13.2, whilst the black line shows the ERF for aviation noise and being highly annoyed for all the studies considered in the meta-analysis, examination of the individual estimates from each of the studies contributing to the ERF shows huge variation, with differences in percent varying from as little as 10% between the studies at the lower noise exposures and up to 50% for some of the higher noise exposures. This in itself illustrates the differences between studies of annoyance and also how uncertainty can be introduced into assessments, where such variation is present. Figure 13.1 illustrates how the ERF for being highly annoyed can be increased or decreased by a substantial percentage according to how noise sensitive the individuals considers themselves to be to noise and also by how much they expect their exposure to noise to increase over the next 12 months. Whilst it is usual practice to apply the 'average' ERF plotted across studies to an assessment, individual variation for the local context being assessed may introduce uncertainty.

A further challenge involves how the research evidence is used to guide the setting of LOAEL and SOAEL values for an assessment, now a requirement of UK policy (see section 13.2.2). As previously described, SOAEL values indicate the level above which significant adverse health effects on health and quality of life occur. However, many, if not most, ERFs do not indicate a clear threshold where a significant adverse health effect begins which is implied by the definition of a SOAEL. Indeed many ERFs are linear which further inhibits the identification of where a significant adverse health effect begins, as if effects increase linearly with exposure, what do we consider to be a significant adverse health effect? For example, if we are considering annoyance, is a rise of 10% in the population being highly annoyed significant or is an absolute threshold of 30% or 50% being highly annoyed significant? Answers to these questions will vary by client and stakeholder. There is no agreement or guidance defining significant adverse health effects. There are some publications addressing how to identify LOAEL and SOAEL levels from research evidence [35,36], but there is little agreement among researchers on this issue, and different assessments can, and should, set different LOAEL and SOAEL levels. This introduces further uncertainty into the assessment. The nature of the information available from research studies can also be limited, impacting on the identification of a LOAEL – that is, the level above which adverse effects on health and quality of life can be detected. Studies use different ranges of noise exposures, and many do not go down to very low levels, which inhibits identification of where the effects on health and quality of life can start to be detected. For example, if a noise and health study assesses the relationship between exposure and health starting the assessment at 50 dB, we are likely to assume no effect below 50 dB, which may not be the case.

Further uncertainty may be introduced into the assessment by the use of steady-state relationships to evaluate the effect of a change in noise on people's health and quality of life. Most research on the noise effects on health to date, whether cross-sectional or longitudinal in study design, demonstrate the steady-state relationships, that is, the relationship between current noise exposure and health, and not the effect of a change in noise exposure on health.

In the 1980s and 1990s, a series of studies evaluating interventions for road traffic noise was carried out in the United Kingdom, evaluating the effect of a decrease in road traffic noise associated with a reduction in trafficvolume associated with the opening of new relief roads [37–39]. The studies found a reduction in reports of being bothered by noise after the new roads opened and that this decrease was in excess of (i.e. larger than) that predicted from the steady-state ERF. This excess response in reports of dissatisfaction with or being bothered by road traffic noise was still evident two years and even seven to nine years later [40]. These studies suggest that where there is an excess response in relation to a change in noise exposure, such effects may be observed many years later, perhaps indicating a real change in a nuisance that persists for several years and not just in the short term [7]. Similar excess responses to changes in aircraft noise exposure have been observed [41].

The 2011 version of the DMRB (now superseded) used the data from the UK road studies to estimate an ERF showing the change in 'nuisance' (being bothered very much or quite a lot by traffic noise) for an increase or a decrease in L_{A1018h} [7] (see Figure 13.3).

The new version of the DMRB [8] no longer recommends use of the nuisance curves. This change was perhaps motivated by the uncertainty associated with the extrapolation of the before and after studies carried out for bypass schemes to other types of scheme (e.g. road widening, new alignment. The new version of the DMRB noise and vibration guidance still retains guidance on the magnitude of a noise change [8] (see Table 13.1).

However, the guidance is now more nuanced by the introduction of a number of additional factors (see Table 13.2) that need to be considered

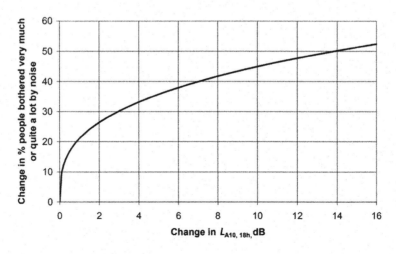

Figure 13.3 Estimation of traffic noise nuisance (change in percent bothered very much or quite a lot by traffic noise) [7].

Table 13.1 Classification of magnitude of noise impacts [8]

Short-term magnitude	Short-term noise change, dB $L_{A10,18h}$ or L_{night}
Major	≥5.0
Moderate	3.0–4.9
Minor	1.0–2.9
Negligible	<1.0

Long-term magnitude	Long-term noise change, dB $L_{A10,18hr}$ or L_{night}
Major	≥10.0
Moderate	5.0–9.9
Minor	3.0–4.9
Negligible	<3.0

in addition to the magnitude of the noise change. The introduction of these new factors allows the noise change criteria to be adapted to reflect different situations and circumstances in recognition of the fact that there will be no universal response to health and quality of life associated with a particular noise change.

The approach now advocated in the DMRB is based on the cumulative experience of promoting and operating numerous road schemes. Therefore, there can be a reasonable level of confidence that the approach will not significantly under- or over-estimate the effects of the scheme. However, it must be recognised that the approach set out in the DMRB represents a pragmatic compromise reflecting the current state of the scientific evidence. Ideally, assessors would not have to rely on noise exposure assessments as a proxy for estimating effects on people, and different ERFs would be available to directly quantify the effects of the scheme and the proposed mitigation measures for a range of circumstances. However, we are a long way from this ever likely being a reality.

The lack of ERFs that will enable changes in health and quality of life to be assessed for changes in the sound environment is not limited to road schemes. On the contrary, the lack of evidence on the effects of a change in noise exposure [41] represents a huge problem in the assessment of the noise effects resulting from the promotion of noise-producing schemes. Until the evidence base on interventions or change situations improves, there will continue to be significant uncertainty resulting from the continuing reliance on the use of noise exposure assessments and steady-state exposure-response relationships that may under-estimate (or over-estimate) impacts for schemes

Table 13.2 Determining final operational significance on noise sensitive buildings [8]

Local circumstance		Influence on significance judgement
Noise level change (is the magnitude of change close to the minor/moderate boundary?)	1.	Noise level changes within 1 dB of the top of the 'minor' range can indicate that it is more appropriate to determine a likely significant effect. Noise level changes within 1 dB of the bottom of a 'moderate' range can indicate that it is more appropriate to consider that a change is not a likely significant effect.
Differing magnitude of impact in the long term and/or future year to the magnitude of impact in the short term	1.	Where a greater impact in the long term and/or future year is predicted, it can be more appropriate to consider that a smaller change is a likely significant effect. A lower impact in the long term and/or future year over the short term can indicate that it is more appropriate to consider that a larger change is not significant.
	2.	A similar change in the long term and non-project noise change can indicate that the change is not due to the project and not an indication of a likely significant effect.
Absolute noise level with reference to LOAEL and SOAEL (by design this includes the sensitivity of the receptor)	1.	A noise change where all do-something absolute noise levels are below SOAEL requires no modification of the initial assessment.
	2.	Where any do-something absolute noise levels are above the SOAEL, a noise change in the short term of 1.0 dB or more results in a likely significant effect.
Location of noise-sensitive parts of a receptor	1.	If the sensitive parts of a receptor are protected from the noise source, it can be appropriate to conclude that a moderate or major magnitude change in the short and/or long term is not a likely significant effect.
	2.	An example of this would be where no windows of sensitive rooms face the road, and outdoor spaces are protected from the road by buildings.
	3.	Conversely, if the sensitive parts of the receptor are exposed to the noise source, it can be more appropriate to conclude that a minor change in the short and/or long term is a likely significant effect.
	4.	An example of this would be where a house has many windows of sensitive rooms and outdoor spaces facing the road.
	5.	It would only be necessary to look in detail at individual receptors in terms of this circumstance where the decision on whether the noise change gives rise to a significant environment effect is marginal.
Acoustic context	1.	If a project changes the acoustic character of an area, it can be appropriate to conclude that a minor magnitude of change in the short and/or long term is a likely significant effect.

(*Continued*)

Table 13.2 (Cont).

Local circumstance	Influence on significance judgement
Likely perception of change by residents	1. If the project results in obvious changes to the landscape or setting of a receptor, it is likely that noise level changes will be more acutely perceived by the noise-sensitive receptors. In these cases, it can be appropriate to conclude that a minor change in the short and/or long term is a likely significant effect. 2. Conversely, if the project results in no obvious changes for the landscape, particularly if the road is not visible from the receptor, it can be appropriate to conclude that a moderate change in the short and/or long term is not a likely significant effect.

that change noise exposure. In the meantime, a number of approaches must be adopted to reflect and acknowledge the limitations associated with these shortcomings.

13.4 Assessment Principles and Approaches

The over-riding aim of the assessment must be to carry out the assessment and report the findings in a way that supports good decision making. There are no universal methods for dealing with the uncertainty associated with predicting the effects of noise on humans. However, a number of principles and approaches can be employed to manage the uncertainty.

13.4.1 Evidence-Led Approaches

Having identified the aspects of the environment potentially affected by the development, it is important to use the best available scientific evidence to undertake the assessment. An evidence-led approach can be used in which the best available scientific evidence on the relevant effects is employed to establish the assessment methodology.

Section 13.3 provided an overview of scientific evidence on the effects of noise and highlighted the limitations of the current state of knowledge in this area. This section explains that there is a high degree of variability in individual and community responses to noise such that it is not possible to reliably predict effects on individuals or specific communities. Moreover, there are significant gaps in our knowledge of the effects of noise on people, especially in relation to how people respond to a change in the acoustic environment. One of the main failings of noise assessments is that they do not assess the effects of noise on people or that the findings of the assessment are reported without properly reflecting the limitations of the assessment. All too often noise exposure is used as a proxy for assessment of the effects of noise on people without proper consideration for how the noise exposure parameter correlates with the human response to noise.

The scientific evidence on the effects of noise should be used to scope the assessment. The potential impacts should be identified at the beginning. If the proposed scheme has the potential to affect health and quality of life, then each aspect of health and quality of life should be identified and properly assessed. This is necessary because different aspects of health and quality of life, such as sleep disturbance and annoyance, would normally justify different methods and approaches. When each aspect of health and quality of life has been identified, the scientific evidence should be used, as far as possible, to define the assessment methodologies and identify the noise assessment parameters that best correlate with each aspect of interest. The type of assessment and the choice of parameter should be appropriate in relation to the type of effect. It may be that in complex situations, a range of parameters, including numbers of events, should be considered.

Steady-state ERFs are available for sources of transportation noise and other sources of noise. In an ideal situation, ERFs for each type of effect should be used to evaluate the prevalence of an effect with and without the proposed development. This will be practical where steady-state ERFs exist and are relevant to the type of scheme and situation that is being assessed. Where ERFs do not exist or are not ideally suited to the situation under consideration, such limitations should be acknowledged in the ES with an explanation of how the proposed methodology was developed and any limitations addressed. In some cases, other types of evidence may be used. For example, it might be possible to use an analysis of complaint data or case studies. For example, this approach was used to assess the impact of noise from engine testing at Cambridge airport. In some cases it will be appropriate to use standards such as BS 4142[11], which was developed by an expert group that used case studies and other scientific evidence to inform its development.

So far we have discussed the use of steady-state ERFs. However, as we have explained, some abrupt changes in the noise environment can provoke changes in annoyance in excess of that which would be predicted by steady-state relationships [38,41,42]. Consequently, reliance on steady-state relationships can under- or over-predict annoyance in the period following a change. It may be legitimate to assume that, all else being equal, the general annoyance response of a community will revert back to a steady-state relationship after a relatively long period of time. This is reasonable because there is likely to be a change in the composition of the population over time. In addition, it is also possible that self-selection may occur and that people who are highly sensitive to noise may move away from an area if they feel that the area has become too noisy. Equally, people sensitive to noise may move to another area if they feel that is has become quieter over time. Even so, the evidence from before and after studies shows that the change in annoyance can last for several years and possibly in excess of 10 or 20 years. If a change in the acoustic environment could cause excess annoyance and ERFs are not available to estimate a change in annoyance, then the assessment should acknowledge this as a limitation, and the ES should explain how short-term changes in effects are being addressed in the assessment.

Typically, noise assessments will consider absolute levels of noise in relation to defined thresholds or criteria (e.g. LOAELs and SOAELs). The best available scientific evidence should be used to define or inform absolute thresholds when they are used. Wherever appropriate, consensus opinion from expert groups, such as the World Health Organization (WHO), on LOAELs should be used. Again, any limitations in the assessment should be identified and reported.

Even with well-established exposure-response relationships, significant limitations can occur when attempting to apply ERFs at a project level. For example, exposure-response studies provide an indication of average community annoyance and may be poor predictors of individual or even community response. Any limitations and shortcomings should be suitably identified and reported so that the decision maker is informed about the level of risk of under- or over-estimating effects.

In the majority of cases, it should be possible to refer to and rely on consensus expert opinions. For example, the WHO *Environmental Noise Guidelines for the European Region* [43] are based on the consensus view of a panel of experts. It must be recognised, however, that such expert reviews are not exhaustive and that new information becomes available on a regular basis. Consequently, it may be necessary to update the reviews and to focus in greater depth on particular aspects relevant to the development. In some cases, therefore, it may be necessary to undertake a bespoke review of the current scientific literature relating to schemes that are particularly contentious or unusual.

13.4.2 Statistical, Contextual and Numerical Approaches

If noise exposure is established for a population within a defined study area, then it will be possible to apply the ERFs to estimate the overall scale of an effect. For example, steady-state ERFs can be used to estimate the number of people highly annoyed or the number of people who will be at risk of being sleep disturbed. It will also be possible to estimate the scale of an effect or effects at different levels of confidence. Then, if that population was exposed to a change in the acoustic environment, it should be possible to estimate the change in effect.

An assessment that is undertaken to estimate the scale of an effect over a large area could be described as a 'statistical' or 'prevalence-based approach', which may be appropriate and sufficient for some situations. Although we can estimate the overall scale of an effect and the change in an effect, it is not possible to identify specific individuals or communities that are likely to experience a significant effect on health. We can only estimate risk or the incidence of an effect in a sample population. For example, we could estimate, within a sample population, the increase in the number of people who might be at risk of stress-related effects, but it will not be possible to identify households or communities within that population that are likely to experience a stroke or a heart attack.

A statistical approach can be used to assess the overall impact of the scheme. A statistical or area-based approach may also be sufficient by itself if the mitigation envisaged to prevent, reduce or offset any significant adverse effect was universal (provided equal benefit to all receptors) and was not spatially relevant. For example, noise control applied to trains, aircraft or other sources would provide an equal or similar noise benefit for each of the receptors within the assessment area. However, for a lot of noise assessments, the envisaged mitigation is spatially relevant or location specific. For example, the noise mitigation envisaged might include measures that reduce noise during transmission (e.g. noise barriers) or at the receptor itself (off-site protection measures such as noise insulation packages). It then becomes necessary to undertake assessments for different receptors, small groups of receptors or defined community areas within an overall study area.

A numerical or noise exposure approach is one in which numerical pass or fail criteria are used to identify the likely significant effects. For example, if the noise level at an identified receptor exceeds a specified noise level and is predicted to receive an increase in noise level equal to or greater than 3 dB when rounded to the nearest 0.1 decibel, it is identified as a significant adverse effect. Anything exposed to an increase in noise level of 2.9 dB or less is not identified as a significant effect. It should be evident from the description of effects given in Section 13.3 that the human response to noise is complex and that such simplistic approaches are not evidence based. Guidance documents, such as the IEMA Guidelines, now recommend that simplistic numerical approaches should be avoided and that a range of factors relating to particular effects on specific receptors should be considered. Consideration of the impact of sound within the context of the particular circumstances of the receiver environment is also a feature of BS 4142 [11] relating to sound generated by commercial and industrial uses. Now noise assessments tend to consider the context in which a noise change will occur by consideration of a range of factors. An example of this is the scope and methodology report for HS2 Phase 2B [44], which defines the LOAELs and SOAELs for operational noise. Where the predicted construction or operational noise level exceeds the relevant SOAEL values, then a likely significant adverse effect is reported for each receptor affected.

It then goes on to explain that for residential receptors, likely significant adverse effects from operational noise (positive from noise reductions and negative from noise increases) will also be determined on a community basis where the calculated noise level exceeds the relevant LOAEL but is less than the relevant SOAEL by taking into account the following factors:

- type of effect being considered (e.g. annoyance);
- the magnitude of the predicted noise level compared to the relevant LOAEL and SOAEL values and available dose-response information;
- for operation of the proposed scheme, the predicted change in noise level (day or night);
- the existing sound environment in terms of the absolute level and the character of the existing environment;
- the number and grouping of receptors subject to noise effect and noise change;
- any unique features of the proposed scheme or the receiving environment;
- the potential combined impacts of sound and vibration; and
- the effectiveness of mitigation through design or other means.

This example shows how numerical criteria can be used as the basis of a noise assessment whilst also giving flexibility to the assessors to consider other relevant factors. This is in contrast to a more numerical approach that only considers numerical criteria.

The main point we are making here is that simple numerical approaches will rarely reflect the available scientific evidence and will be rarely justified. There is an increasing trend towards a combination of statistical and contextual approaches, underpinned by evidence. Calculating the prevalence of an effect using ERFs with confidence intervals will allow the uncertainty in the assessments to be considered and reported.

13.4.3 The Precautionary Principle

The 'precautionary principle' was adopted by the UN Conference on the Environment and Development in 1992 [45], suggesting that a precautionary approach should be widely applied to protect the environment, meaning that where there are threats of serious or irreversible damage to the environment, lack of full scientific certainty should not be used as a reason for postponing cost-effective measures to prevent environmental degradation. This principle can be extended to the protection of people's health and well-being. In essence, it means that assessors should err on the side of caution where uncertainties exist. This does not, however, mean that all risk should be avoided or that the assessment must be entirely confident that all the significant effects are identified. Rather, risks and probabilities should be properly balanced, having regard for the seriousness of the risk and the implications of any decision if the environmental information over- or under-estimated the significant effects.

Dealing with vulnerable or highly sensitive groups represents an example of how a precautionary approach could be applied to noise assessment. Any assessment of health effects on the general population could under-estimate the effects on vulnerable or highly sensitive groups or individuals. It is generally considered best practice, therefore, where it is relevant to the scheme, to assess the effects of a scheme on vulnerable groups as well as the general population.

13.4.4 Transparency

Transparency helps interested parties to properly access, understand and scrutinise the results of the assessment. The methods used in the assessment should be sufficiently described. It is also a requirement of the EIA regulations that any limitations or uncertainties that may affect the outcome of the assessment should be acknowledged. This will enable decision makers to understand the level of risk of under-estimating the likely significant effect.

13.4.5 Proportionality

The assessment should produce information that is sufficient, relevant and reliable to support effective decision making. Uncertainty should not be used as an excuse for driving unrealistic or disproportionate demands to either eliminate or minimise risk. All forecasts or assessments of impacts on the environment have an element of uncertainty, and this is something that needs to be practically managed so as to focus efforts on areas where there is the greatest level of risk. In this context, the level of rigour applied should reflect the level of risk. For example, where the level of uncertainty can materially affect the outcome of the assessment and where the effects may be irreversible, the assessment should be achievable within the limits of available time and resources.

13.4.6 Participative

Consulting and engaging with stakeholders on the data and assessment methods as far as is reasonable will increase the robustness of the assessment because it will provide opportunities for stakeholders to voice their concerns and, where appropriate, scrutinise and challenge the assessment methodologies. Promoters of schemes can go a long way towards addressing concerns about the assessment by dealing with them openly and effectively. In many cases, concerns can be addressed through the provision of information. However, where concerns are valid, they can be overcome by considering and agreeing to alternative approaches or undertaking sensitivity analyses (see Section 13.4.10).

As well as helping to ensure that the ES is compliant with the EIA regulations, engaging and consulting on the proposed methodology provide opportunities to incorporate feedback and suggestions on the methodology which, if carefully managed, will improve the objectivity of the assessment. Agreeing with the methodology as far as possible in advance of the EIA means that there is less opportunity to carry out a post hoc analysis of the results or adapt the methodology in light of the emerging results in favour of the developer. This is why it is considered good practice to prepare and consult on the scope and methodology of the EIA, for example, through consultation on a 'scope and methodology report', at the start of the EIA process.

Effective consultation and engagement are now integral to the planning process. However, for consultation and engagement to be effective, it has to be meaningful and fair. If it is not, there is a risk that the credibility of the assessment will be undermined.

13.4.7 Expert Review

The planning and legal processes allow for independent review and challenge, especially if an objector to a scheme or an operation feels so strongly that the assessment methods are not adequate and wishes to present evidence on what they consider to be deficiencies in the EIA. However, there is growing recognition that conflict and challenge between technical experts are not helpful to the decision-making process, and as a consequence, planning and legal processes are becoming less adversarial, with increasing emphasis on consultation and reaching agreement between experts. But this means that more responsibility is being placed on technical assessors, acting on behalf of scheme promoters, to proactively engage and reach agreement on technical matters as far as it is possible to do so.

A number of nationally important infrastructure schemes have used expert peer review to improve the objectivity and robustness of assessments. Recently, the application of expert review has become more prominent on nationally important schemes. For example, HS2 set up an Acoustics Review Group to review and challenge the approach to sound and vibration assessment, management, control, policy and planning matters. The terms of reference state that its purpose was

> [t]o provide an independent and highly experienced perspective on the development of sound and vibration assessment methods and criteria and proposals for effective management and control of noise and vibration.

The intention of review groups such as these is not to replace external consultation or engagement but instead to provide an additional layer of advice and expertise to maximise the objectivity and credibility of the assessment.

IEMA used to review environmental statements and give them ratings. IEMA's review and rating mechanism has been replaced with their EIA Quality Mark [46]. The EIA Quality Mark registrants have to adhere to seven key commitments. These commitments underpin and keep the high standard of the scheme:

1. EIA Management – We commit to using effective project control and management processes to deliver quality in the EIA we co-ordinate and the Environmental Statements we produce.
2. EIA Team Capabilities – We commit to ensuring that all our EIA staff have the opportunity to undertake regular and relevant continuing professional development.
3. EIA Regulatory Compliance – We commit to delivering Environmental Statements that meet the requirements established within the appropriate UK EIA Regulations.
4. EIA Context and Influence – We commit to ensuring that all EIAs we coordinate are effectively scoped and that we will transparently indicate how the EIA process, and any consultation undertaken, influenced the development proposed and any alternatives considered.
5. EIA Content – We commit to undertaking assessments that include: a robust analysis of the relevant baseline; assessment and transparent evaluation of impact significance; and an effective description of measures designed to monitor and manage significant effects.
6. EIA Presentation – We commit to deliver Environmental Statements that set out environmental information in a transparent and understandable manner.
7. Improving EIA practice – We commit to enhance the profile of good quality EIA by working with IEMA to deliver a mutually agreed set of activities, on an annual basis, and by making appropriate examples of our work available to the wider EIA community.

13.4.8 Consideration of Alternatives and Optioneering

The assessment should be objective, impartial, sufficiently robust, balanced and fair. Consideration of alternatives and mitigation options can represent an important means of identifying the most sustainable solution, namely the solution that gives rise to the least net adverse environmental impacts whilst meeting other economic and social objectives.

Options appraisal is a key component for many infrastructure schemes. The government's Transport Analysis Guidance (TAG) provides guidance on, among other things, environmental impact appraisal [47]. TAG places emphasis on appraisal as a continuous process to encourage an interactive design and management process. It provides a framework for developers and designers to promote better balanced projects that support a wide range of social, economic and environmental objectives. Options appraisal is therefore seen as a means of promoting better design and achieving wider acceptance of the scheme.

The importance of optioneering is not limited to the promotion of physical infrastructure schemes and is seen as a necessary step or steps in the development and design of other proposals. For example, the CAA's guidance on the process for changes to airspace design [48] requires options to be developed, appraised and consulted on as an integral step in the design process and the identification of air change proposals.

13.4.9 Reasonable Worst-Case Assumptions

It is important to consider the level of uncertainty associated with the assumptions that are used to predict the likely significant effects. This is particularly important for major projects where the environmental assessment needs to be conducted long before any detailed design is carried out and before contractors are engaged. Consequently, it is often necessary to make a number of assumptions about the operations of the scheme and about the construction arrangements so as to be able to conduct the EIA.

The DMRB (volume 11, section 3, part 7: HD213/11) [7] recognised that uncertainty decreases over time and specifically recommended the following:

> During an assessment of the impacts from noise and vibration, the uncertainty associated with input data is an important factor in determining how confident the Overseeing Organisation's supply chain can be with the assessment results. As the road project progresses, the quality and accuracy of the assessment should normally improve. This in turn will influence the accuracy of designed mitigation measures, for example the height and positioning of any barriers. The most up to date scheme design and traffic flow information should be used in the final assessment.

The level of uncertainty associated with construction assumptions can be particularly important, to the extent that the final construction methods may be materially different to those assumed if the assessment is undertaken well in advance (in some cases several years) of the start of the works and well in advance of when detailed geotechnical and other information is made available. In such instances, reasonably pessimistic assumptions in terms of geotechnical information, construction methods, percentage on-times for plant and equipment and location are typically used.

The assessment should not assume an unrealistic or unlikely worst-case situation. For example, it would not be reasonable to assume that a scenario represents a worst-case simply because it is physically possible or technically feasible. Rather, the worst case to be assessed should be reasonably likely. For example, a railway may have a theoretical capacity of so many trains per hour over an operating period and a theoretical maximum design speed. However, there is no point in assuming a maximum theoretical capacity throughout the full operational period if there is less demand for trains outside the peak hours of demand or assuming a maximum speed if the trains are unlikely to reach the maximum design speeds for most of the time. For the noise assessment, it would be reasonable to expect a relatively pessimistic forecast to be used (e.g. a higher level of operational train frequency than anticipated) but not one that is improbable. In any event, any forecasts used in the assessment should be described and justified. Good justifications will provide sufficient context. For example, it may be said that the forecasts are 10% below the operating capacity because the demand forecasts predict that peak demand will only occur between specified hours.

For noise predictions, reasonably worst-case meteorological assumptions are often used. For example, the ISO 9613 general prediction method [49], the Calculation of Railway Noise [50], and the Calculation of Road Traffic Noise [51] all predict the propagation of sound under favourably mild downwind conditions. Any technical difficulties and limitations associated with the assumptions should be identified and reported within the framework for addressing uncertainty.

13.4.10 Sensitivity Testing

Sensitivity analysis can be used to assess and communicate the sensitivity of different assumptions and the importance of different factors or the interrelationship between different factors. For example, a number of sensitivity analyses were undertaken and reported for the proposed fifth terminal at Heathrow to consider and understand the influence of different forecasts for aircraft fleet mix and numbers of aircraft. The EIA base-case assumptions were 50 million passengers per annum (mppa) and 425,000 air transport movements (ATMs) for the without situation and 80 mppa and 458,00 ATMs with the proposed terminal. A range of sensitivity tests was then carried out for different passenger throughput (up to 100 mppa), movements (up to 530,000 ATMs), different modes of operation (segregated mode and mixed mode) and different fleet-mix assumptions (BAA/66). These sensitivity tests were used to assess and demonstrate how different assumptions might affect the results of the noise assessment.

13.4.11 Predictions

Where it is practical to do so, the level of uncertainty in calculations and measurements that are relied on should be quantified. For calculation of vibration impacts, there is a reasonable expectation that any prediction models/methods used to predict noise and vibration levels will be suitably validated and verified. It is worth noting that BS ISO 14837–1 provides guidance on the validation and verification of prediction models and the derivation of confidence limits [15]. Among others, the following recommendations are made:

- The accepted level of accuracy will depend upon the different stages of assessment (scoping, environmental assessment and detailed design) and planning sensitivity (see figure 13.5 of BS ISO 14837–1)[15].
- Models should be validated by comparing the output of the model with an independent set of measurements.
- Model accuracy may also be estimated by comparison, for a defined test case, against other validated models.
- Improvement of a prediction model at a specific site could be assisted with measurements at that site.

13.4.12 Measurements

Uncertainty in measurement may occur as a result of several factors: error and uncertainty in the instrument chain (including coupling of transducers), temporal and spatial sampling issues associated with the measurement method and the influence of environmental factors. The evaluation of uncertainty associated with the measurement of noise is hampered by the fact that little information has hitherto been published evaluating the errors and variability associated with such measurements. There still remains little available information regarding the repeatability or reproducibility of tests carried out in the field. However, there are some examples where uncertainty has been estimated. For example, the level of error was estimated for the measurements taken as part of the Defra study on human response to vibration in residential environments [52]. These are reproduced in Table 13.3.

In some cases, it may be appropriate to conduct repeatability and reproducibility tests as part of the EIA. Clearly, any opportunities to undertake such tests should reflect the scale and nature of the project under consideration. The North London Waste Power Project [53] provides an

Table 13.3 Measurement uncertainty estimates for the measurement methods used on the Defra vibration dose-response study [52]

Vibration source	Measurement type	Exposure uncertainty, dB
Railway	Internal	±2.2
Railway	Extrapolation	±6.2
Construction	Extrapolation	±8.6
Internal	Internal	±2.2

example where an assessment was undertaken in accordance with BS 4142. The standard requires uncertainty to be considered. As part of the evaluation of uncertainty, two baseline surveys, each lasting a week, were carried out, and the repeatability of the survey results was analysed. For most of the measurement locations, the data were fairly stable, and the differences in the background sound levels between the two surveys were within the standard deviation of the dataset. However, where there were significant differences this led to recommendations to use lower values for the background sound levels to be used in the assessment. This is an example of a reasonably precautionary approach that was proportionate to the circumstances of the development, with it being a major development for a waste-to-energy plant that could potentially affect several communities.

13.5 Reporting of Effects

It is worth reiterating that the ES or noise assessment report is a communicative tool to aid good decision making. It is important therefore that the results of the findings are reported in a way that enables all stakeholders to understand and access the findings of the EIA. To be fair to all stakeholders any uncertainty or limitations in the assessment methods should be properly reflected in the approach used to reporting the findings of the EIA. For example, it would not be appropriate to identify and report significant effects arising from the construction at specific receptors and at particular times if the level of uncertainty associated with the construction assumptions is high. If so, it may be more appropriate to adopt a risk assessment approach to indicate the general geographical extent and to estimate overall number of dwellings which may be at risk of significant effects with an overall description of the duration of the impacts and an indication of when the impacts are likely to arise in the programme. Predicting at individual properties and within specific months or even weeks may suggest a level of precision or accuracy that is not justified.

The DMRB [8] summary noise tables provides an example of how information can be presented transparently to give the decision maker and other interested parties the main findings of an assessment and also providing context. In this example, noise level changes are presented for each type of receptor of relevance within different noise change bands (see Table 13.4).

Advances in mapping and GIS tools clearly aid the ability of the practitioner to report findings of noise assessments. The HS2 environmental statement, which presented contours of noise levels, provides a useful example [54].

Ideally, the results should be reported so that people living or working in specific areas can easily identify the predicted impacts on them. In some instances, it may be possible to report findings of the EIA on individual receptors. However, the desirability of reporting effects at a local or even individual level must be balanced against the level of certainty in them. It may only be possible to report on a broad or large area basis. Reporting effects at specific receptors or small areas should be avoided if the level of uncertainty inherent in the assessment does not justify reporting to this level of detail. This is especially the case for construction noise and vibration.

Table 13.4 Operational noise reporting table for noise assessment [8]

Project:					
Scenario/comparison:					
		Daytime		Night-time	
Change in noise level, dBA		Number of dwellings	Number of other noise-sensitive receptors	Number of dwellings	Number of other noise-sensitive receptors
Increase in noise level, dB $L_{A10,18h}$ /L_{night}	<1.0				
	1.0–2.9				
	3–4.9				
	>5				
No change	0				
Decrease in noise level, dB $L_{A10,18h}$ /L_{night}	<1.0				
	1.0–2.9				
	3.0–4.9				
	>5				

13.6 Assurance Processes after Consent

So far we have talked about the EIA in the context of seeking and obtaining planning permission. However, for many major applications assurance processes require assessments and uncertainty to continue to be considered beyond the time when the consent is granted. There are numerous examples of assurances that relate to the design and commissioning of the permitted scheme. There are also examples that relate to the ongoing operation and maintenance of the permitted scheme. We explained earlier that uncertainty is time related and will decrease during the design development of a scheme. It follows that post-consent assurance processes can provide significant opportunities to manage and reduce uncertainty in noise assessment.

The Crossrail *Environmental Minimum Requirements* [55] provide a good example of ongoing assurance processes as they relate to noise. The controls contained in *Environmental Minimum Requirements: General Principles* (when considered along with the powers contained in the Crossrail Act and the Undertakings) are intended to ensure that the impacts assessed in the ES will not be exceeded. They also require the nominated undertaker to 'use reasonable endeavours to adopt mitigation measures that will further reduce any adverse environmental impacts caused by Crossrail, insofar as these mitigation measures do not add unreasonable costs to the project or unreasonable delays to the construction programme'. In other words, reasonable steps must be taken to reduce the impact of the scheme by comparison to the impacts reported in the ES.

Through the ongoing design and construction of the scheme, Crossrail adopted a comprehensive package of measures to evaluate the impact of the scheme by comparison with that reported in the ES to ensure that the impacts were not exceeded and to make improvements where it was reasonable to do so. In the case of the noise impacts, uncertainty was considered part of the ongoing design and assurance processes (see Table 13.5). For example, Information Paper IPD10, relating to ground-borne noise and vibration [56], requires

Table 13.5 Operational ground-borne noise criteria. (Rreproduced with permission from IPD10 [56])

Building	Level/measure
Residential buildings, offices, hotels, schools, colleges, hospitals, laboratories, libraries	40 dB $L_{Amax,S}$
Theatres	25 dB $L_{Amax,S}$
Large auditoria/concert halls	25 dB $L_{Amax,S}$
Sound recording studios	30 dB $L_{Amax,S}$
Courts, lecture theatres, places of meeting for religious worship, small auditoria/halls	35 dB $L_{Amax,S}$

The nominated undertaker will be required to design the permanent track system so that the level of groundborne noise arising from it near the centre of *any noise-sensitive room* is predicted in all *reasonably foreseeable circumstances* not to exceed the levels in Table 1. The nominated undertaker will be required to install the permanent track using a standard track system for the Crossrail tunnel sections. In any location where the standard system is predicted during detailed design to cause levels of groundborne noise exceeding the relevant assessment criterion an enhanced track support system will be installed [emphasis added].

IDP10 suggests that the method for demonstrating compliance with the requirements is by design, not measurement, after the opening of the railway. This is for very good reason, as it will be unlikely, if not impossible, to take remedial action after the track is installed. Predictions used in the demonstration of compliance must be carried out using a model complaint with ISO 14837–1: 2005 [15]. It can be seen that the assurance requires the acoustic criteria, specified for different types of use, to be achieved with confidence.

The track was designed by the Arup Atkins joint venture, and the ground-borne noise and vibration modelling was carried out using the Arup calculation procedures, which have been validated on a wide range of railways worldwide (Figure 13.4) [56].

This enabled the 95% confidence limit for a ground-borne noise prediction to be estimated at +8 dB. On this basis, Crossrail decided to add a conservative +10-dB allowance for prediction uncertainty to all predictions to demonstrate that the design criteria would be met in *all reasonably foreseeable circumstances* inside all the noise-sensitive receptors located above the tunnels.

So far we have described assurances that apply during the design stage or up to the point where the scheme becomes operational. However, there are some assurances that apply throughout the operational life of the scheme. For example, several railway schemes have committed to maintenance regimes, often defined by reference to objective noise criteria, which aim to maintain the condition of train wheel sets and the rail head in good condition. Examples include the Thameslink Programme, Crossrail and HS2.

At the time when construction is about to commence, for many types of schemes, far more information is likely to be known about the construction methods, ground information, plant and equipment by comparison to the time when the findings of the initial EIA are reported. As a consequence, construction noise levels can be predicted with more confidence at the start of construction by comparison with the time when earlier assessments may have been made. This provides opportunities to manage the uncertainty in

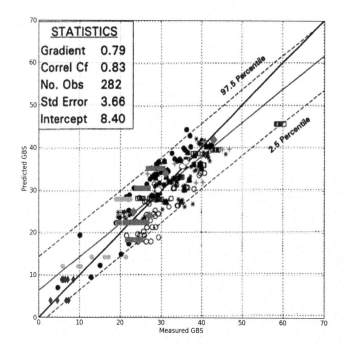

Figure 13.4 Regression analysis of predicted versus measured ground borne noise levels [57].

the construction noise assessment. A number of major projects commit to a 'construction code of practice' or similar arrangements that require further assessments to be carried out before the start of construction or at the time shortly before the start of the works. These assessments may lead to an increase in protection measures. For example, many public funded, major infrastructure schemes provide noise insulation or temporary rehousing when noise levels are likely to be acute e.g. in excess of 75 dB during the day. Fresh assessments are then undertaken about 9 months ahead of the start of construction to determine eligibility in accordance with defined criteria. For many projects this re-assessment can be undertaken in excess of a year after the initial EIA. There is also an increasing trend for major projects to require applications for prior consent (under Section 61 of the Control of Pollution Act 1974) [58] to be prepared and submitted shortly (typically six weeks) before the start of construction. This often requires further and up-to-date assessments to be carried out, which are then used to check whether additional protection measures, including noise insulation and temporary re-housing, may be necessary.

13.7 Monitoring

Until recently monitoring has not been required by the EIA Regulations and was not a matter of routine. Consequently, feedback on the effectiveness of assessments and the effectiveness of mitigation or protection methods has been lacking. However, the assessment regulations now require monitoring to be carried out and there are a number of recent examples where monitoring will be required.

For example, the A14 Cambridge to Huntingdon Improvement Scheme Development Consent Order (DCO) 2016 requires a post-construction monitoring and mitigation plan to be implemented [59]. The DCO requires the following:

(1) No part of the authorised development within the area of South Cambridgeshire District Council is to be opened for public use until a post-construction noise monitoring plan for that part complying with this requirement has been submitted to and approved in writing by the Secretary of State, following consultation with South Cambridgeshire District Council ("the monitoring plan").

(2) The monitoring plan must make provision for the monitoring of traffic flows with reference to the Important Areas identified within the area of South Cambridgeshire District Council in the environmental statement.

(3) The monitoring plan must provide that

   (a) during the 12 month period after the authorised development has been opened for public use, and during the 12 month period after the authorised development has been opened for public use for 4 years, traffic monitoring must be undertaken for the locations referred to in sub-paragraph (2) in accordance with the Post Opening Project Evaluation procedure operated by the undertaker;

   (b) if following analysis by the undertaker of the monitoring data derived from the monitoring mentioned in sub-paragraph (a), in consultation with South Cambridgeshire District Council, it reasonably appears to the undertaker that as a result of the authorised development traffic flows are materially greater than those predicted in the environmental statement, the assessment of noise effects at the locations where those materially greater flows are identified is to be re-calculated utilising the monitored data and using the methodology set out in the environmental statement; and

   (c) if it reasonably appears to the undertaker from the re-calculations mentioned in sub-paragraph (b) that the noise effects of the authorised development are materially greater than those predicted in the environmental statement, the undertaker, in consultation with South Cambridgeshire District Council, must develop a scheme of reasonable and sustainable mitigation at each relevant location, which the undertaker must submit to the Secretary of State for approval.

(4) Post-construction noise monitoring must be carried out by the undertaker in accordance with the monitoring plan and the results of the monitoring must be submitted to South Cambridgeshire District Council.

(5) Before considering whether to approve any scheme of mitigation submitted by the undertaker to the Secretary of State, the Secretary of State must consult South Cambridgeshire District Council.

(6) Any scheme of mitigation approved by the Secretary of State must be implemented by the undertaker.

The purpose of the requirement is to establish whether the actual change in road traffic noise levels resulting from the scheme, which was used as the basis of the environmental assessment, is materially greater than predicted and reported in the ES. The requirement is carefully worded to reflect the significant uncertainty that exists in undertaking measurements at different points in time to establish a change in noise levels, especially when the measured noise levels may be significantly influenced by: extraneous noise (not directly attributable to the scheme); meteorological conditions affecting the propagation of sound from the source to the receiver location; and other factors that may influence the variability in sound emissions. As a consequence, the post-construction plan requires traffic flow to be monitored and for the road traffic noise levels to be recalculated if the actual traffic flows are materially greater than assumed. The Post Opening Project Evaluation procedure, referred to in the requirement, also requires the performance of the noise control measures, such as barriers, to be checked and assured. This form of monitoring, within the context of this scheme, represents a more reliable approach than comparing the results of noise measurements taken before construction and post-construction of the scheme. This is because the calculations utilise identical attenuation losses between source and receiver for the before and after cases and assume the same meteorological conditions assumed for the EIA. In this way, potential variability and uncertainty resulting from meteorological variability are eliminated. Thus, the monitoring requirement will establish whether a material change in road traffic noise levels occurs as a result of a significant increase in traffic flows, which are attributable to the scheme.

HS2 Information Paper F4 [60] relates to the requirements for monitoring the performance of noise and vibration control measures applied to the operational phase of the proposed scheme, namely:

Noise and vibration monitoring will be carried out at different times during the lifetime of the Proposed Scheme at a combination of carefully selected monitoring locations including: adjacent or attached to moving vehicles, at fixed positions or in the vicinity of individual assets; and locations within the surrounding areas and communities alongside the railway corridor.

A wide range of noise and vibration related data for assets such as trains, tracks, noise fence barriers, earthworks, fixed installations and how track and overhead catenary systems interact with the rolling stock will be collected. These data, together with noise and vibration measurements will be used to monitor the operational noise and vibration performance of the Proposed Scheme. Where noise and vibration performance deviates from expected conditions, the following actions will be undertaken:

If the measured performance is better than the expected conditions:

- A study to document the reasons why assets are achieving a higher performance than expected
- A review of further improvements to other assets that could potentially benefit from the technology transfer of the high performing assets

If the measured performance is worse than the expected conditions:

- A study to identify the root cause and all possible solutions to the low performance
- An investigation of other similar assets that could also be underperforming
- Corrective action to improve existing performance and prevent future loss of performance so far as this may be required to achieve the objectives set out in the [assurances relating to noise]

The expected conditions will be determined with reference to: predictions during the project's development; related noise and vibration data of the Proposed Scheme's assets; baseline noise and vibration monitoring information gathered prior to construction of the Proposed Scheme; laboratory test information; and previous in situ noise and vibration measurements.

In many respects, the HS2 assurance is similar to the A14 requirement in that it requires monitoring and may lead to further intervention to either enhance the performance of the noise and vibration control measures or address under-performance of the measures. However, this example is also useful in that it demonstrates the potential legacy value of such monitoring that could be of benefit to future railway projects. This offers a potentially powerful means of improving knowledge and dealing with uncertainty into the future. This is perhaps, until recently, one of the greatest failings in the planning system but is one that is being addressed in the guidance and some major projects, as we have seen. For example, the *Professional Practice Guidance: Planning and Noise* [61] encourages developers to obtain post-occupancy feedback from new residents on acoustic design issues and that the lessons learned from such surveys should inform future good practice.

So far the focus has been on monitoring the performance of the asset or the noise control measures. However, valuable information could be obtained if health and quality of life were monitored during construction or operation of the scheme.

The consortium of local authorities that objected to Heathrow Terminal 5 recommended that aircraft noise annoyance studies be carried out every five years around Heathrow. Unfortunately, the government at the time did not accept the proposals and instead commissioned a one-off study (ANASE [62], which has proved to be contentious). However, had these recommendations been accepted, we would now have far better information about how annoyance and other effects may change over time and how well those changes in effects correlate with changes in noise descriptors or indices. The most recent study reported by the government does indeed suggest – consistent with other EU studies – that people are now more sensitive to aircraft noise than they were in the 1980s [34].

13.7.1 Progressive Validation and Verification of Predictions

Wherever possible, noise predictions should be carried out at the time of the EIA using validated noise prediction methods and models (see Section 13.6), but this is not always possible. However, some schemes may provide opportunities to validate or verify predictions and assessments during the implementation or operational stage.

Again, Crossrail provides a useful example. As well as controlling operational noise, Crossrail Information Paper IPD10, relating to ground-borne noise and vibration [56], required the temporary construction railway to meet the specified noise and vibration criteria. However, prior to Crossrail, very little information was available about ground-borne noise and vibration levels that could be used to validate the noise predictions. To address this uncertainty, both the contractors responsible for the east and west tunnelling contracts, extending from Westborne Park in the west and Pudding Mill Lane in the east of London, undertook validation measurements at the beginning of the tunnel drives. This information was then used to refine the designs and extents of the resiliently supported temporary rails. These validation studies are described on the Crossrail Legacy website [63].

An example of the results of the validation analysis, carried out in accordance with ISO 14837–1: 2005, is provided in Table 13.6 and Figure 13.5.

It is also worth adding that both contractors designed modular track designs that could be interchanged, and there were a number of examples where, in response to complaints, non-resilient track was substituted with resilient track. Although there were few cases where this happened, it provides a useful example of how flexibility can be introduced into the design to reduce noise levels if complaints and other information suggest that the scheme is having a greater impact than was predicted.

Construction noise assessment provides an example where ongoing validation and verification can be incorporated into normal practice. Noise

Table 13.6 Results of the validation analysis [63]

Data	Prime A	Prime C
Gradient	0.85	0.66
Standard deviation	3.32	2.80
Offset	4.20	5.53
Random error	±3.32	±2.80

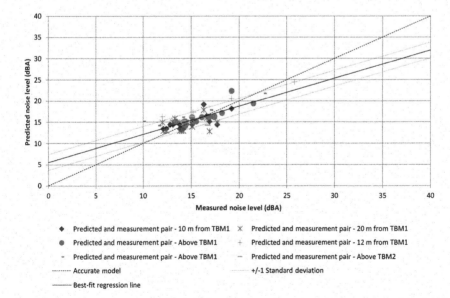

Figure 13.5 Track type C ISO 14837–1: 2005 model validation analysis [63].

monitoring is often required as part of the ongoing management of construction noise. The results of noise and vibration monitoring can then be used to check and verify any noise calculations presented in the applications for prior consent, carried out under Section 61 of the Control of Pollution Act [58], or other assessments. This information can be used to modify the predictions used in subsequent calculations and assessments. This approach has been used successfully on a number of construction sites. For example, Crossrail's supplier performance matrix for noise and vibration included requirements for exemplary practice demonstrated on validation and verification of noise predictions [64].

13.8 Conclusions

A major failing of many EIAs and environmental statements is that

- there is little or no systematic evaluation or consideration of uncertainty; and
- findings are reported to a level of accuracy that is not justified.

Noise assessments are often poor at considering – if not neglecting – and reflecting the scientific evidence on the effects of noise and vibration on human health and quality of life.

A number of principles and approaches have been described herein that will encourage uncertainty to be properly managed in the EIA process. Several examples of best practice are given. Continuing processes for assessing and esnuring that the environmental effects are no greater, and, where sustainable, better, than those reported in the ES during construction, design and operation of schemes can help to overcome uncertainty and any risk that the adverse effects of a scheme are greater than assessed. Finally, monitoring provisions and arrangements can be used to significantly improve the quality and robustness of noise assessments that are undertaken.

References

1. Department for Communities and Local Government, *The Planning Act 2008: Guidance on Nationally Significant Infrastructure Projects and Housing*, March 2017.
2. *Town and Country Planning (Environmental Impact Assessment) Regulations 2017: Statutory Instrument.*
3. *Directive 2014/52/EU of the European Parliament and of the Council dated 16 April 2014 amending Directive 2011/92/EU on the Assessment of the Effects of Certain Public and Private Projects on the Environment.* European Parliament, Brussels.
4. Environmental Impact Assessment (EIA), available at http://ec.europa.eu/environment/eia/eia-legalcontext.htm.

5. Department for Environment, Food and Rural Affairs, 'Noise policy statement for England', in *Secondary Noise Policy Statement for England*. DEFRA, London, 2010.
6. Ministry of Housing, Communities and Local Government, *National Planning Policy Framework: Noise*, Ministry of Housing, London, 2019.
7. *Design Manual for Roads and Bridges*, Volume 11, Section 3, Part 7, Annex 6: *Assessing Traffic Noise and Vibration Nuisance*. 2011.
8. Highways England, *Sustainability and Environment Appraisal: LA111 Noise and Vibration* (formerly HD 213/11, IAN 185/15). Highways England, London, 2019.
9. Institute of Environmental Management and Assessment, *IEMA Guidelines for Environmental Noise Impact Assessment*. IEMA, Lincoln, UK, 2014.
10. Acoustics and Noise Consultants, *Red Book: Measurement and Assessment of Groundborne Noise and Vibration*. ANC, Northallerton, UK, 2012.
12. Ministry of Housing, Communities and Local Government, *Planning Practice Guidance: Minerals*. Ministry of Housing, London, 2014.
13. Ministry of Housing, Communities and Local Government, *Planning Practice Guidance: Noise*. Ministry of Housing, London, 2014.
14. British Standards Institution, *Guide to Evaluation of Human Exposure to Vibration in Buildings: Vibration Sources Other than Blasting*, BS 6472: 2008. BSI, London, 2008.
15. International Organization for Standardization, *Mechanical Vibration: Ground-Borne Noise and Vibration Arising from Rail Systems*, Part 1: *General Guidance*, ISO 14837-1: 2005. ISO, Geneva, 2005.
16. British Standards Institution, *Code of Practice for Noise and Vibration Control on Construction and Open Sites: Noise*, BS 5228-1: 2009+A1: 2001. BSI, London, 2014.
17. Basner M, et al., 'Auditory and non-auditory effects of noise on health', *Lancet* 383(9925)(2014), 1325–1332.
18. Basner M, et al., 'Aviation noise impacts: State of the science', *Noise and Health* 19(87)(2017), 41–50.
19. Clark C, et al., 'Association of long-term exposure to transportation noise and traffic-related air pollution with the incidence of diabetes: A prospective cohort study', *Environ Health Perspect* 125(8)(2017).
20. Eze IC, et al., 'Association between ambient air pollution and diabetes mellitus in Europe and North America: Systematic review and meta-analysis', *Environ Health Perspect* 123(5)(2015), 381–389.
21. Gehring U, et al., 'Impact of noise and air pollution on pregnancy outcomes', *Epidemiology* 25(3)(2014), 351–358.
22. Nieuwenhuijsen MJ, Ristovska G, Dadvand P, 'WHO environmental noise guidelines for the European region: A systematic review on environmental noise and adverse birth outcomes', *Int J Environ Res Public Health* 14(10)(2017), 1252.
23. Roswall N, et al., 'Modeled traffic noise at the residence and colorectal cancer incidence: A cohort study', *Cancer Causes Control* 28(7)(2017), 745–753.
24. Clark C, Paunović K, 'Systematic review of the evidence on the effects of environmental noise on quality of life, wellbeing and mental health', *Int J Environ Res Public Health* 15(11)(2018), 2400.

25. Babisch W, 'Updated exposure-response relationship between road traffic noise and coronary heart diseases: A meta-analysis', *Noise and Health* 16(68) (2014), 1–9.

26. Guski R, Schreckenberg D, Schuemer R, 'WHO environmental noise guidelines for the European region: A systematic review on environmental noise and annoyance', *Int J Environ Res Public Health* 14(12)(2017), 1539.

27. Basner M, McGuire S, 'WHO environmental noise guidelines for the European region: A systematic review on environmental noise and effects on sleep', *Int J Environ Res Public Health* (in press).

28. Clark C, Paunović K, 'WHO environmental noise guidelines for the European region: A systematic review on environmental noise and cognition', *Int J Environ Res Public Health* 15(2018), 285.

29. Clark C, et al., 'Exposure-effect relations between aircraft and road traffic noise exposure at school and reading comprehension: The RANCH project', *Am J Epidemiol* 163(1) (2006), 27–37.

30. International Organization for Standardization *Acoustics: Assessment of Noise Annoyance by Means of Social and Socio-acoustic Surveys*, ISO/TC 43/SC 1 N 1313: 2003. ISO Geneva, 2003.

31. European Commission, 'Position paper on dose-response relationships between transportation noise and annoyance',European Commission, Brussels, 2002.

32. Janssen SA, et al., 'Trends in aircraft noise annoyance: The role of study and sample characteristics', *J Acoust Soc Am* 129(4) (2011), 1953–1962.

33. Hill AB, 'The environment and disease: Association or causation?', *Proc R Soc Med* 58(2) (1965), 295–300.

34. Civil Aviation Authority, *CAP1506 Survey of Noise Attitudes 2014*. CAA, London, 2017.

35. European Environment Agency, 'Good practice guide on noise exposure and potential health effects', EEA Technical Report No 11/2010.EEA, Brussels, 2010.

36. AECOM, Ltd. *Possible Options for the Identification of SOAEL and LOAEL in Support of the NPSE*.AECOM, Ltd., Glasgow, 2013.

37. Langdon F, Griffiths I, 'Subjective effects of traffic noise exposure. II. Comparisons of noise indices, response scales and the effects of changes in noise levels', *J Sound Vib* 83(2)(1982), 171–180.

38. Griffiths I, Raw G, 'Community and individual response to changes in traffic noise exposure', *J Sound Vib* 111(2)(1986), 209–217.

39. Huddart L, Baughan CJ, 'The effects of traffic change on perceived nuisance', Transport Research Laboratory Research Report RR363. Transport Research Laboratory, Crowthorne, UK, 1994.

40. Griffiths I, Raw G, 'Adaptation to changes in traffic noise exposure', *J Sound Vib* 132(2)(1989), 331–336.

41. Brown AL, van Kamp I, 'WHO environmental noise guidelines for the European region: A systematic review of transport noise interventions and their health effects', *Int J Environ Res Public Health* 14(8)(2017), 873.

42. Brink M, et al., 'Annoyance responses to stable and changing aircraft noise exposure', *J Acoust Soc Am* 124(5)(2008), 2930–2941.

43. World Health Organization, *The World Health Organization Guidelines for Environmental Noise Exposure for the European Region*, WHO, Geneva, 2018.

44. High Speed Two (HS2), Ltd., 'HS2 Phase 2b: Crewe to Manchester and West Midlands to Leeds. Environmental Impact Assessment Scope and Methodology Report' (draft for consultation HS2, Ltd., London, 2017.

45. United Nations, *The Rio Declaration on Environment and Development*, UN Conference on the Environment and Development. United Nations, New York, 1992.

46. Institute of Environmental Management and Assessment, *EIA Quality Mark*. IEMA, Lincoln, UK.

47. Department for Transport, 'TAG Unit A3 Environmental Impact Appraisal, December 2015', Department for Transport, London, 2015.

48. Civil Aviation Authority, *Airspace Design: Guidance on the Regulatory Process for Changing Airspace Design Including Community Engagement Requirements*, CAP1616. CAA, London, 2018.

49. International Organiztion for Standardization, *Acoustics: Attenuation of Sound during Propogation Outdoors*, Part 2: *General Method of Calculation*, ISO 9613–2: 1996. ISO, Geneva, 1996.

50. Department for Transport, *Calculation of Railway Noise*. Department for Transport, London, 1995.

51. Department for Transport, *Calculation of Road Traffic Noise*. Department for Transport, London, 1988.

52. Department for Environment Food and Rural Affairs, *Human Response to Vibration in Residential Environments*, Defra NANR209. Defra, London, 2012, available at http://hub.salford.ac.uk/vteam/publications/defra-nanr209-human-response-to-vibration-in-residential-environments.

53. https://infrastructure.planninginspectorate.gov.uk/projects/london/north-london-heat-and-power-project.

54. High Speed Two (HS2), Ltd., *HS2 Phase One Environmental Statement*, Volume 5: *Sound, Noise and Vibration*. HS2, Ltd., London, 2013, available at www.gov.uk/government/publications/hs2-phase-one-environmental-statement-volume-5-sound-noise-and-vibration/hs2-phase-one-environmental-statement-volume-5-sound-noise-and-vibration.

55. Crossrail Ltd., *Environmental Minimum Requirements: General Requirements*, CR/HB/EMR/0001. Crossrail Ltd., London, 2008, available at www.crossrail.co.uk/about-us/crossrail-act-2008/environmental-minimum-requirements-including-crossrail-construction-code.

56. Crossrail Ltd., 'Groundborne noise and vibration', Crossrail Information Paper D10. Crossrail Ltd., London, 2008.

57. Crossrail Ltd, 'Control of railway induced groundborne noise and vibration from the UK's Crossrail Project'. Crossrail Ltd., London, 2018, available at: https://learninglegacy.crossrail.co.uk/documents/control-of-railway-induced-groundborne-noise-and-vibration-from-the-uks-crossrail-project/.

58. Control of Pollution Act, Section 61, 1974.

59. Correction Notice: The A14 Cambridge to Huntingdon Improvement Scheme Development Consent Order 2016 (S.I. 2016/547)', 2017, available at: https://infrastructure.planninginspectorate.gov.uk/wp-content/ipc/uploads/projects/TR010018/TR010018-004535-171113%20FINAL%20CORRECTION%20NOTICE.pdf.

60. High Speed Two (HS2) Limited, *F4: Operational Noise and Vibration Monitoring Framework*. HS2 Limited, London, 2017.
61. ANC, CIEH and IOA, *ProPG: Planning and Noise: Professional Practice Guidance on Planning and Noise. New Residential Development*. ANC, CIEH and IOA, London 2017, available at www.ioa.org.uk/publications/propg.
62. Department of Transport, *ANASE Attitudes to Noise from Aviation Sources in England*. Department of Transport, London, 2007, available at www.dft.uk/pgr/aviation/environmentalissues/ANASE.
63. Crossrail Ltd. 'Noise and vibration controls for the TBM and temporary construction railway'. Crossrail Ltd., London, 2016,available at https://learninglegacy.crossrail.co.uk/documents/noise-vibration-controls-tbm-temporary-construction-railway.
64. https://learninglegacy.crossrail.co.uk/wp-content/uploads/2017/03/ENV26-01_Draft-Noise-matrix-Supplier-Performance-feb-16.pdf.

Chapter 14

Uncertainty in International Acoustics Standards

Douglas Manvell

Introduction

Head pounding, I return to my hotel in Le Mans in June 2005. In the midst of a heat wave, without the assistance of air conditioning, a large number of acousticians had presented and discussed uncertainty in acoustics. A common understanding of the terminology – uncertainties, probabilities, measurands, coverage factors and the like – was beginning to form. As I downed a litre of water in a split second, I wondered whether the sensation of my brain cooking was caused by the fact that as a session chair and presenter, I didn't get to any of the limited supplies of bottled water in time and how much this new subject matter and the various views and inputs presented contributed to this. The Institute of Noise Control Engineering (INCE)–Europe Uncertainty Seminar, co-organized by Jean Tourret, held in le Mans, France, is one of my earliest and strongest surviving memories of my first experiences with uncertainty in acoustics.

I had joined the International Organization for Standardization (ISO) Technical Comittee 43 Sub Comitte 1 on noise in 1998 as a member of Working Group 45, which was in the process of revising the ISO 1996 environmental noise assessment standard – a process that continues today. In addition to comparing assessment methodologies from around the world and working to create universal concepts to cover the many sources of noise affecting communities, a major part of the workload was to provide guidelines on the determination of the uncertainty of the assessment at a time where little documentation and applied research were available. In addition to this, I became involved in standards on airport noise monitoring (ISO 20906), on the quality assurance of calculation software (the ISO 17534 series) and, by invitation from the working group on occupational noise exposure (ISO 9612), where one of the biggest challenges we faced was determining uncertainty and documenting how to do this.

Now, 20 years later, as chairman of ISO TC 43 and of European Committee for Standardization (CEN) TC 211,[1] both covering acoustics standardization, the topic of uncertainty in standards remains challenging despite the progress made.

ISO and ISO TC 43

Before we progress, a few words about ISO. The International Organization for Standardization (ISO) is the world's largest developer of voluntary international standards. As a consequence of a meeting in London comprising delegates from 25 countries, ISO was founded in 1947 with 67 technical committees, groups of experts focusing on a specific subject. Since its inception, over 20,000 standards covering almost all aspects of technology and business have been published. ISO is an independent, non-governmental international organization with a membership of over 160 national standards bodies and is based in Geneva, Switzerland [1].

Through its members, ISO brings together experts to share knowledge and develop voluntary, consensus-based market relevant international standards that support innovation and provide solutions to global challenges. International standards provide internationally accepted and understood specifications for products, services and systems to ensure quality, safety and efficiency. As a result, they are instrumental in facilitating international trade.

One of ISO's initial technical committees, ISO TC 43 (acoustics), deals with standardization in the field of acoustics, including methods of measuring acoustical phenomena – their generation, transmission and reception – and all aspects of their effects on humans and their environment. It has sub-committees dealing with noise (SC1), building acoustics (SC2) and underwater acoustics (SC3) and has published over 200 standards on these topics.

The latest strategic business plan of ISO/TC 43, *Acoustics*, from July 2016 [2] includes the improvement in dealing with uncertainty in noise measurement standards as one of its expanding work areas. It also continues to identify measurement uncertainty as one of the areas needing further research.

The technical committee's operating plan states that each revised or newly developed standard or technical specification concerning acoustic measurements or sound predictions must include valid statements on uncertainty in accordance with its policy for the treatment of measurement uncertainty in standards. From 2010 on, each newly published relevant standard/technical specification must contain an uncertainty evaluation fully complying with the *Guide to the Expression of Uncertainty in Measurement* (GUM).

---

1 The European standardisation body which is, for acoustics, basically identical to ISO TC 43.

Uncertainty in ISO

The management and reporting of uncertainty in ISO is built up around the principles outlined in ISO/IEC Guide 98, *Uncertainty of Measurement* [3], which consists of three parts:

- Part 1 from 2009: *Introduction to the Expression of Uncertainty in Measurement.*
- Part 3 from 2008 with supplements and corrections published between 2008 and 2011: *Guide to the Expression of Uncertainty in Measurement* (GUM1995).
- Part 4 from 2012: *Role of Measurement Uncertainty in Conformity Assessment.*

Part 3 is commonly referred to as the GUM.

The GUM was first published in 1993 by ISO in the name of the seven international organizations that supported its development, including, among others, the International Electrotechnical Commission (IEC) and the International Organization of Legal Metrology (OIML). It is a general document that can be applied to several fields of application [4].

The focus of the GUM is the establishment of "general rules for evaluating and expressing uncertainty in measurement that can be followed at various levels of accuracy and in many fields – from the shop floor to fundamental research." As a consequence, the principles of the GUM are intended to be applicable to a broad spectrum of measurements. It is widely accepted by regional and national bodies including the European Union (EU) and, in the United States, the National Institute of Standards and Technology (NIST) and the American National Standards Institute (ANSI), as well as the National Research Council (NRC) in Canada.

Moreover, the GUM has been adopted by national metrology institutes throughout the world, such as NIST in the United States, the NRC in Canada, the National Physical Laboratory (NPL) in the United Kingdom, and the Physikalisch-Technische Bundesanstalt (PTB) in Germany, to name but a few.

A new international organization, the Joint Committee for Guides in Metrology (JCGM), with members from the seven original international organizations together with the International Laboratory Accreditation Cooperation (ILAC), has been formed to assume responsibility for the maintenance and revision of the GUM.

Uncertainty in Acoustics in ISO

As mentioned earlier, ISO TC 43, *Acoustics*, has defined a policy for the treatment of measurement uncertainty in standards [5]. This was prepared by ISO/TC 43, *Acoustics*, and ISO/TC 43/SC 1, *Noise*, and is referred to in the ISO TC 43 business plan. Thus, it is a central document for all its sub-committees.

The need to put more emphasis on the issue of measurement uncertainty when developing new or revised measurement standards was recognized by ISO/TC 43/SC 1 (*Noise*) already in 1999 when "ISO/TC 43/SC 1 requests each of its Working Groups to consider the 'Guide to the expression of uncertainty in measurement' in the preparation of documents and, if appropriate, include a statement of measurement uncertainty" [6]. However, there turned out to be insufficient knowledge about the proper treatment of measurement uncertainty in the working groups. Therefore, in Berlin in September 2003, a brief technical seminar was held, attended by around 100 experts, with the aim of providing more information and some direct guidance on the principles of the GUM and its application to measurements in acoustics. As a result, a policy document was issued in 2004 (Document 43 N 1023).

Nevertheless, in the ISO TC43 business plan of May 2005, the committee's chairman, Professor Dr. Klaus Brinkmann, stated that expanding work areas include "improvement in dealing with uncertainty in noise measurement standards," that further research was needed and that "the uncertainty issue which is becoming increasingly important."

There was continued discussion on the policy document, and in fact, it was first published as an official guide of ISO TC43 in 2012 as edition 1 of the *Guide for Policy on Treatment of Measurement Uncertainty in Standards Prepared by ISO/TC 43/SC 1, "Noise."* In this document, guidelines were given to working groups in "preparing clauses of standards for which a certain degree of harmonization within the standards from the committee is desirable." Convenors, project leaders and member bodies were invited to forward comments and experience with this guide and, as a result, edition 2 of our Guide 2 on Measurement Uncertainty was published in April 2018.

Edition 2 of the guide

- States that every time a measurement standard is elaborated, the working group in charge should

  - Study ISO/IEC Guide 98–3 (the GUM),
  - Choose one of the following approaches that all are GUM compatible:

    - The modelling approach that identifies and quantifies all major sources of uncertainty (the so-called uncertainty budget). This is the preferred method.
    - The inter-laboratory approach: round-robin tests in accordance with ISO 5725–2 to determine reproducibility of the method.

- The hybrid approach: using the inter-laboratory approach for components of the uncertainty budget where the technical knowledge to model contributions is lacking

- Provides default content of a normative clause on measurement uncertainty:

  - Mandatory text: "The determination of the uncertainty of results obtained from measurements according to this International Standard comply with ISO/IEC Guide 98–3." Or indicating deviations from this and providing the reason why such deviations were introduced.
  - Coverage factor to be stated with reference to a specified coverage probability p (preferably 95%)

- Demands mandatory reporting of measurement uncertainty unless there are good reasons for leaving it optional.
- States that the evaluation of measurement uncertainty is the responsibility of each person/team performing the measurement.
- Provides guidance on implementing the modelling approach, including

  - Standardised structure of the uncertainty on annex.
  - Working groups are to decide on whether quantitative information on the various uncertainty contributions can be given and whether this information is representative of a typical situation or a worst case.
  - Allows an approach to uncertainty based on solid and reliable information on reproducibility data from inter-laboratory comparisons and provided that no sufficient information is available to apply the modelling approach. An explanation of why the modelling approach was not chosen shall be given.

The latest strategic business plan of ISO/TC 43, *Acoustics*, from July 2016 includes the improvement in dealing with uncertainty in noise measurement standards as one of its expanding work areas. It also continues to identify measurement uncertainty as one of the areas needing further research and that "the uncertainty issue … is becoming increasingly important". In addition, each revised or newly developed standard or technical specification concerning acoustic measurements or sound predictions must include valid statements on uncertainty in accordance with its policy for the treatment of measurement uncertainty in standards. Since 2010, each newly published relevant standard/technical specification must contain an uncertainty evaluation fully complying with the GUM.

Examples of Uncertainty in Standards

ISO 1996–2, Environmental Noise Assessment

When, in 1998, I joined the ISO TC 43 SC1 Working Group 45 on revising the ISO 1996 environmental noise assessment standard, the need and desire to determine how to determine and document uncertainty had just been raised within ISO.

At the same time, the working group was faced with the challenge of revising a standard that was, at that time, over 20 years old. The initial focus was on collating and comparing assessment methodologies from around the world and then working to create and secure in our minds some universal concepts to cover the many sources of noise affecting communities. Many countries had and still have relatively diverse national methodologies linked into legal frameworks, with strong opinions from national experts on the choice of assessment method. In addition, the relatively uncontrolled nature of the topic, with varying residual sound and meteorological conditions, requires the use of one of a range of methodologies to enable the determination of the specific sound under investigation. This resulted in the standard developing into an umbrella document that provides guidance and good practice rather than explicit step-by-step instructions on assessing environmental noise. The development of the revision of ISO 1996 Part 2 [7] covering the actual determination of sound pressure levels by measurement, calculation and/or post-processing continued until 2007 with a relatively simple and easy-to-understand model to determine the major contributors to uncertainty and to create an uncertainty budget. The major factors to consider were the sound source and the measurement time interval, the weather conditions, the distance from the source and the measurement method and instrumentation. Some guidelines on how to estimate the measurement uncertainty were given (see Table 14.1), based on a coverage probability of approximately 95%.

This table refers to A-weighted equivalent continuous sound pressure levels only measured using IEC 61672 Class 1 instrumentation from at

Table 14.1

Standard uncertainty				Combined standard uncertainty σ_t, dB	Expanded measurement uncertainty, dB
Due to instrumentation, dB	Due to operating condition, dB	Due to weather and ground conditions, dB	Due to residual sound, dB		
1.0	X	Y	Z	$\sqrt{1.0^2 + X^2 + Y^2 + Z^2}$	$\pm 2\sigma_t$

least three measurements under repeatability conditions. Uncertainty due to weather and ground conditions came from a simplified table of possible values. It was noted that the uncertainty due to residual sound will vary depending on the difference between measured total values and the residual sound.

Higher uncertainties can be expected on maximum levels, frequency band levels and levels of tonal components in noise. It was noted that insufficient information was available and that, in many cases, it is appropriate to add more uncertainty contributions (e.g. the one associated with the selection of microphone location).

Some of the uncertainty content was educational regarding concepts, terminology and the determination of uncertainty budgets. During its development, much discussion was focused on the impact of measurement instrumentation and weather conditions.

New or revised standards have a timeline within which they need to progress and be published or the work abandoned. This factor, together with the desire to publish a standard that was a significant revision of the previous version, resulted in the standard's publication in 2007. During the approval process up to publication, a strong desire to further develop information and guidelines on uncertainty assessment was universally supported. As a result, ISO approved that the working group continue work with the primary purpose of refining and extending guidance on uncertainty assessment. A revised method, based on reports from the EU Imagine project [8], was introduced, and an Excel spreadsheet template was made available in the subsequent revision published in 2017 [9].

The uncertainty of sound pressure levels depends on the sound source and the measurement time interval, the meteorological conditions, the distance from the source and the measurement method and instrumentation. The measurement uncertainty must be determined in compliance with ISO/IEC Guide 98-3 (GUM). The modelling approach is suggested to be used, identifying and determining the impact of all significant sources of uncertainty. Systematic effects must be eliminated or reduced by the application of corrections. The expanded measurement uncertainty is to be reported, usually with a coverage probability of 95%, such that the result becomes $L \pm 2\,u$.

For environmental noise measurements, uncertainty is extremely complicated, and with different assessment methodologies sometimes required, the standard does not provide an exact formula for determining uncertainty. Instead, the principle of identifying the most important sources of uncertainty and creating an appropriate uncertainty budget is described, and a default assessment table is provided (see Table 14.2).

The numbers given in this table refer to A-weighted equivalent-continuous sound pressure levels only. Higher uncertainties are to be expected on maximum levels, frequency-band levels and levels of tonal components in noise.

Table 14.2

Quantity	Estimate, dB	Standard uncertainty u_i, dB	Magnitude of sensitivity coefficient c_i	Clause for guidance
$L' + \delta_{slm}$	L'	$u(L')$ 0.5^a	$\dfrac{1}{1-10^{-0.1(L'-L_{res})}}$	Annex F
δ_{sou}	0	u_{sou}	1	7.2–7.5, annex D
δ_{met}	0	u_{met}	1	Clause 8, annex A
δ_{loc}	0.0–6.0	u_{loc}	1	Annex B
$L_{res} + \delta_{res}$	L_{res}	u_{res}	$\dfrac{10^{-0.1(L'-L_{res})}}{1-10^{-0.1(L'-L_{res})}}$	Annex F

a Here 0.5 dB refers to a Class 1 sound level meter. A Class 2 sound level meter would have the standard uncertainty of 1.5 dB.

In many cases, the measured values must be corrected to other source operating conditions, such as the yearly average level. Similarly, other measurements may be corrected to other meteorological conditions in order to make L_{den} calculations possible. Some examples, including a link to a spreadsheet (http://standards.iso.org/iso/1996/2/), of complete uncertainty calculations are given in the annexes. Further work to provide more worked examples that ISO or national standard bodies could publish as guidance for the general practitioner is encouraged.

ISO 9612-2, Occupational Noise Exposure

This standard [10] is slightly simpler than ISO 1996 in that is describes a more limited number of valid assessment. Here the assessment method may be based on person-borne noise dose meters or on manned sound level meter assessments at operator locations. In addition, the standard describes how to apply assessments of individual noise exposure to determine how representative they are and the associated uncertainty when reporting larger groups of employees. The uncertainty assessment, where relevant, includes an assessment of the representativeness of the working duration for each operating condition in determining the noise exposure.

Importantly, occupational noise exposures can have high standard deviations, that is, greater than 3 dB, and as such are not explicitly covered under Guide 98–3. However, Guide 98–3 Supplement 1 (Monte Carlo methods) provides a practical alternative to the original guide when linearization of the model provides an inadequate representation or for when the probability density function (PDF) for the output quantity "departs appreciably from a Gaussian distribution or a scaled and shifted *t*-distribution, e.g. due to marked asymmetry". Thus, as indicated in Supplement 1 to the GUM, it can be applied under a range of conditions which are valid for occupational noise exposures, as well as for several other standards. Rather than use Monte Carlo methods, ISO 9612–2 tabulates the properties of their probability distribution.

Similar to many acoustical standards, the standard uses the arithmetic average of the squared pressure values that were obtained from measured decibel values. Assuming that the decibel values have a normal distribution, the squared values have a skewed distribution. In this case, using mathematics based on squared pressure, then, as the standard deviation becomes larger (e.g. >3 dB), the uncertainty becomes disproportionately larger, and the average of the squared values shifts farther away from the peak (i.e. the mean or median) of the normally distributed decibel values. Note that the assumption of a normal distribution does not cause problems where small standard deviations occur.

ISO 20906, Airport Noise Monitoring

For airport noise monitoring (ISO 20906 [11]), the uncertainty assessment concerned the uncertainty associated with a noise monitoring system based on noise monitors at suitable acoustically robust locations, where residual sound has little impact on measured levels. Here, however, the bias in uncertainty introduced by residual sound was dealt with in a simple manner as there were no studies available on the impact of this on the uncertainty.

Other standardization bodies and their approach to uncertainty

The ISO/IEC GUM is widely accepted by regional and national bodies, including the European Union (EU) and, in the United States, NIST and ANSI, as well as the NRC in Canada [4]. Moreover, the GUM has been adopted by NIST and most of NIST's sister national metrology institutes throughout the world, such as the NRC in Canada, the NPL in the United Kingdom and the PTB in Germany.

The GUM is adopted by ANSI as the *American National Standard for Expressing Uncertainty: U.S. Guide to the Expression of Uncertainty in Measurement*, ANSI/NCSL Z540-2-1997. The American Society of Mechanical Engineers (ASME) publishes a number of uncertainty standards and guides [12].

ANSI (United States)

In many cases, ANSI standards are identical to ISO standards and thus inherit the uncertainty assessments, where available. In addition, work is ongoing to update existing ANSI standards to include uncertainty assessment. Differences, where present, are mainly due to the identification and quantification of the sources of the component uncertainties. Whilst some are well known and clearly defined, some are estimated, some are measured and others are modelled (calculated within the method itself).

One example is ANSI/ASA S12.42 [13], *Measurement of Insertion Loss of Hearing Protection Devices in Continuous or Impulsive Noise Using Microphone-in-Real-Ear or Acoustic Test Fixture Procedures.* Here annexes F and I show uncertainty calculations for the insertion loss measurement and calculations for the continuous noise and impulsive methods, respectively. Although calculation of the overall uncertainty generally follows the ISO/IEC 98 GUM, calculation of the acoustic test fixture active and passive insertion loss uncertainties are calculated as linear sums of their respective component uncertainties.

Canada

Canada, particularly in acoustic terms, is a small country with more interest in using and participating in the standards used by its larger trading partners than in developing national standards. Hence, there is significant interest in ISO standards. In addition, there is also a great deal of overlap in the United States and Canada, with ANSI standards being used in Canada. Also, Canadians work on ANSI standardization and US citizens work on Canadian Standards Association (CSA) committees. Thus, there are only five Canadian acoustical standards left (i.e. CSA Z94.2, Z107.6, 56, 58 and Z1007), and unless they refer to ISO standards, they are basically at the same state of development concerning uncertainty as ISO standards from 1998. However, CSA Z107.56–06 (R2013), *Procedures for the Measurement of Occupational Noise Exposure* [14], uses a similar approach to ISO 9612 for determining how many measurements of group members are required to obtain a given precision.

Australia

The general approach from Standards Australia is to seek adoption of ISO standards where relevant and applicable and to incorporate any uncertainty sections in the ISO in national standards. For example, when considering uncertainty in building acoustics measurements of sound reduction and sound absorption both in the laboratory and in the field, Australia adopts the ISO approach, albeit at a later date. AS/NZS 3817, 1998, on measurement of single impulses or series of impulses [15], has a section on uncertainty based on the adoption of ISO 10843:1997.

However, some purely Australian standards are still being developed, and uncertainty is treated differently. AS/NZS 2107, 2016, on recommended design sound levels and reverberation times for building interiors [16], has no ISO equivalent and has no text on uncertainty. By contrast, AS 4959, on wind turbine generator noise [17], states that documentation must provide an "estimation of prediction uncertainty" and an "estimation of measurement uncertainty" but does not provide a method to do so. It also states that "the accuracy of such (predictive) models is heavily dependent on the assumptions used within them, and such assumptions should be reported and justified together with an estimation of predictive accuracy". AS 1269, on occupational noise [18], has the statement, "When carrying out any form of measurement there is always an uncertainty in the value of the reading obtained" but does not have a special section on uncertainty.

Most Australian standards do not have a formal section on uncertainty, and in some cases, the topic is described in terms of "accuracy". The uncertainty, accuracy or precision of measurement equipment is often given in terms of compliance with IEC 61672. However, it may be referred to in general terms as in AS 2377 on measurement of rail-bound vehicle noise [19] and AS 2702 (20) on measurement of road traffic noise. Here, exceeding calibration deviation limits of the sound measuring equipment used may render results invalid. In addition, there is general guidance regarding checking and ensuring that accuracy is accounted for.

There is currently an update in process on AS 1055 on the description and measurement of environmental noise. The revision introduces, compared with the current published standard, a section on uncertainty that is similar to that of ISO 1996-2: 2007, rather than the more recent 2017 version, as the earlier version is considered a more practical for inclusion in the standard at this point in time. The current draft of this standard states

"in presenting the results of environmental noise monitoring, the uncertainty and variability in the results shall be stated. The methods for assessing the uncertainty shall be described. Appendix F provides an informative explanation of methods to assess uncertainty and the components of measurements to be considered."

Appendix F, which is informative, is based on the principles of ISO IEC Guide 98-3 and describes the total uncertainty as the sum of the square root of the squares of five uncertainty components – instrument, location, source, meteorological conditions and residual sound. It describes what these are related to and gives typical values for the errors associated with some of them.

Concluding Remarks

We have come a long way since the publication of the GUM 25 years ago. Further research and work to document best available practices in dealing with uncertainty in noise measurement standards continues, as it must. The benefits of determining the impact of different factors on the quality of one's assessment are obvious. Applying quantitative uncertainty to results and ensuring a better and clearer understanding of what this means are important not only for standards writers and the scientific community but also for the community in which we live and work. The inclusion and refinement of the treatment of measurement uncertainty in new and revised ISO standards on acoustics will continue.

It is not feasible for one person, even with expert assistance, to describe the uncertainty assessment in all international and national standards within the scope of this chapter. However, I hope that this chapter gives you some insight.

My head no longer pounds in quite the way it did at the INCE–Europe Uncertainty Seminar in 2005. But the challenges of uncertainty and the work to progress this discipline still occupy my mind.

Acknowledgements

Many thanks go to Chris Struck of CJS Labs for his input regarding US ANSI standards and to Stephen Keith of the National Research Center of Canada for his input regarding Canadian standards and his review of the manuscript. In addition, thanks also goes to the Australian contingent of Marion Burgess of the University of New South Wales, Peter Teague of Vipac and Colin Tickell, Chair Standards Australia Technical Committees for "Community and Environmental Noise" and "Human Vibration", for their input.

References

1. www.iso.org/the-iso-story.html.
2. Strategic Business Plan of ISO/TC 43, *Acoustics*, July 2016, https://isotc.iso.org/livelink/livelink/fetch/2000/2122/687806/ISO_TC_043__Acoustics_.pdf?nodeid=1161983&vernum=-2, can be found under www.iso.org/committee/48458.html.
3. International Organization for Standardization, *ISO/IEC Guide 98: Uncertainty of Measurement*. ISO, Geneva.
4. NIST, https://physics.nist.gov/cuu/Uncertainty/international2.html.
5. International Organization for Standardization, *Guide for Policy on Treatment of Measurement Uncertainty in Standards Prepared by ISO/TC 43/SC 1, Noise, Edition 2, 2018*. ISO, Geneva.
6. Brinkmann K, Higginson R, Nielsen L, "Treatment of measurement uncertainties in international and European standards on acoustics", in *Proceedings of the INCE Europe Symposium on Uncertainty*, Le Mans, France, 2005.
7. International Organization for Standardization, *Acoustics: Description, Measurement and Assessment of Environmental Noise*, Part 2: *Determination of Environmental Noise Levels*, ISO 1996–2. ISO, Geneva, 2007.

8. IMAGINE, Technical Report IMA09TR-040830-dBA01 on Measurements of Road, Rail and Air Traffic Noise.
9. International Organization for Standardization, *Acoustics: Description, Measurement and Assessment of Environmental Noise*, Part 2: *Determination of Sound Pressure Levels*, ISO 1996–2. ISO, Geneva, 2017.
10. International Organization for Standardization, *Acoustics: Determination of Occupational Noise Exposure – Engineering Method*, ISO 9612. ISO, Geneva, 2009.
11. International Organization for Standardization, *Acoustics: Unattended Monitoring of Aircraft Sound in the Vicinity of Airports*, ISO 20906:2009, incl Amd 1. ISO, Geneva, 2013.
12. American Society of Mechanical Engineers, *Measurement Uncertainty: Understanding and Economic Benefits*, ASME B89.7. ASME, New York, http://files.asme.org/Catalog/Codes/PrintBook/25586.pdf.
13. American National Standards Institute, *American National Standard Methods for the Measurement of Insertion Loss of Hearing Protection Devices in Continuous or Impulsive Noise Using Microphone-in-Real-Ear or Acoustic Test Fixture Procedures*, ANSI/ASA S12.42: 2010. ANSI, New York, 2010.
14. Canadian Standards Association, *Procedures for the Measurement of Occupational Noise Exposure*, CSA Z107.56-06 (R2013). CSA, Ontario, 2006.
15. Standards Australia and Standards New Zealand, *Acoustics: Methods for the Description and Physical Measurement of Single Impulses or Series of Impulses*, AS/NZS 3817: 1998. AS, Sydney, 1998.
16. Standards Australia and Standards New Zealand, *Acoustics: Recommended Design Sound Levels and Reverberation Times for Building Interiors*, AS/NZS 2107: 2016. AS, Sydney, 2016.
17. Standards Australia, *Acoustics: Measurement, Prediction and Assessment of Noise from Wind Turbine Generators*, AS 4959: 2010. AS, Sydney, 2010.
18. Standards Australia, *Occupational Noise Management*, AS 1269, 2005-14. AS, Sydney, 2005.
19. Standards Australia, *Acoustics: Methods for the Measurement of Railbound Vehicle Noise*, AS 2377: 2002. AS, Sydney, 2002.
20. Standards Australia, *Acoustics: Methods for the Measurement of Road Traffic Noise*, AS 2702: 1984. AS, Sydney, 1984.

Index